Series in BioEngineering

The Series in Bioengineering serves as an information source for a professional audience in science and technology as well as for advanced students. It covers all applications of the physical sciences and technology to medicine and the life sciences. Its scope ranges from bioengineering, biomedical and clinical engineering to biophysics, biomechanics, biomaterials, and bioinformatics.

More information about this series at http://www.springer.com/series/10358

Dyah Ekashanti Octorina Dewi ·
Yuan Wen Hau · Ahmad Zahran Mohd Khudzari ·
Ida Idayu Muhamad · Eko Supriyanto
Editors

Cardiovascular Engineering

Technological Advancements, Reviews, and Applications

Springer

Editors
Dyah Ekashanti Octorina Dewi
School of Biomedical Engineering
and Health Sciences, Faculty of Engineering
Universiti Teknologi Malaysia
Johor Bahru, Johor, Malaysia

Yuan Wen Hau
School of Biomedical Engineering
and Health Sciences, Faculty of Engineering
Universiti Teknologi Malaysia
Johor Bahru, Johor, Malaysia

Ahmad Zahran Mohd Khudzari
School of Biomedical Engineering
and Health Sciences, Faculty of Engineering
Universiti Teknologi Malaysia
Johor Bahru, Johor, Malaysia

Ida Idayu Muhamad
School of Chemical and Energy
Engineering, Faculty of Engineering
Universiti Teknologi Malaysia
Johor Bahru, Johor, Malaysia

Eko Supriyanto
School of Biomedical Engineering
and Health Sciences, Faculty of Engineering
Universiti Teknologi Malaysia
Johor Bahru, Johor, Malaysia

ISSN 2196-8861 ISSN 2196-887X (electronic)
Series in BioEngineering
ISBN 978-981-10-8404-1 ISBN 978-981-10-8405-8 (eBook)
https://doi.org/10.1007/978-981-10-8405-8

This Springer imprint is published by the registered company Springer Nature Singapore Pte Ltd.
The registered company address is: 152 Beach Road, #21-01/04 Gateway East, Singapore 189721, Singapore

Contents

Part I Cardiovascular Imaging

Tissue-Mimicking Materials for Cardiac Imaging Phantom—Section 1: From Conception to Materials Selection 3
Dyah Ekashanti Octorina Dewi and Nurul Shafiqa Mohd Yusof

Tissue-Mimicking Materials for Cardiac Imaging Phantom—Section 2: From Fabrication to Optimization 35
Nurul Shafiqa Mohd Yusof and Dyah Ekashanti Octorina Dewi

Computational Human Models in Cardiovascular Imaging: From Design to Generations 65
Nurulazirah Md Salih and Dyah Ekashanti Octorina Dewi

Multimodal Medical Image Fusion as a Novel Approach for Aortic Annulus Sizing 101
N. A. Hamzah, Z. Omar, M. Hanafi and L. I. Izhar

Part II Cardiovascular Informatics

Introduction of Telecardiology and Challenges in Developing Countries 125
Kui Lin Kam, Hoe Tung Yew and Yuan Wen Hau

Clinical Pathway in Cardiovascular Disease Management 143
Rania Hussien Al-Ashwal and Eko Supriyanto

Square Root Design for Natural Frequency Module of Dynamic ECG Features—a Preliminary Study 155
Nurul Ashikin Abdul-Kadir, Norlaili Mat Safri, Mohd Afzan Othman and Fauzan Khairi Che Harun

Part III Cardiovascular Biomechanics

Mock Circulatory Loop for Cardiovascular Assist Device Testing 177
A. Z. M. Khudzari, M. R. A. Kadir, K. Osman and A. H. M. Hudzari

Computational Fluid Dynamics Application in Reducing Complications of Patent Ductus Arteriosus Stenting 201
M. I. Kori, K. Osman, A. Z. M. Khudzari and I. Taib

Part IV Cardiovascular Biomaterials

Biopolymers as Potential Carrier for Effervescent Reaction Based Drug Delivery System in Gastrointestinal Condition 221
I. I. Muhamad, S. Selvakumaran, M. A. M. Sis, N. Pa'e and M. H. Salehudin

3D Bioprinting of a Tissue Engineered Human Heart 243
Khalida Fakhruddin, Saiful Izwan Abd Razak, Nadirul Hasraf Mat Nayan and Mohammed Rafiq Abdul Kadir

Part I
Cardiovascular Imaging

Tissue-Mimicking Materials for Cardiac Imaging Phantom—Section 1: From Conception to Materials Selection

Dyah Ekashanti Octorina Dewi and Nurul Shafiqa Mohd Yusof

Abstract The effectiveness of cardiac imaging system is valued from its reliability, reproducibility, accuracy, and usefulness in the clinical settings. These parameters are often evaluated, validated, and justified during the optimization process of the system. A large number of calibration techniques have been used in this process to provide values that specify how standardized the imaging system is. The most common technique is using physical imaging phantom. This device can clarify the degree of image quality and object detectability produced by the imaging system. However, even various imaging phantoms have been widely available, it is still difficult to obtain the phantoms that mimic the realistic biological tissues and functions, particularly for cardiac imaging applications. As cardiac imaging systems capture and analyse dynamic cardiac morphology and function in motions, the main issue in cardiac imaging phantoms is how close the phantom properties to those of realistic biological tissues so that the phantom can guarantee for a reproducible measurement. As cardiac imaging phantom materials play vital roles in the standardized validation for cardiac imaging systems, it is important to study Tissue Mimicking Materials (TMMs) for cardiac imaging systems, materials, and their properties that build the phantom structures. This review study is divided into two parts. Part 1 highlights on preparation processes in phantom development that consist of conception, design, simulation, and materials selection stages, while part 2 concentrates on realization processes from fabrication to optimization stages. This part 1 is aimed to briefly review the current state of knowledge regarding TMMs and their uses for cardiac imaging phantoms. Introduction to systematic processes in the phantom development is also presented to provide an understanding on how to generate the physical phantom step by step.

Keywords Tissue mimicking materials · Cardiac imaging phantom · Systematic processes · Conception · Design · Simulation · Materials selection

---

D. E. O. Dewi (✉) · N. S. M. Yusof
IJN-UTM Cardiovascular Engineering Center, Institute of Human Centered Engineering, Universiti Teknologi Malaysia, Johor Bahru, Malaysia
e-mail: dyah@utm.my

School of Biomedical Engineering and Health Sciences, Faculty of Engineering, Universiti Teknologi Malaysia, Johor Bahru, Malaysia

D. E. O. Dewi et al. (eds.), *Cardiovascular Engineering*, Series in BioEngineering,
https://doi.org/10.1007/978-981-10-8405-8_1

1 Introduction

Recent advancements in biomedical imaging systems that emphasize on precision medicine based on quantitative imaging biomarkers necessitate accurate and proper tools for validation and evaluation. The conventional technique for validating the imaging systems using image data directly collected from human subjects or patients has some restrictions in clinical ethics aspects [1]. Also, this technique is often limited by difficulties in obtaining standardized quantification of human image data [2]. At the moment, imaging models and phantoms in computational and physical means have been widely used to perform validation and evaluation of biomedical imaging system [3]. Compared to the computational imaging models that require extensive data simulations, the physical imaging phantoms are mostly used directly on the imaging machine during data acquisition to justify the image quality and object detectability, as it may simply represent the actual status of the imaging system when scanning the phantom. With these physical imaging phantoms, a large number of measurements and tests can be performed in a more measurable, simpler, and safer way [4].

In cardiac imaging systems, the main concern in physical cardiac imaging phantoms is that the heart organ has its specialties. The heart has dynamically beating and pumping mechanism with complex morphological tissue, valvular, and vasculature that differ from other organs [5, 6]. Therefore, the basic issues in the development of physical cardiac imaging phantoms are that the material properties must to meet the standardized characteristics to that of actual biological heart. In this way, the phantom materials must be composed and fabricated in such a way that it can mimic the heart structure [7]. Also, due to the dynamic properties, the phantom materials must be durable and elastic enough to perform beating and pumping mechanism in a large number of experimental cycles [8]. Due to these special requirements, the study on Tissue Mimicking Materials (TMMs) for cardiac imaging models is becoming more challenging. Besides, as technological advancements are directed into quantitative cardiac imaging [9, 10], image-guided cardiac drug delivery, intervention, and surgery [11], and so on, fusion with other imaging modalities or navigation systems, also with robotic surgery, is needed to assist cardiac procedures, from diagnosis, therapy, to surgery [11, 12]. With these various cardiac alternatives, TMMs need to be advanced and customized with distinctive design, materials composition, fabrication, and testing for specific physical cardiac imaging phantom application. Consequently, the TMMs development turns out to be more challenging and is important to be studied further in details.

In this chapter, our review on the current state of TMMs and their uses for cardiac imaging phantoms is started with the role of imaging modalities in cardiovascular (CV) applications. In this part, overview of cardiac imaging modalities is described together with the needs of physical imaging phantoms. Then the systematic processes in the physical cardiac imaging phantom development, from concept to materials selection, are portrayed.

2 Cardiac Imaging Systems

Cardiovascular diseases (CVDs) and their related forms, as the leading cause of increased morbidity and mortality rates nowadays, have become main concerns to be solved through prevention, management, and monitoring strategies in diagnosis and treatment. A large number of actions have been made to solve these epidemic diseases, from screening [13, 14], diagnostic [15–17], monitoring [18, 19], treatment [15, 20], prevention [21, 22], to rehabilitation [21–23]. Numerous technologies have been developed and implemented as efforts in tackling these health problems.

Cardiac imaging, as one type of these sophisticated favorable technologies in assisting diagnosis and treatment, has its interactive visualization and multimedia data representation [24], morphological and functional measurements, localization parameters [25, 26], and wide integration with various clinical procedures and other healthcare technologies [27]. From these wide ranges of contributions, cardiac imaging has been endorsed in guidelines of CVDs. With this importance, high quality assurance of cardiac imaging is a must.

2.1 *Imaging Modalities*

X-rays as conventional imaging modality in cardiac imaging management have been widely used to acquire heart radiographic image and reveal basic cardiac morphological parameters measurement, such as size and shape of the heart that may indicate specific heart diseases, such as cardiac hypertrophy [28] and congenital heart disease [29]. Though, the common issue in conventional X-rays in image quality has limited its uses in high resolution cardiac applications. Besides, it also only produces the two-dimensional (2D) projection image and still suffer from organ overlapping appearances that may complicate the analysis.

With the same source as 2D X-rays, angiography imaging captures visualization of coronary arteriography in real-time mode during catheterization procedure [30]. Uses of contrast injection has also enhanced the visibility of coronary arteries, providing better observation of coronary diseases. In spite of that, flow characteristics are still poorly identified in this way [31].

A more advanced radiation-based imaging, Computed Tomography (CT), provides three-dimensional (3D) volumetric representation that makes it suitable for quantification and visualization of cardiovascular morphology and function [32, 33]. Electrocardiography (ECG)-gating system is also usually attached to the cardiac CT system during scanning to observe functional parameters of the heart during systole and diastole phases [34–36]. These functional parameters can identify left ventricular aneurysms, enhancement of scars, and thrombus [37]. Moreover, CT developments with multi-slice and multi-detector technologies for higher spatial and temporal resolutions have provided physiological myocardial perfusion from stenosis [38]. Yet, the main problem in cardiac CT is about excessive radiation exposure for maintaining image quality [39]. A number of low-dose imaging systems with compensated image quality have been built [40]. However, the implementation of these low-dose imaging system into various cardiac diagnosis and therapeutic monitoring necessitates optimal protocol design to achieve balance between proper image quality and radiation dose reduction.

Alternatively, echocardiography or ultrasound (US) imaging technology as the first-line non-invasive imaging tool has allowed for a broader range of cardiac quantification parameters in the heart chambers, wall, valves, pericardium and intracardiac shunts. Various echocardiographic modes, such as tissue Doppler, strain imaging, speckle tracking, and so on, have also been widely implemented for a more comprehensive observation [41, 42]. In order to solve limitations in 2D representation and operator dependency, conventional echocardiography has moved towards 3D transducer technology and more automatic quantification system [43]. Also, due to freehand transducer mechanism, the skill of the operator in positioning the transducer plays an important role in generating the anatomically proper observation. Still, with real-time visualization and non-radiation exposure, echocardiography is preferred to be used to assist during cardiac intervention or surgery, together with navigation devices or other imaging modalities in the image-guided cardiac systems [44]. In this regard, the accuracy of the system highly relies on how the echocardiography image frames are interconnected and synchronized with the navigation devices or fused with multidimensional images from other imaging modalities.

Optical imaging techniques have provided new options for evaluation of the various factors associated with atherosclerosis, plaque stability, and treatment of coronary artery disease [45]. By using multiple movable light sources and detectors attached to the tissue surface, light attenuation characteristics can be obtained and reconstructed into the internal absorption and scattering distributions. Unusual growths inside the tissue may be discerned from the recovered optical densities due to different scattering and absorption properties in layer architectures [46, 47]. The challenge in optical cardiac imaging is more to image reconstruction from the track that the optical imaging device has pass through.

Molecular characterization in cardiac imaging can be obtained from Cardiac Magnetic Resonance (CMR) and Nuclear Medicine (NM). With the strength in soft tissue characterization, CMR has demonstrated excellent capability in portraying the spectrum of cardiovascular anatomy and physiology [48]. The ECG-gating setting in the CMR has also provided improved image quality and visualization that may reveal cardiomyopathy abnormalities in pericardial thickening and constriction [49]. The use of contrast medium highlights the blood supply mapping in heart [49]. For aortic dissection diagnosis, not only cardiac CT and CMR that are known for their ability in several aortic dissection types of observation, transesophageal echocardiography also has been used for this purpose in high sensitivity [50].

Compared to CMR, NM reveals better functional, metabolic, and molecular information of the cardiac tissues [51]. Myocardial perfusion defects and coronary artery diseases can also be detected and characterized more accurately using this modality [52, 53]. The use of certain radiopharmaceuticals, for instance technetium 99 m stannous pyrophosphate, may visualize the accumulation of calcium ions within mitochondria of damaged myocardial cells in myocardial infarction [53, 54]. Yet, the main issue in NM is that the image quality is considerably low, so that anatomical details are blurred [55]. The alternative of this problem is performing image fusion between NM and CT or CMR for obtaining both anatomical and metabolic representation in one image [56, 57]. However, this procedure imposes hardware and software modification.

2.2 *Issues in Imaging Applications*

Broad uses of cardiac imaging cover from conventional single imaging modality, in combination with diagnostic or therapeutic devices, to multimodality imaging modes. Depending on the procedure, cardiac imaging can be performed in one single scanning mode, multiple scanning for treatment planning in the pre-intervention and pre-surgery stages, monitoring the medication progression in periods of time, or in synergy during intra-intervention and intra-surgery procedures for visualization assistance. In this way, all cardiac imaging modalities are interconnected, enabling for multimodality data fusion [12]. With these roles, imaging accuracy is uncompromised for a precise data merging.

In contrast, some imaging factors, from system faulty to human errors, have been reported to influence the quality and safety of cardiac imaging fusion in clinical settings [58, 59]. If standardized reference in image acquisition is not provided, cardiac imaging may suffer from inter- and intra-viewer variability [60]. In addition, various cardiac procedures, including cardiac catheterization, open heart surgery, coronary artery bypass surgery, and heart transplants, have high variability in the procedures. This variability occurs due to uniqueness of every case and experiences, specifically when dealing with the heart complex structure and motion compensation in the cardiac deformation and respiration. Furthermore, introduction of various multimodality imaging techniques, such as automatic quantification, image fusion, robotic navigation, and so on, have brought positive impacts in these cardiac procedures. But the differences from the conventional procedures have occasionally triggered any errors and imposed the clinicians to equip with more specific skills [61]. In radiation safety, along with greater needs of image quality, radiation dose from cardiac multimodality imaging tends to be higher if not appropriately maintained. An accurate radiation dose monitoring system may help controlling the dose. Therefore, to accommodate the accuracy issues in cardiac multimodality imaging, a translational method to bridge the gap among research, quality assurance, and clinical implementation of cardiac multimodality imaging is vital. In this way, standardization and calibration mechanisms can be developed in the research and industrial bases as imaging quality assurance for good trade-offs between clinical accuracy and safety. Figure 1 illustrates these interconnections.

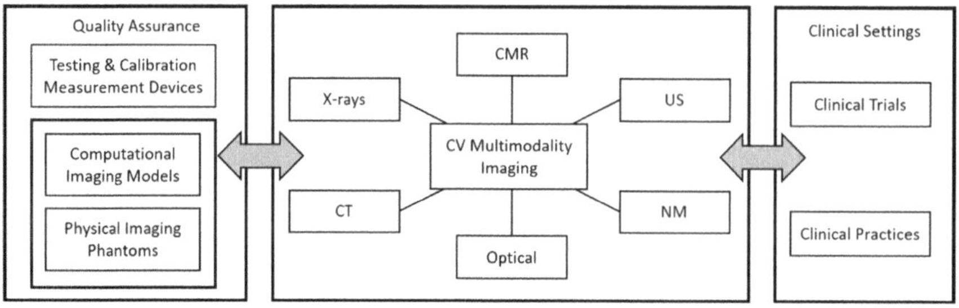

Fig. 1 Interconnection among imaging modalities, quality assurance, and clinical implementation in assisting cardiac diagnosis and treatment management

3 Cardiac Phantoms and Tissue Mimicking Materials

3.1 Physical Imaging Phantoms

Physical imaging phantoms are devices utilized for validation and evaluation of many processes in biomedical imaging systems. Its direct implementation like patient dummy placed into the imaging machine makes the scanning validation and evaluation more standardized, measurable, and safer. Physical imaging phantoms are mostly used for image quality verification and calibration during acquisition processes [62, 63]. These phantoms are designed with standardized measurement scales that signify how good the imaging modality capture particular objects of interest in the scanned phantom with specified imaging settings [64, 65]. Also, physical imaging phantoms can be used to identify any functional defects in the instrumentation of the imaging modality and related systems by analysing the quality of the image output or detecting from special sensors in the phantom [65, 66]. With some adjustments, radiation safety during imaging and radiotherapy can also be measured and controlled through phantoms [64, 66].

In advanced applications for quantitative imaging and image analysis, physical imaging phantoms can play role as in vitro testing tools representing objects of interest with certain size, shape, and properties. The performance of the applications is determined by how accurate, fast, and robust the objects can be detected and measured [67]. Besides quantitative imaging and image analysis, the imaging technology is also progressing into an integrated system that combines imaging modalities with positioning techniques for navigation in interventional and surgical tools. With this integration, planning procedures in medical intervention and surgery have become standardized protocols to encounter any obstacles and minimize errors during the actual intervention and surgery, especially when the procedures implement minimally invasive method with imaging guidance. In this way, phantom is specifically used as planning model replacing the actual patient during pre-intervention or pre-surgical planning stage [68, 69]. In a more basic procedure, phantoms have also been used in many medical trainings substituting biological subjects, real patients, or cadavers as learning objects for the trainees in handling incisions, insertion of sutures, catheterization, removal of unwanted cells, therapeutic penetration, and open surgery in the training [70].

Additionally, along with fast growth of sensing technology for biophysical measurements, the application of physical imaging phantoms is then growing wider. Integration is not only for navigation, but also for synchronization with other measurements. The implementation of this synchronization has been made in, image-guided hyperthermia therapy, ECG-gating cardiac imaging, and so forth [69, 71]. With these important roles, it can be seen that phantom does not function as imaging calibration tool per se, but it has expanded as quality assurance in image quality, quantitative imaging system, image navigation, biophysical measurement integration, and many more. Therefore, research in this area is one of important parts in the in vitro stage of imaging system development. Not only research, a number of industries have also concentrated in this business since long time ago.

This physical model strategy is chosen due to issues on durability, preservation, availability, cost, and ethics that come up in biological subjects. This strategy has also been widely commercialized and available in the markets. These phantom products have also been endorsed in a lot of medical education and trainings for skill improvements in anatomy understanding, imaging navigation for intervention and surgery, visual-tactile coordination in handling intervention and surgical devices, image quality calibration and adjustment, and other technical assessments. With this model, learning process can be developed in a more standardized and structured way. Even so, due to materials characteristics limitation, not all models can be made truly realistic. Therefore, research in modifying these materials characteristics is highly needed to improve the phantoms performance.

3.2 Cardiac Tissue Mimicking Materials

In general biomedical applications, TMMs can be defined as materials that mimic the actual tissues and their properties in physical imaging phantoms [72]. Physical imaging phantoms for cardiac imaging applications are quite specific because of complex cardiac morphological structures, motion behaviours, imaging characteristics, electrical and chemical activities, and mechanical properties in interaction with medical devices [5–8]. Also, when pathophysiological aspects are included, the cardiac phantoms are becoming more complicated. Hence, cardiac TMMs must be carefully studied, selected, fabricated, characterized, and matched in order to correctly simulate the actual cardiac tissues.

In developing TMMs, knowledge about physical properties of materials is essential, as it determines the outcomes in the applications. Once the appropriate TMMs are obtained, the design process is conducted to determine the most optimum phantom features and its fabrication technique that fits with the main uses of the physical imaging phantom. In this stage, it is important to define which parts that need to be emphasized to fit with the application. Similar to development of other biomedical products and devices, development of any type of physical imaging phantoms using TMMs requires systematic processes to generate the best phantom prototype with high efficiency, accuracy, and fitting with the needs in biomedical imaging application [73, 74].

4 Systematic Processes in Phantom Development

Adapted from prototyping theory in generic biomedical product design method [73], phantom prototype can be developed iteratively through seven stages, from concept, design, simulation, material selection, fabrication, testing and analysis, to optimization, as in Fig. 2. In this chapter, our discussion is focused from conception to materials selection. Design, simulation, and materials selection are made in a triangle reverse closed loop. This explains that this looping enables for developers to repetitively utilize interconnecting parameters, such as phantom morphology, that circulate among these processes.

4.1 Concept

Concept is the first stage of systematic processes of physical imaging phantom development. The concept stage is aimed to understand the underlying aspects in TMMs applications and deliver into ideas and prototype design [73, 74]. The concept is built by elaborating questions in problem analysis and synthesis like so:

1. Identify the core problems using analysis process
2. Establish target specifications

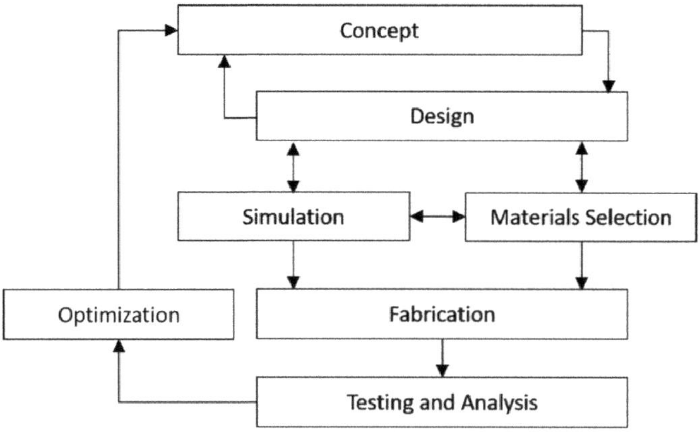

Fig. 2 Seven systematic processes in physical imaging phantoms development

3. Develop list of requirements
4. Translate into several systematic ideas with features
5. Study the feasibility
6. Develop plans to realize the ideas
7. Address some common challenges to be anticipated.

One example is in physical cardiac imaging phantoms development, the critical issue is cardiac TMMs need to mimic biological cardiac structures under imaging modalities. As the purposes can vary, from image quality assurance, quantitative imaging system, radiation dose safety control, to interventional and surgical planning and learning models, it is important to narrow down the target into one specific idea through analysis process before the conception. The analysis process generate list of phantom requirements, then determines the priorities to obtain the main purpose. Once one purpose is taken, then several ideas can be generated.

For the concept on cardiac learning model, two ideas can be proposed. The first idea is to develop cardiac model based on 3D printing or molding system with single homogenous material. With this technique, the anthropomorphic phantom can function to visualize the basic morphology of the heart, but producing phantom that mimics the functional characteristics of the heart is not easy. However, homogeneity of the phantom is one important material in imaging. Yet, homogenous tissue is rarely found in real biological situation.

The second idea can be made by developing more complex cardiac tissue structures. With this choice, it consequently needs more advanced techniques to combine various materials with different size and shape. As cardiac structures are composed of multiple layers with various thickness, shape, size, and elasticity, the concept can be made more innovative by creating the tissues layer by layer, then combining the layers into one united cardiac structure. The development of each tissue layer can be performed similarly to that of in homogenous tissue. Once all separated tissues have been developed, the layers unification can be made by using special sticking material. With this technique, the complex layers can be maintained more properly, producing in a more realistic physical cardiac imaging phantom. This phantom can be used as cardiac patient specific model for a learning and planning model [74, 75].

Above all, TMMs concept needs to consider about some hindering aspects in phantom development to meet with requirements. Unmatched properties due to improper materials interaction of TMMs frequently occur in the development of physical imaging phantoms for image quality testing in imaging system [76, 77]. This common problem can be anticipated with good observation and understanding on materials mixing and property alteration. Especially for cardiac phantom in multimodality imaging, mixing the materials with some contrast additives has been renowned to improve imaging quality and object detectability in specific imaging modalities [77]. For phantoms with repetitive mechanical uses, such as for needle puncture, stitching, or physical examination, study on durability and consistency of the materials is also important [70]. Additionally, in order to produce imaging phantoms for motion analysis, some TMMs can be designed to have certain elasticity properties that can be actuated electrically or mechanically, generating specified motions that can be captured using dynamic imaging modalities [72]. However, this motion elasticity properties can only be obtained if the selected materials have high level of elasticity and durability. Adding additives in the base material may also alter the imaging or mechanical properties.

4.2 Design

From the concept stage, design of TMMs for physical imaging phantoms can be described. Adapted from the principles in biomedical product design process, design stage in the physical phantom development is aimed to translate the ideas into the real phantom prototype by specifying plans for synthesizing the ideas and anticipating the challenges to find the appropriate structures [73]. In this way, expansion of the synthesis process can be detailed in these work activities:

1. Identify the morphological and functional schemes
2. Detail out the schemes and parameters
3. Develop the technical drawing of the design
4. Deliver the prototype
5. Generate some scenarios for simulation and materials selection.

Identification of morphological and functional parts of the phantom, together with parameters specification of the phantom, need to be supported by good understanding and analysis of anatomy and physiology of the body, organ, or tissue. Once the details of the phantom morphology and function are identified, technical drawing of the design is needed to equip with comprehensive volumetric scaling and visualization, functional feasibility, and scenario developments for the next steps of phantom development [73]. In this way, prototyping software, for instance Solidworks, in combination with 3D volumetric medical image, can be used for designing the 3D phantom shape. Alternatively, some ready-to-use biological organ drawings from open source or paid system may also provide the instantly available designs that cut down the design process time and efforts. Lastly, some scenarios need to be planned for simulation, materials selection, and fabrication stages.

In the design process of phantom, characterizing the best type of phantom is necessary to focus on the main requirements and organize the systematic processes more efficiently. As simply described in [71], physical phantoms can be grouped based on the type of measurement. The phantoms can be in the form of liquid, semi-solid, and solid forms. The liquid phantoms are mostly meant to represent and mimic the fluid mechanisms in the biological settings and imaging systems that employ and interact with liquid materials. With certain solution and concentration designs, liquid phantoms can be varied and implemented in wide range of applications. The semi-solid phantoms are designed to model the physical characteristics of materials in between liquid and solid, such as absorption, diffusion, perfusion, and so on. The properties of these semi-solid phantoms can also be tuned either more to liquid or solid forms, depending on the applications. The solid phantoms, as are usually used as tissue-equivalent models, act in two-sides. The phantoms can simulate not only about the changes in the phantoms due to external stimulations, such as imaging, interventional, or surgical applicators, but also the detectability of the phantoms in measurement devices, such as imaging.

In the context of phantom design, the phantom can also be simply classified into three types, anthropomorphic, functional, and combination of both types of phantoms. Since each of these types have their own purposes and specific applications, the simulation, materials, and fabrication technique in phantom development are also following its purposes.

4.2.1 Anthropomorphic Phantom

Phantom that mimics the shape, regions, and structures of human anatomy is called anthropomorphic phantom. The anthropomorphic phantom emphasizes on morphological structure to mimic the exterior shape and size of the organ as close as possible. Although it does not always provide functional features, this phantom is mainly utilized for physical anatomical model that provides volumetric and structural visualization with tactile impression for clinical modelling, consultation, planning, and learning to for better observation about the overall representation of the organ [73].

A term called reference man is developed as a reference model based on recommendation of International Commission on Radiological Protection (ICRP) about standardized anthropometry of the organs (height, weight, volume) and represented in an anthropomorphic phantom [78, 79]. With this reference, anthropomorphic phantoms can be used to test imaging techniques and evaluate the structure and geometry. For ICRP reference man, phantoms are mainly aimed to measure radiation dose and predict the effective dose.

Since human geometry and structure vary, it is no surprising that anatomical geometry and structure of phantoms can and should vary as well. Even provided with the ICRP specifications, customizations of anatomical details for physical phantoms are still valid and always needed, especially to reflect certain population group, such as in Asian and Caucasian people that differ in anatomical geometry [78, 79]. Besides races, phantom variations can also be made based on gender, age, anatomical region, special clinical situations, like pregnancy and abnormality. However, although some differences may affect some parts in the phantom design, the general description of the reference model remains the same, as the reference model plays role as the ground truth. Hence, with this standardized anthropometry, a proper phantom can be built. In this way, fabrication technique also needs to be adapted to produce the anatomical structures of the phantoms as in the design.

In the context of measurement type as in [70], phantoms are mostly developed as modification and combination of semi-solid or solid phantoms that are the representation of body, organ, or tissue. However, as liquid is also a major part of the human body, including liquid phantoms as parts of anthropomorphic phantoms can improve the performance of the phantom, in some extent. This can be realized through designing the suitable solution and concentration of the liquid phantom that can describe the biological liquid components in microscopic and functional scales [70, 80]. Then the liquid phantom can be added up in the anthropomorphic phantom component. Additionally, it is possible to do on the other way around by including the semi-solid and solid phantoms in the liquid phantom systems as an integral mechanism to model the biological system in the body or organ.

To fulfil the needs of diversification in sophisticated imaging systems that undergo for full body observation, specific organ only, molecular imaging, or cellular imaging, the phantom development can also be expanded for the full body, partial organ, tissue structures, or even in the cellular levels [78, 81, 82, 83]. In the full body phantom design, the purpose is more to systemic modelling that considers all parts of the organ as one big interconnected system. In the partial organ phantom, the development of the phantom is more concentrated on the details of organ of interest only, such as heart with its compartments, levels of tissues, valves, vessels, and all parts. In the tissue structure modelling, the phantom can be developed to mimic the structures and functions of the tissue. In this way, the measurement is more to characterizing the structure patterns, composition, elasticity, and many more. In the cellular levels, the phantom design is more complicated as this cellular part mostly imposes functional representation of each components that build the phantom. Furthermore, the size of phantoms in cellular levels are mostly up scaled to ease the modelling mechanism.

In the anthropomorphic cardiac imaging phantom, the general dimensions also need to follow the sizes provided from the ICRP recommendation. However, due to the high need of customization in cardiac imaging applications, some cardiac imaging phantoms have been modified into specific sizes, shapes, and structures [84]. Some innovations, such as adding some features into the phantom, like heartbeat, blood flow, and so on, have made the anthropomorphic cardiac phantom closer to the real function of the heart [84, 85]. Some of the commercialized examples are presented in Table 1.

4.2.2 Functional Phantom

Functional phantom highlights on how to mimic physical, electrical, chemical, or mechanical functions of biological subject with some adjustable parameters to mimic certain mechanism of the organ. The functional phantom applications range from pre-clinical research, clinical modelling, consultation, planning, to learning system to analyze functions in the organ [88]. With these advantages, functional phantoms have also been widely used in testing and validation of imaging modalities. Through 3D printing, functional phantom can also become proof of concept in early development of artificial organs [89]. Also, some functional phantoms also may have more than one function to accommodate the industrial needs. Even not fully following the anthropometry standard from the ICRP recommendation, the functional phantom still accommodate certain basic functions as recommended, especially in providing a reliable quantitative measurement.

The exterior appearance of this type of phantom is mostly not similar to that of anthropomorphic phantom. The phantom is more developed in a simplified version

Table 1 Examples of commercialized anthropomorphic cardiac phantoms

Phantoms	Manufacturers	Application
Static heart phantom	CIRS Inc. [86]	– Heart tissue equivalent material – Contrast agent features – Scanned for US and MRI modalities
Dynamic cardiac phantom	CIRS Inc. [86]	– Simulates realistic motion of the heart – Complex heart motion with respiratory motion – Correlated with ECG signal gating – Training for radiation dosimetry
Dynamic heart and lung phantom	Kyoto Kagaku Co.Ltd [87]	– Represents 3D movement of the heart, lungs and pulmonary nodule with respiratory cycle and abdominal movement – Elastic heart with systolic and diastolic motions – Coronary arteries with stenotic examples – Connectable to ECG for ECG gating
Myocardial phantom HL	Kyoto Kagaku Co.Ltd [87]	– Simulates high radio accumulation interference in the liver with myocardial SPECT images – Allows for setting defects in LV muscle with background in each lung field, mediastinum, and RV – Scanned for SPECT and CT

describing a certain function in imaging system, either it is in round or box shapes, or arrangement of simple shapes, or only arrangements of interconnected compartments that represent a function, as it is mainly used only as model and simulation of several strategies in an imaging system or application with high precision and reproducibility. In this way, liquid phantoms with their derivatives are also largely used in this type, especially in the combination with semi-solid and solid phantoms as an integrated system. However, this simplification can also limit the detailed measurement, especially for structural analysis.

As most functional phantom represents physical, electrical, chemical, or mechanical activities, phantom does not only consist of TMMs to represent the organ or body, but it is also connected to devices that deliver physical, electrical, chemical, or mechanical effect to TMMs, such as sensor and electronic components, chemical reactor, or mechanical actuator [90, 91]. With these devices, the functional phantom can work based on its function. Also, as this functional phantom is working closely with imaging modalities, one important concern about the use of devices for functional phantom in imaging modalities is about imaging safety. For instance in the MRI systems, it needs to ensure that the materials, devices, and protocols for the MRI functional phantom are compatible to magnetism [92].

Similar to anthropomorphic phantoms, functional phantoms can vary from whole body systemic mechanism into specific organ levels. The whole body phantom mostly has quite extensive design, as it also considers functional mechanism and relationship of multi organ. For specific organ phantom, some connections related to multi

Table 2 Examples of commercialized functional cardiac phantoms

Phantoms	Manufacturers	Applications
Dynamic multimodal heart phantom	Shelley medical imaging technologies [96]	– Simulates realistic beating motion of the heart – Uses air pump attached to cardiac structure to inflate and deflate LV chamber – Scanned for multimodality imaging system (US, MRI, CT, PET) – Allows for motion rate controlling application
Cardiac Doppler flow ultrasound phantom ATS 523 and 523A	ATS Laboratories (CIRS Inc.) [86]	– Simulates deep vasculature of cardiac and abdominal vessels – Rubber-based TMM for Doppler flow phantoms – Contains 4 flow channels with 2, 4, 6, and 8 mm diameters – Scan surfaces at 2 fixed angles of 45° and 60° with depths from 5 to 18 cm – Scanned for US modality
11347-210 ultrasound training block (Real Vessel)	Kyoto Kagaku Co.Ltd [87]	– Consumable TMM consisting of skin, muscle, and skelton layers – Training to identify injection spots with accuracy indicator
HM4.0 reusable 4 vessel ultrasound phantom training model	Humimic Medical [97]	– Simulates blood vessels with the consistency of human soft tissues – Scanned for US modality – Training reusable block model for ultrasound-guided vascular and arterial interventions to recognize a bifurcation and locate the correct position of needle insertion

organ is usually simplified or assumed to be default. The development of specific organ phantom is also more concentrated on function of certain organ, such as cardiac function. Not only that, due to complexity of the organ and its function, some functional phantoms even restrict their scope by only modelling certain part of the organ, such as the function of left ventricle region in the heart [93].

Cardiac functional phantoms can also focus on modelling one of the physical, electrical, chemical, or mechanical activities of the heart. Some of the examples of functional cardiac phantoms are presented in Table 2. From these examples, it can be inferred that combination of more than one function is also possible, however, the complexity is also higher, following the features provided in the measurement.

In cardiac functional phantom for radiation dose safety control applications, special radiation sensor systems are used to detect the degree of radiation exposure in the phantom captured from image acquisition [64–66]. In this way, the phantom is functionally designed by inserting small model dosimeters in different locations of the phantom to measure the doses from external irradiation in the specified points during imaging procedure [94, 95]. From these dosimeters, the organ dose and effective dose can be quantified to evaluate the radiation risk to the body or organ.

4.2.3 Combination of Anthropomorphic and Functional Phantom

This hybrid version between anthropomorphic and functional phantom concentrates on compensating the morphological structure of the phantom together with its function as the actual organ. With these complete features, this phantom may work as an organ simulator. Some of the examples in cardiac imaging phantom are beating heart phantoms [64, 98].

The functional design can be verified by using simulation stage, while the morphological design can be detailed by using materials selection stage. Although these two stages have different direction, they are synchronizing each other and can work concurrently.

4.3 Simulation

Once TMMs design process is established, one of the next steps is to perform simulation on the proposed design as a part of initial functional testing and verification before TMMs are fabricated. A simulation is a computer calculation performing a program that implements a mathematical model for the designed physical system [99]. This TMMs simulation is aimed as an efficient experimental way to obtain the predicted outcomes of the design computationally without working on the real TMMs fabrication process. The simulation results can help understand how the effect of body composition in the TMMs, particularly, in physical properties parameters. By employing several strategies of experimental protocols, simulation can suggest for making decision about the most optimum design. With this way, any inefficiency or errors can also be anticipated and solved in the design optimization process [99]. A number of numerical simulations have been used in simulation of any design to produce estimation values. The two popular numerical simulation methods, Finite Element Method (FEM) and Monte Carlo (MC), have been widely used for TMMs simulation [100–102]. The choice of method highly depends on the applications and needs of the TMMs. Table 3 shows comparison between these two simulation techniques in several imaging systems.

In this initial testing and verification, performing simulation is mainly meant to answer the research questions and objectives of the study to define the proper experimental protocol. These research questions and objectives also justify which numerical simulation method is suitable to describe the model. Before the simulation process is started, the following list of requirements are needed to ensure that the simulation is running properly

1. Model or prototype to be simulated
2. Simulation or computational modeling software/tool
3. Working principle of the model or prototype
4. Measurement parameters
5. External data to run the simulation
6. Experimental protocol of the simulation.

Table 3 Comparison of numerical simulations in the tissue properties study

	FEM	MC
Purpose	FEM is used to model complicated problems involving differential equations which generally have no analytical solution [104]	MC simulations are computer algorithms to calculate radiation interactions and energy deposition in the body [109]
Process	Define load, constraints or materials based on the analysis type (thermal, structural, or fluid). Then, boundary conditions are applied to represent the part will be analyzed	A random value is selected for each tasks based on the range of estimates. Then, it is calculated based on this random value. The result is recorded, and the process is repeated
Software	COMSOL, Solidworks, ABAQUS, ANSYS	C and Matlab programming
Application	– Layered soft tissue estimation using MRI [105] – Biomechanical estimation in soft tissue in US [106] – Pinch elastomer estimation for soft tissues [107] – Estimation of hyperelasticity soft materials [108]	– Patient dose estimation and imaging phantom [110] – Radiation dose estimation in CT and X-rays [111]

The model or prototype is the main aspect, as it defines the suitable tool, working principle, and parameters. Determination of physical properties in the measurement parameter and appropriate external data from the reference man may provide a more realistic basis for numerical simulations of biological structures. Also, the use of realistic 3D models from the references of real tissue can be beneficial in generating a more comprehensive and accurate solution [103].

4.4 Materials Selection

4.4.1 Synthesis Mapping Process

When the simulation process works to provide the best and optimum design for the phantom, the materials selection process works to determine the best materials for the prepared phantom design. Basically, during the simulation process, this materials selection process can be combined all together with the simulation process by including the material properties as part of the parameters in the simulation process. However, separating these two processes independently can also be considered, especially when some non-technical aspects, such as material safety, cost, fabrication complexity, preservation, clinical applicability, market availability, and so on, need to be taken into account in choosing the best materials. Furthermore, in order to obtain a more comprehensive materials selection outcome, it is advisable to develop a synthesis mapping based on the observation, materials nomination, and selection

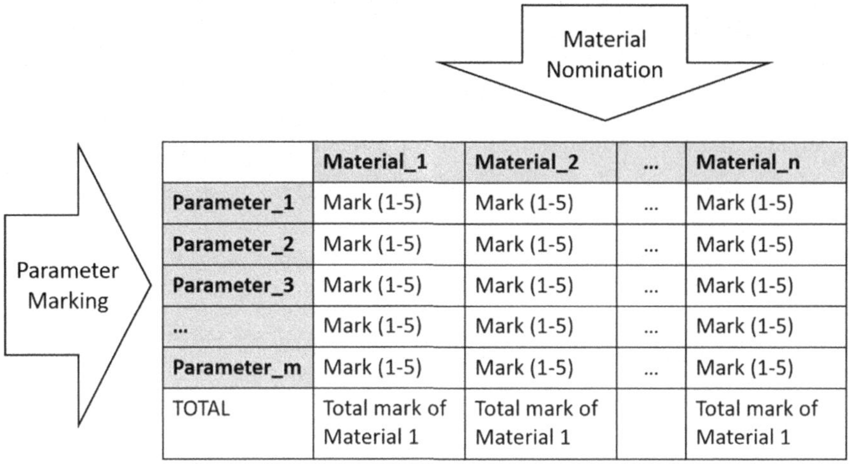

	Material_1	**Material_2**	**...**	**Material_n**
Parameter_1	Mark (1-5)	Mark (1-5)	...	Mark (1-5)
Parameter_2	Mark (1-5)	Mark (1-5)	...	Mark (1-5)
Parameter_3	Mark (1-5)	Mark (1-5)	...	Mark (1-5)
...	Mark (1-5)	Mark (1-5)	...	Mark (1-5)
Parameter_m	Mark (1-5)	Mark (1-5)	...	Mark (1-5)
TOTAL	Total mark of Material 1	Total mark of Material 1		Total mark of Material 1

Fig. 3 Synthesis mapping process in physical imaging phantoms development

method. In this way, the observed resources and selection method are mapped relative to the materials nomination. Thus, the contents of both parts are scaled accordingly.

The synthesis mapping is a part of design methodology that is used for understanding the problem by collecting various resources to determine the best option in a semi-quantitative manner [112]. This technique is applicable for materials selection and performed by generating a table consisting of selection method strategies in the row part, matched with the material nomination in the column part. Figure 3 describes a simple schematic development of synthesis mapping process. For defining the selection method, the strategies can be divided into two main parts, the material nomination and parameter marking strategies. It is suggested to work on material nomination first, then continued by parameter marking part.

The material nomination strategy is aimed to provide list of nominated materials that have been prioritized in the analysis process in the conception. Mostly, the analysis process can generate for more than ten materials. The steps can be described as follows:

1. *Observation*: Elaborate nominated materials from the analysis process.
2. *Review*: Collect information and analyze their material properties, such as types and characteristics of materials, composition of materials, and possibility for combining more than one materials.
3. *Pre-materials selection*: Narrow down the nominated material into only maximum three main materials.
4. *Plotting*: Map the three nominated materials into synthesis table in the column section. This is described by *Material_1*, *Material_2*, *Material_n*, and so on.
5. Go for parameter marking steps, then continue to the next steps.

6. *Synthesizing*: Fill up the synthesis table by connecting nominated materials in the column section with the parameters and scaling in the row section based on the available sources.
7. *Marking*: Calculate the total mark for each column. The material that has the highest mark is considered to be the best selected material.

The parameter marking strategy is aimed to generate the parameters used to assess the performance of the materials in a scaled marking. As illustrated in Fig. 3, components in the row section, from *Parameter_1*, *Parameter_2*, to *Parameter_m*, are the parameters produced from parameter marking steps as listed in this fashion:

1. *Characteristics analysis*: Define the requirement list about the phantom. Perceive the underlying characterization techniques and its clinical applications through literature study. Then correlate between basic requirement list and materials characteristics, covering biological, mechanical, chemical, and physical aspects.
2. *Parameterization*: From step 1, extract at least ten important parameters. Use simple words and include technical and non-technical aspects that can be used for selection method.
3. *Plotting*: Map the parameters into synthesis table in the row section.
4. *Scaling*: Determine marking value scale and describe the marking definition to be used in the synthesis for materials selection, for instance, 1–5 of mark value range, where 1 is poor, 2 is almost adequate, 3 is adequate, 4 is good, and 5 is excellent. Even this scaling is subjectively developed, each mark in each material needs to be scientifically and practically proven and verified.
5. Return back to Step 6 in the material nomination strategy.

From synthesis mapping process, it is important to note that this process requires extensive literature study and observation about the suitable materials in the analysis process in the conception. The analysis process produces requirement list about the imaging application. From this list, possible materials can be chosen and numerous characteristics of the materials can be defined accordingly. To ease the selection, the availability of databank for materials characterization is highly needed, especially for TMMs. This databank is very helpful in supporting the analysis and synthesis processes in the phantom design [112]. Besides databank for TMMs characteristics, databanks for the biological tissue characteristics are also needed. These biological tissue characteristics databanks can be used as the material characteristics reference that can be applied for both computational human model and physical human phantoms. Some efforts actually have been initiated, however, comprehensive information about tissue characteristics is still difficult to access, in some extent.

Even certain phantom materials have been preferably used in various studies, it does not always guarantee that these selected materials can be directly apt for all purposes. Some significant biases may possibly occur obscuring the material matching process. These biases can be due to the nature of the materials that have their own specific characteristics, or due to specific treatments, for instance physical exposures or preservation processes that may cause material degradation [113, 114]. The obvious impact of these biases may affect the TMMs application, such as unstandardized image quality in TMMs or unrealistic mechanical properties. Some physical, chemical, or electrical processes can be used as material restoration techniques to re-adjust the imaging properties of the materials. However, it strongly depends on the material capabilities [113, 114].

Material restoration techniques are basically comparable to those of properties alteration. Some processes have shown important impacts. Physical exposures, such as heating and cooling, can be used to modify imaging characteristics of the materials [113]. Chemical treatments, for instance mixing with additives, have also been found to improve visibility of base material under imaging system [115]. Also, adding some semiconductor additives to the base material can induce electrical processes, such as conductivity, that mimic dielectric properties to those of biological tissues [7]. However, defining the best process to alter the material properties needs to be done carefully and accurately, as some processes may be successful in adjusting the properties, but some others have some effects on other part of the TMMs, causing erroneous characteristics in the TMMs. In some ways, trade-offs in TMMs are inexorable. In one side, TMMs can be designed accurately to simulate the outer part of organ or tissue, but then the detailed internal anatomy must be sacrificed, and vice versa. Above all, materials selection in TMMs does not only need careful procedures, but also requires good knowledge background of materials to support the synthesis mapping process.

In accommodating these needs, based on the fabrication composition TMMs can be grouped into two parts, base material and additive material. The base material is the main material that builds the main component of TMMs and its characteristics dominate the TMMs. Additive material is specific material that is added to the base material to alter the characteristics of the base material so that it fits with the TMMs requirements or closes to that of the biological tissues. The uses of additive materials may range from improving the electrical, physical, chemical, and mechanical properties of the base material, material preservatives, to dramatically changing the characteristics of the base material, depending on the applications [116].

4.4.2 Polymers for TMMs

Polymers, chemical compounds consisting of repeated units of hydrocarbons and other materials, have been commonly used in all parts of daily life. Due to the ease of processing, high resistance to chemical, physical, and electrical penetrations, flexibility in forming and mixing with other materials, and recycle facilitation, polymers are preferred for TMMs since long time ago. Based on their source and content, polymers can be categorized into biological and synthetic materials. The biological or natural polymers mostly have high water content characteristics, while the synthetic polymers have low water content characteristics [117, 118]. Summary of polymers for TMMs is given in Table 4.

Even most TMMs are made from synthetic polymers, TMMs can also be produced from biological polymers or even biological tissues, especially when there is a specific need to have high degree of closeness to the biological tissues. However, some

Table 4 Summary of polymers for TMMs based on the source of materials related to their fabrication characteristics

Category	Materials	Advantages	Disadvantages	Stability
Biopolymer	Agarose [124, 134, 135]	Well-characterized performance	Limited to damage of structure	Days to weeks
	Gelatin [134, 136]	High stability and scatter uniformity	Instability with temperature variations	Days to weeks
	Gellan gum [137, 138]	Can be adjusted to different levels of elastic	Complex preparation	Years, depend on storage
Synthetic polymer	Polyurethane [139, 140]	Adjustable elasticity Long lasting	Challenging standardization technique	Years
	Silicone rubber [141–144]	Long lasting Easy to de-mold	Limited high attenuation and low speed sound	Years
	Polyvinyl alcohol [145–148]	Good elasticity	Challenging in preparation	Years, depend on storage

concerns need to be taken into account when implementing the biological TMMs, especially in material preservation and stability for long term uses. Another type of phantom uses combination of both synthetic and biological materials for the reasons of simplicity and closeness to the real object. Plastination, technique to preserve biological tissue, organ, or body, can be used as TMMs by undergoing mixing process with synthetic polymers layer by layer [119, 120].

Synthetic polymers have gained high attentions as TMMs in technical as well as medical applications. This is because a range of physical and chemical properties, such as monomer units, polymerization reaction, and formation of co-polymers at some concentrations, can be achieved, so that these polymers fulfil structural and mechanical properties for TMMs applications [121, 122]. Structurally, most synthetic polymers have more stable matrix composition [123]. Hence, these materials are easy to characterize. Highly advanced mechanical properties of synthetic polymers also enable for easy processing, flexible mixing with other materials, and shape deformation for adjustments into various characteristics, such as free shape deformation and plasticity upon special stimulus [124].

Specifically for cardiac imaging applications, cardiac TMMs must have imaging properties that are mimicking to those of biological cardiac tissues under certain imaging modality observation, particularly in the structural visibility of the cardiac tissues and morphological characteristics of the cardiac parts [125, 126]. In order to obtain the appropriate imaging properties, it is important to observe the underlying imaging physics of the imaging modality, the type of properties that can be captured, and how the materials are characterized in this imaging modality. Similar to imaging applications, TMMs for verifying cardiac mechanical properties also must be treated to mimic mechanical properties to those of biological cardiac tissues. Databanks of material property characterization may provide references for suitable materials for this purpose [112]. Similar to that of imaging properties, in order to improve the mechanical properties of TMMs, modification can also be made through physical, chemical, or electrical processes [127]. Recently, due to their advantages, there is growing interest in the use of materials from Silicone Rubber (SR) and polyvinyl alcohol cryogel (PVA-C) for cardiac TMMs.

Silicone Rubber

The molecular structure of silicone rubber (SR) consists of silicone, hydrogen and oxygen atoms. The SR structure is from organic groups (methyl, vinyl or phenyl groups) attached to silicon. SR is favored in many high technological applications, especially in phantom development, as it provides combination of thermal and chemical resistance with low risk to body reactions and elastomer properties, and also long-term stability [128].

Silicones can be divided into two categories, high-temperature vulcanizing types (HTV) and room-temperature vulcanizing types (RTV) [129]. Two-part RTV in SR are common materials that are commonly used in phantom. This type of silicone can be prepared from a two-component poly-dimethylsiloxane (PDMS) composition in platinum or catalyst, and cross-linked with materials by the addition hydrosilation mechanism [129]. Another kind of RTV can be prepared using linear hydroxyl terminated PDMS macromolecules and cross-linked with materials by low molecular weight tetra (alkyloxysilane) cross-linking agent, catalysed by stannous octoate on the basis of a condensation cure mechanism. Another method for cross-linking elastomers is irradiation curing. The procedure of mechanism for cross-linking SR reaction is based on the scission of C–H and C–Si bonds in the side chains [130].

Polyvinyl Alcohol (PVA) Cryogel

PVA is derived from the hydrolysis of poly(vinyl acetate), resulting in a final product that is a mixture of PVA and poly(vinyl acetate). An aqueous solution of high-grade PVA can be gelled into a cross-linked hydrogel by the formation of crystallites during repeated freeze/thaw cycles [131]. The stiffness properties of poly(vinyl alcohol) cryogel (PVA-C) can be manipulated by processing the material through a number of freeze/thaw cycles, varying the PVA concentration and by controlling the rate of thawing [132].

Although going into very lengthy processes, commercialized PVA powder has been widely used to make PVA liquids and solid by combining the PVA powder and de-ionized water through magnetic stirring, heating, cooling, molding, resting, freeze/thaw, de-frosting, and de-molding processes. However, with these long processes, it can only be kept for a maximum of few weeks. In order to avoid any biological contamination, a preservation technique is added [133, 134].

5 Documentation in Synthesis Processes

Although this documentation part is not a specific process in the systematic processes of physical imaging phantoms, documentation is crucially required in all research and also imperatively applicable in this study. As systematic processes involve extensive studies and experiments with a lot of scenarios, it is important to develop a good documentation management for recording all information produced along the processes. The uses of conventional methods, such as log book, labels, stationeries, and so on, are flexible and practical enough in documenting any ideas, conception process, design sketch, experimental schedules, list of questions, categorizing the materials, authentic research proof, and so forth. However, some issues, like environment, mobility, filing size, life time, reproducibility, and so forth, have limited the usages in wider scale and longer term research. Therefore, as digital facilities are nowadays becoming common, accessible, and considerably low cost in daily life, most detailed documentation is easily made and recorded through computer, cameras, videos, digital storage media, and many more. Certain software is also dedicated for this documentation purpose.

6 Summary

As a part of the big process in the quality assurance of cardiac imaging system, physical imaging phantom has progressed significantly together with the intensive developments of hardware and software in cardiac imaging system. Based on accuracy problems that occur in cardiac imaging systems, physical imaging phantom seems to be an effective solution in supporting the system in maintaining the quality of the cardiac procedures. With the feasibility of customizing the phantom design, various applications can be accurately verified in a standardized way. Through the proposed synthesis processes, the phantom development can be performed in a more standardized, measurable, simpler, and easier way. The preparation part of the synthesis processes, from conception to materials selection, results in the details of materials and strategic planning for the next steps of phantom development in fabrication, testing and analysis, and optimization. With these processes, the mimicking properties can be more ensured and any unexpected errors can be minimized.

Acknowledgements The authors are grateful for funding supports by Universiti Teknologi Malaysia and Ministry of Higher Education Malaysia under FRGS Grant R.J130000.7845.4F764 and GUP Tier 1 Grant Q.J130000.2545.20H36.

References

1. Emanuel, E.J.: Ethical and Regulatory Aspects of Clinical Research: Readings and Commentary. philpapers.org (2003)
2. Budoff, M., Cohen, M., Garcia, M., et al.: ACCF/AHA clinical competence statement on cardiac imaging with computed tomography and magnetic resonance. Circulation **112**(4), 598–617 (2005)
3. Hubbard, P.L., Zhou, F.-L., Eichhorn, S.J., Parker, G.J.M.: Biomimetic phantom for the validation of diffusion magnetic resonance imaging. Magn. Reson. Med. **73**(1), 299–305 (2015)
4. Koonce, J.D., Vliegenthart, R., Schoepf, U.J., et al.: Accuracy of dual-energy computed tomography for the measurement of iodine concentration using cardiac CT protocols: validation in a phantom model. Eur. Radiol. **24**, 512 (2014). https://doi.org/10.1007/s00330-013-3040-6
5. Hill, A.J., Iaizzo, P.A.: Comparative cardiac anatomy. In: Iaizzo, P. (ed.) Handbook of Cardiac Anatomy, Physiology, and Devices. Springer, Cham (2015)
6. Mathur, A., Ma, Z., Loskill, P., Jeeawoody, S., Healy, K.E.: In vitro cardiac tissue models: current status and future prospects. Adv. Drug Deliv. Rev. **96**(15), 203–213 (2016)
7. Garrett, J., Fear, E.: Stable and flexible materials to mimic the dielectric properties of human soft tissues. IEEE Antennas Wirel. Propag. Lett. **13**, 599–602 (2014)
8. Vannelli, C., Moore, J., McLeod, J., Ceh, D., Peters, T.: Dynamic heart phantom with functional mitral and aortic valves. In: Proceedings, Medical Imaging 2015: Image-Guided Procedures, Robotic Interventions, and Modeling, vol. 9415, p. 941503 (2015). https://doi.org/10.1117/12.2082277
9. Avendi, M.R., Kheradvar, A., Jafarkhani, H.: A combined deep-learning and deformable-model approach to fully automatic segmentation of the left ventricle in cardiac MRI. Med. Image Anal. **30**, 108–119 (2016)
10. Shahzad, R., Bos, D., Budde, R.P.J., Pellikaan, K., Niessen, W.J., van der Lugt, A., van Walsum, T.: Automatic segmentation and quantification of the cardiac structures from non-contrast-enhanced cardiac CT scans. Phys. Med. Biol. **62**(9)
11. Xu, R., Athavale, P., Nachman, A., Wright, G.A.: Multiscale registration of real-time and prior MRI data for image-guided cardiac interventions. IEEE Trans. Biomed. Eng. **61**(10), 2621–2633 (2014)
12. Abi-Jaoudeh, N., Kruecker, J., Kadoury, S., et al.: Multimodality image fusion-guided procedures: technique, accuracy, and applications. Cardiovasc. Intervent. Radiol. **35**, 986 (2012). https://doi.org/10.1007/s00270-012-0446-5
13. Turner, L.R., et al.: Cardiovascular disease screening in general practice: general practitioner recording of common risk factors. Prev. Med. **99**, 282–285 (2017)
14. Duffy, J.Y., et al.: Cardiovascular disease screening. Semin. Perinatol. **39**(4), 264–267 (2015)
15. Bekar, L., et al.: The preference of the physicians in diagnosis and treatment of cardiovascular diseases. Int. J. Cardiovasc. Acad. **3**(1–2), 11–15 (2017)
16. Liu, Y., et al.: Epidermal mechano-acoustic sensing electronics for cardiovascular diagnostics and human-machine interfaces. Sci. Adv. **2**(11), e1601185 (2016)
17. Danad, I., et al.: Diagnostic performance of cardiac imaging methods to diagnose ischaemia-causing coronary artery disease when directly compared with fractional flow reserve as a reference standard: a meta-analysis. Eur. Heart J. **38**(13), 991–998 (2017)

18. Bountry, C.M., et al.: A sensitive and biodegradable pressure sensor array for cardiovascular monitoring. Adv. Mater. **27**, 6954–6961 (2015)
19. Sun, J., et al.: Carotid magnetic resonance imaging for monitoring atherosclerotic plaque progression: a multicenter reproducibility study. Int. J. Cardiovasc. Imaging **31**, 95 (2015)
20. Noc, M., et al.: Invasive coronary treatment strategies for out-of-hospital cardiac arrest: a consensus statement from the European Association for Percutaneous Cardiovascular Interventions (EAPCI)/Stent for Life (SFL) groups. EuroIntervention **10**, 31–37 (2014)
21. Piepoli, M.F., et al.: 2016 European guidelines on cardiovascular disease prevention in clinical practice. Atherosclerosis **252**, 207–274 (2016)
22. Catapano, A.L., et al.: 2016 ESC/EAS guidelines for the management of dyslipidaemias. Atherosclerosis **253**, 281–344 (2016)
23. Arena, R., et al.: Healthy lifestyle interventions to combat noncommunicable disease—a novel nonhierarchical connectivity model for key stakeholders: a policy statement from the American Heart Association, European Society of Cardiology, European Association for Cardiovascular Prevention and Rehabilitation, and American College of Preventive Medicine. Mayo Clin. Proc. **90**(8), 1082–1103 (2015)
24. Kersten-Oertel, M., Jannin, P., Collins, D.L.: The state of the art of visualization in mixed reality image guided surgery. Comput. Med. Imaging Graph. **37**(2), 98–112 (2013). https://doi.org/10.1016/j.compmedimag.2013.01.009
25. Badano, L.P., Miglioranza, M.H., Edvardsen, T., Colafranceschi, A.S., Muraru, D., Bacal, F., Nieman, K., Zoppellaro, G., Marcondes Braga, F.G., Binder, T., Habib, G., Lancellotti, P., Document reviewers Sicari, R., Cosyns, B., Donal, E., Lombardi, M., Sarvari, S.: European Association of Cardiovascular Imaging/Cardiovascular Imaging Department of the Brazilian Society of Cardiology recommendations for the use of cardiac imaging to assess and follow patients after heart transplantation. Eur. Heart J. Cardiovasc. Imaging **16**(9), 919–948 (2015). https://doi.org/10.1093/ehjci/jev139
26. Swift, A.J., Rajaram, S., Condliffe, R., Capener, D., Hurdman, J., Elliot, C.A., Wild, J.M., Kiely, D.G.: Diagnostic accuracy of cardiovascular magnetic resonance imaging of right ventricular morphology and function in the assessment of suspected pulmonary hypertension results from the ASPIRE registry. J. Cardiovasc. Magn. Reson. **14**, 40 (2012). https://doi.org/10.1186/1532-429X-14-40
27. Safavi, K.C., Li, S., Dharmarajan, K., et al.: Hospital variation in the use of noninvasive cardiac imaging and its association with downstream testing, interventions, and outcomes. JAMA Int. Med. **174**(4), 546–553 (2014). https://doi.org/10.1001/jamainternmed.2013.14407
28. Hasan, M.A., Lee, S.-L., Kim, D.-H., Lim, M.-K.: Automatic evaluation of cardiac hypertrophy using cardiothoracic area ratio in chest radiograph images. Comput. Methods Programs Biomed. **105**(2), 95–108 (2012)
29. Dimopoulos, K., Giannakoulas, G., Bendayan, I., Liodakis, E., Petraco, R., Diller, G.-P., Piepoli, M.F., Swan, L., Mullen, M., Best, N., Poole-Wilson, P.A., Francis, D.P., Rubens, M.B., Gatzoulis, M.A.: Cardiothoracic ratio from postero-anterior chest radiographs: a simple, reproducible and independent marker of disease severity and outcome in adults with congenital heart disease. Int. J. Cardiol. **166**(2), 453–457 (2013). https://doi.org/10.1016/j.ijcard.2011.10.125
30. Venugopal, A.N., Koshy, R.C., Koshy, S.M.: Role of chest X-ray in citing central venous catheter tip: a few case reports with a brief review of the literature. J. Anaesthesiol. Clin. Pharmacol. **29**(3), 397–400 (2013). https://doi.org/10.4103/0970-9185.117114
31. Ubeda, C., Vano, E., Gonzalez, L., Miranda, P.: Influence of the antiscatter grid on dose and image quality in pediatric interventional cardiology X-ray systems. Cathet. Cardiovasc. Intervent. **82**(1), 51–57 (2013). https://doi.org/10.1002/ccd.24602
32. Schoenhagen, P., Halliburton, S.S., Stillman, A.E., Kuzmiak, S.A., Nissen, S.E., Tuzcu, E.M., White, R.D.: Noninvasive imaging of coronary arteries: current and future role of multi–detector row CT. Radiology **232**(1) (2004). https://doi.org/10.1148/radiol.2321021803
33. Wilson, J., Saremi, F., Narula, J., Narayan, S.M.: CT in the management of cardiac arrhythmias. In: Budoff, M., Achenbach, S., Hecht, H., Narula, J. (eds.) Atlas of Cardiovascular Computed Tomography. Springer, London (2018)

34. Sun, Z., Ng, K.H.: Diagnostic value of coronary CT angiography with prospective ECG-gating in the diagnosis of coronary artery disease: a systematic review and meta-analysis. Int. J. Cardiovasc. Imaging **28**, 2109 (2012). https://doi.org/10.1007/s10554-011-0006-0
35. Morin, R.L., Gerber, T.C., McCollough, C.H.: Radiation dose in computed tomography of the heart. Circulation **107**, 917–922 (2003). https://doi.org/10.1161/01.CIR.0000048965.56529.C2
36. Ulzheimer, S., Kalender, W.A.: Assessment of calcium scoring performance in cardiac computed tomography. Eur. Radiol. **13**, 484–497 (2003). https://doi.org/10.1007/s00330-002-1746-y
37. Sun, Z., Al Moudi, M., Cao, Y.: CT angiography in the diagnosis of cardiovascular disease: a transformation in cardiovascular CT practice. Quant. Imaging Med. Surg. **4**(5), 376–396 (2014). https://doi.org/10.3978/j.issn.2223-4292.2014.10.02
38. Scholtz, J.-E., Ghoshhajra, B.: Advances in cardiac CT contrast injection and acquisition protocols. Cardiovasc. Diagn. Ther. **7**(5), 439–451 (2017). https://doi.org/10.21037/cdt.2017.06.07
39. Baker, J.E., Moulder, J.E., Hopewell, J.W.: Radiation as a risk factor for cardiovascular disease. Antioxid. Redox Signal. **15**(7), 1945–1956 (2011). https://doi.org/10.1089/ars.2010.3742
40. Thai, W., Wai, B., Lin, K., et al.: Pulmonary venous anatomy imaging with low-dose, prospectively ECG-triggered, high-pitch 128-slice dual source computed tomography. Circ. Arrhythm. Electrophysiol. **5**(3), 521–530 (2012). https://doi.org/10.1161/CIRCEP.111.968313
41. Muraru, D., Niero, A., Rodriguez-Zanella, H., Cherata, D., Badano, L.: Three-dimensional speckle-tracking echocardiography: benefits and limitations of integrating myocardial mechanics with three-dimensional imaging. Cardiovasc. Diagn. Ther. **8**(1), 101–117 (2018). https://doi.org/10.21037/cdt.2017.06.01
42. Armstrong, A.C., Ricketts, E.P., Cox, C., et al.: Quality control and reproducibility in M-mode, two-dimensional, and speckle tracking echocardiography acquisition and analysis: the CARDIA study, year-25 examination experience. Echocardiography (Mount Kisco, NY) **32**(8), 1233–1240 (2015). https://doi.org/10.1111/echo.12832
43. Badano, L.P.: The clinical benefits of adding a third dimension to assess the left ventricle with echocardiography. Scientifica **2014**, 897431 (2014). https://doi.org/10.1155/2014/897431
44. Bencsik, G.: Novel strategies in the ablation of typical atrial flutter: role of intracardiac echocardiography. Curr. Cardiol. Rev. **11**(2), 127–133 (2015). https://doi.org/10.2174/1573403X10666141013121843
45. Soloperto, G., Casciaro, S.: Progress in atherosclerotic plaque imaging. World J. Radiol. **4**(8), 353–371 (2012). https://doi.org/10.4329/wjr.v4.i8.353
46. Vignali, L., Solinas, E., Emanuele, E.: Research and clinical applications of optical coherence tomography in invasive cardiology: a review. Curr. Cardiol. Rev. **10**, 369–376 (2014). https://doi.org/10.2174/1573403X10666140604120753
47. Suter, M.J., Nadkarni, S.K., Weisz, G., et al.: Intravascular optical imaging technology for investigating the coronary artery (2011). https://doi.org/10.1016/j.jcmg.2011.03.020
48. Peterzan, M.A., Rider, O.J., Anderson, L.J.: The role of cardiovascular magnetic resonance imaging in heart failure. Card. Fail. Rev. **2**(2), 115–122 (2016). https://doi.org/10.15420/cfr.2016.2.2.115
49. Parsai, C., O'Hanlon, R., Prasad, S.K., Mohiaddin, R.H.: Diagnostic and prognostic value of cardiovascular magnetic resonance in non-ischaemic cardiomyopathies. J. Cardiovasc. Magn. Reson. **14**(1), 54 (2012). https://doi.org/10.1186/1532-429X-14-54
50. Sherrah, A.G., Grieve, S.M., Jeremy, R.W., Bannon, P.G., Vallely, M.P., Puranik, R.: MRI in chronic aortic dissection: a systematic review and future directions. Front. Cardiovasc. Med. **2**, 5 (2015). https://doi.org/10.3389/fcvm.2015.00005
51. Gulenchyn, K., McEwan, A., Freeman, M., Kiess, M., O'Neill, B., Beanlands, R.: Treating the right patient at the right time: access to cardiovascular nuclear imaging. Can. J. Cardiol. **22**(10), 827–833 (2006)

52. Boogers, M.J., Fukushima, K., Bengel, F.M., Bax, J.J.: The role of nuclear imaging in the failing heart: myocardial blood flow, sympathetic innervation, and future applications. Heart Fail. Rev. **16**, 411–423 (2011). https://doi.org/10.1007/s10741-010-9196-0
53. Merhige, M.E., Breen, W.J., Shelton, V., et al.: Impact of myocardial perfusion imaging with PET and (82)Rb on downstream invasive procedure utilization, costs, and outcomes in coronary disease management. J. Nucl. Med. **48**, 1069–1076 (2007). https://doi.org/10.2967/jnumed.106.038323
54. Kircher, M., Lapa, C.: Novel noninvasive nuclear medicine imaging techniques for cardiac inflammation. Curr. Cardiovasc. Imaging Rep. **10**(2), 6 (2017). https://doi.org/10.1007/s12410-017-9400-x
55. Li, T., Ao, E.C.I., Lambert, B., Brans, B., Vandenberghe, S., Mok, G.S.P.: Quantitative imaging for targeted radionuclide therapy dosimetry—technical review. Theranostics **7**(18), 4551–4565 (2017). https://doi.org/10.7150/thno.19782
56. Pan, J.A., Salerno, M.: Clinical utility and future applications of PET/CT and PET/CMR in cardiology. In: Kjaer, A. (ed.) Diagnostics **6**(3), 32 (2016). https://doi.org/10.3390/diagnostics6030032
57. Zhuang, H., Codreanu, I.: Growing applications of FDG PET-CT imaging in non-oncologic conditions. J. Biomed. Res. **29**(3), 189–202 (2015). https://doi.org/10.7555/JBR.29.20140081
58. Zhenzhen, X., Tao, B., Li, Y., et al.: 3D fusion framework for infarction and angiogenesis analysis in a myocardial infarct minipig model. Mol. Imaging **16**, 1536012117708735 (2017). https://doi.org/10.1177/1536012117708735
59. Woo, J., Stone, M., Prince, J.L.: Multimodal registration via mutual information incorporating geometric and spatial context. IEEE Trans. Image Process. **24**(2), 757–769 (2015). https://doi.org/10.1109/TIP.2014.2387019
60. Nordenfur, T., Babic, A., Bulatovic, I., Giesecke, A., Günyeli, E., Ripsweden, J., Samset, E., Winter, R., Larsson, M.: Method comparison for cardiac image registration of coronary computed tomography angiography and 3-D echocardiography. J. Med. Imaging **5**(1), 014001 (2018). https://doi.org/10.1117/1.JMI.5.1.014001
61. Shrestha, U.M., Seo, Y., Botvinick, E.H., Gullberg, G.T.: Image reconstruction in higher dimensions: myocardial perfusion imaging of tracer dynamics with cardiac motion due to deformation and respiration. Phys. Med. Biol. **60**(21), 8275–8301 (2015). https://doi.org/10.1088/0031-9155/60/21/8275
62. Lopez-Perez, A., Sebastian, R., Ferrero, J.M.: Three-dimensional cardiac computational modelling: methods, features and applications. BioMed. Eng. OnLine **14**, 35 (2015). https://doi.org/10.1186/s12938-015-0033-5
63. Fonseca, C.G., Backhaus, M., Bluemke, D.A., Britten, R.D., Chung, J.D., Cowan, B.R., Dinov, I.D., Finn, J.P., Hunter, P.J., Kadish, A.H., Lee, D.C., Lima, J.A.C., Medrano-Gracia, P., Shivkumar, K., Suinesiaputra, A., Tao, W., Young, A.A.: The Cardiac Atlas Project—an imaging database for computational modeling and statistical atlases of the heart. Bioinformatics **27**(16), 2288–2295 (2011). https://doi.org/10.1093/bioinformatics/btr360
64. Wong, M.D., Wu, X., Liu, H.: Image quality and dose efficiency of high energy phase sensitive X-ray imaging: phantom studies. J. X-ray Sci. Technol. **22**(3), 321–334 (2014). https://doi.org/10.3233/XST-140428
65. Ceh, J., Youd, T., Mastrovich, Z., et al.: Bismuth infusion of ABS enables additive manufacturing of complex radiological phantoms and shielding equipment. In: Choi, J.-W., Engeberg, E.D. (eds.) Sensors (Basel, Switzerland) **17**(3), 459 (2017). https://doi.org/10.3390/s17030459
66. Negron, L.A., Viola, F., Black, E.P., Toth, C.A., Walker, W.F.: Development and characterization of a vitreous mimicking material for radiation force imaging. IEEE Trans. Ultrason. Ferroelectr. Freq. Control **49**(11), 1543–1551 (2002)
67. Rozenkrantz, A.B., Mendiratta-Lala, M., Bartholmai, B.J., Ganeshan, D., Abransom, R.G., Burton, K.R., Yu, J.-P.J., Scalzetti, E.M., Yankeelov, T.E., Subramaniam, R.M., Lenchik, L.: Clinical utility of quantitative imaging. Acad. Radiol. **22**, 33–49 (2015)

68. De Grand, A.M., Lomnes, S.J., Lee, D.S., et al.: Tissue-like phantoms for near-infrared fluorescence imaging system assessment and the training of surgeons. J. Biomed. Opt. **11**(1), 014007 (2006). https://doi.org/10.1117/1.2170579
69. Boltz, T., Pavlicek, W., Paden, R., Renno, M., Jensen, A., Akay, M.: An anthropomorphic beating heart phantom for cardiac X-ray CT imaging evaluation. J. Appl. Clin. Med. Phys. (2010). https://doi.org/10.1120/jacmp.v11i1.3129
70. Nattagh, K., Siauw, T., Pouliot, J., Hsu, I.C., Cunha, J.A.: A training phantom for ultrasound-guided needle insertion and suturing. Brachytherapy **13**(4), 413–419 (2014). https://doi.org/10.1016/j.brachy.2014.01.003. Epub 12 Feb 2014
71. Seegenschmiedt, M.H., et al.: Thermoradiotherapy and Thermochemotherapy, Volume 1: Biology, Physiology, Physics. Springer-Verlag, Berlin (1995)
72. Boutchko, R., Balakrishnan, K., Gullberg, G.T., O'Neil, J.P.: Human torso phantom for imaging of heart with realistic modes of cardiac and respiratory motion. US8535061B2, US Patent 2007
73. Verkerke, G.J., van der Houwen, E.B.: Design of biomedical products. In: Rakhorst, G., Ploeg, R. (eds.) Biomaterials in Modern Medicine: The Groningen Perspective, pp. 23–38. Biomechanical Engineering, World Scientific Publishing (2008)
74. Jones, P., Bowes, J.: Rendering systems visible for design: synthesis maps as constructivist design narratives. She Ji J. Des. Econ. Innov. **3**(3), 229–248 (2017). https://doi.org/10.1016/j.sheji.2017.12.001
75. Garcia, J., Yang, Z., Mongrain, R., et al.: 3D printing materials and their use in medical education: a review of current technology and trends for the future. BMJ Simul. Technol. Enhanc. Learn. (2017). https://doi.org/10.1136/bmjstel-2017-000234
76. Zainon, R.: Design and fabrication of multipurpose smart phantom for positron emission tomography/computed tomography imaging (2008)
77. Shikhaliev, P.M.: Dedicated phantom materials for spectral radiography and CT. Phys. Med. Biol. **57**(6), 1575–1593 (2012). https://doi.org/10.1088/0031-9155/57/6/1575. Epub 7 Mar 2012
78. Park, S., Lee, J.K., Kim, J.I., Lee, Y.J., Lim, Y.K., Kim, C.S., Lee, C.: In vivo organ mass of Korean adults obtained from whole-body magnetic resonance data. Radiat. Prot. Dosimetry **118**(3), 275–279 (2006). https://doi.org/10.1093/rpd/nci340
79. Bosgra, S., van Eijkeren, J., Bos, P., Zeilmaker, M., Slob, W.: An improved model to predict physiologically based model parameters and their inter-individual variability from anthropometry. Crit. Rev. Toxicol. **42**(9) (2012)
80. Del Bianco, S., Martelli, F., Cignini, F., Zaccanti, G., Pifferi, A., Torricelli, A., Bassi, A., Taroni, P., Cubeddu, R.: Liquid phantom for investigating light propagation through layered diffusive media. Opt. Express **12**, 2102–2111 (2004)
81. Fieseler, M., Kugel, H., Gigengack, F., Kösters, T., Büther, F., Quick, H.H., Faber, C., Jiang, X., Schäfers, K.P.: A dynamic thorax phantom for the assessment of cardiac and respiratory motion correction in PET/MRI: a preliminary evaluation. Nucl. Instrum. Methods Phys. Res. A **702**, 59–63 (2013)
82. Boote, E., Fent, G., Kattumuri, V., Casteel, S., Katti, K., Chanda, N., Kannan, R., Katti, K., Churchill, R.: Gold nanoparticle contrast in a phantom and juvenile swine: models for molecular imaging of human organs using X-ray computed tomography. Acad. Radiol. **17**(4), 410–417 (2010)
83. Abdullah, K.A., McEntee, M.F., Reed, W., Kench, P.L.: Development of an organ-specific insert phantom generated using a 3D printer for investigations of cardiac computed tomography protocols. J. Med. Radiat. Sci. (2018). https://doi.org/10.1002/jmrs.279
84. Lubis, L.E., Craig, L.A., Bosmans, H., Soejoko, D.S.: Task-based phantom evaluation of cardiac catheterization imaging modes. Phys. Med. **46**, 114–123 (2018). https://doi.org/10.1016/j.ejmp.2018.02.002. Epub 5 Feb 2018
85. Tavakoli, V., Kendrick, M., Shakeri, M., Alshaher, M., Stoddard, M.F., Amini, A.: A multimodal (MRI/ultrasound) cardiac phantom for imaging experiments. In: Proceedings SPIE 8672, Medical Imaging 2013: Biomedical Applications in Molecular, Structural, and Functional Imaging, p. 867202, 29 Mar 2013. https://doi.org/10.1117/12.2008783

86. http://www.cirsinc.com/
87. https://www.kyotokagaku.com/
88. Thakur, Y., Nikolov, H.N., Gulka, B., Holdsworth, D.W., Drangova, M.: Design and construction of a multipath vessel phantom for interventional training. Br. J. Radiol. **83**(995), 979–982 (2010). https://doi.org/10.1259/bjr/91767642
89. Ventola, C.L.: Medical applications for 3D printing: current and projected uses. P&T **39**(10), 704–711 (2014)
90. Abayazid, M., Kemp, M., Misra, S.: 3D flexible needle steering in soft-tissue phantoms using fiber bragg grating sensors. In: 2013 IEEE International Conference on Robotics and Automation, 6–10 May 2013. https://doi.org/10.1109/icra.2013.6631418
91. Yokoyama, K., Nakagawa, H., Shah, D.C., Lambert, H., Leo, G., Aeby, N., Ikeda, A., Pitha, J.V., Sharma, T., Lazzara, R., et al.: Novel contact force sensor incorporated in irrigated radiofrequency ablation catheter predicts lesion size and incidence of steam pop and thrombus. Circ. Arrhythm. Electrophysiol. **1**, 354–362 (2008)
92. Magnetic free MRI phantom: Madsen, E.L., Fullerton, G.D.: Prospective tissue-mimicking materials for use in NMR imaging phantoms. Magn. Reson. Imaging **1**(3), 135–141 (1982)
93. Cygan, S., Werys, K., Błaszczyk, Ł., Kubik, T., Kałużyński, K.: Left ventricle phantom and experimental setup for MRI and echocardiography—preliminary results of data acquisitions. Biocybern. Biomed. Eng. **34**(1), 19–24 (2014). https://doi.org/10.1016/j.bbe.2013.12.002
94. Stabin, M.G., Xu, X.G., Emmons, M.A., Segars, W.P., Shi, C., Fernald, M.J.: RADAR reference adult, pediatric, and pregnant female phantom series for internal and external dosimetry. J. Nucl. Med. **53**(11), 1807–1813 (2012). https://doi.org/10.2967/jnumed.112.106138
95. Kim, J.I.: Physical phantom of typical Korean male for radiation protection purpose. Radiat. Prot. Dosimetry **118**, 131–136 (2005). https://doi.org/10.1093/rpd/nci338
96. http://www.simutec.com/
97. https://humimic.com/
98. Bolwin, K., Czekalla, B., Frohwein, L.J., Büther, F., Schäfers, K.P.: Anthropomorphic thorax phantom for cardio-respiratory motion simulation in tomographic imaging. Phys. Med. Biol. **63**, 035009 (2018)
99. Jan, S., Benoit, D., Becheva, E., Carlier, T., Cassol, F., Descourt, P., Frisson, T., Grevillot, L., Guigues, L., Maigne, L.: GATE V6: a major enhancement of the GATE simulation platform enabling modelling of CT and radiotherapy. Phys. Med. Biol. **56**, 881 (2011)
100. Veress, A.I., Segars, W.P., IEEE Member, Tsui, B.M.W., IEEE Fellow, Gullberg, G.T.: Incorporation of a left ventricle finite element model defining infarction into the XCAT imaging phantom. IEEE Trans. Med. Imaging **30**(4), 915 (2011)
101. Paganetti, H.: Range uncertainties in proton therapy and the role of Monte Carlo simulations. Phys. Med. Biol. **57**, R99 (2012)
102. Fang, Q.: Mesh-based Monte Carlo method using fast ray-tracing in Plücker coordinates. Biomed. Opt. Express **1**(1), 165–175 (2010). https://doi.org/10.1364/BOE.1.000165
103. Dobre, A.A., Morega, A.M., Morega, M.: The investigation of flow—structural interaction in an arterial branching by numerical simulation. Proc. IEEE/EMBS Reg. 8 Int. Conf. Inf. Technol. Appl. Biomed. ITAB, 4–7 (2010). https://doi.org/10.1109/itab.2010.5687648
104. Kung, E.O., Les, A.S., Figueroa, C.A., et al.: In vitro validation of finite element analysis of blood flow in deformable models. Ann. Biomed. Eng. **39**, 1947 (2011). https://doi.org/10.1007/s10439-011-0284-7
105. Tada, M., Nagai, N., Maeno, T.: Material properties estimation of layered soft tissue based on MR observation and iterative FE simulation. Med. Image Comput. Comput. (2005)
106. Han, L., Noble, J.A., Burcher, M.: A novel ultrasound indentation system for measuring biomechanical properties of in vivo soft tissue. Ultrasound Med. Biol. **29**, 813–823 (2003). https://doi.org/10.1016/S0301-5629(02)00776-7
107. Harrison, S.M., Bush, M.B., Petros, P.E.: A pinch elastometer for soft tissue. Med. Eng. Phys. **29**, 307–315 (2007). https://doi.org/10.1016/j.medengphy.2006.03.011
108. Zhang, M.G., Cao, Y.P., Li, G.Y., Feng, X.Q.: Spherical indentation method for determining the constitutive parameters of hyperelastic soft materials. Biomech. Model. Mechanobiol. **13**, 1–11 (2014). https://doi.org/10.1007/s10237-013-0481-4

109. Dewerd, L.A.: The Phantoms of Medical and Health Physics. M. Kissick (ed.). Springer, Berlin (2014)
110. Johnson, P.B., Geyer, A., Borrego, D., et al.: The impact of anthropometric patient-phantom matching on organ dose: a hybrid phantom study for fluoroscopy guided interventions. Med. Phys. **38**, 1008–1017 (2011). https://doi.org/10.1118/1.3544353
111. Akhlaghi, P., Miri, H., Motavalli, L.R.: Determination of tissue equivalent materials of a physical 8-year-old phantom for use in computed tomography. Radiat. Phys. Chem. **112**, 169–176 (2015). https://doi.org/10.1016/j.radphyschem.2015.03.030
112. Brundle, C.R., Evans, C.A., Wilson, S.: Encyclopedia of Materials Characterization: Surfaces, Interfaces, Thin Films. Elsevier (1992)
113. Zalba, B., Marın, J.M., Cabeza, L.F., Mehling, H.: Review on thermal energy storage with phase change: materials, heat transfer analysis and applications. Appl. Therm. Eng. **23**(3), 251–283 (2003)
114. Madsen, E.L., Zagzebski, J.A., Banjavie, R.A., Jutila, R.E.: Tissue mimicking materials for ultrasound phantoms. Med. phy. **5**(5), 391–394 (1978)
115. Yusof, N.S.M., Dewi, D.E.O., Faudzi, A.A.M., Salih, N.M., Bakar, N.A., Hamid, H.A.: Ultrasound imaging characterization on tissue mimicking materials for cardiac tissue phantom: texture analysis perspective. MJFAS (2017)
116. Vogt, W.C., Jia, C., Wear, K.A., Garra, B.S., Pfefer, T.J.: Biologically relevant photoacoustic imaging phantoms with tunable optical and acoustic properties. J. Biomed. Opt. **21**(10), 101405 (2016). https://doi.org/10.1117/1.JBO.21.10.101405
117. Kawaguchi, Y., Iwazaki, H., Ida, T., Nishi, T., Tanikawa, Y., Nitta, N.: New polymer-based phantom for photoacoustic imaging. In: Proceedings Volume 8945, Design and Performance Validation of Phantoms Used in Conjunction with Optical Measurement of Tissue, vol. VI, p. 89450A (2014) https://doi.org/10.1117/12.2037517
118. Hron, P.: Hydrophilisation of silicone rubber for medical applications. Polym. Int. **52**, 1531–1539 (2003). https://doi.org/10.1002/pi.1273
119. Latorre, R., Bainbridge, D., Tavernor, A., López Albors, O.: Plastination in anatomy learning: an experience at Cambridge University. J. Vet. Med. Educ. **43**(3). https://doi.org/10.3138/jvme.0715-113r1
120. Yoon, S., Henry, R., Bouley, D., Bennett, N., Fahrig, R.: Characterization of a novel anthropomorphic plastinated lung phantom. Med. Phys. **35**, 5934–5943 (2008)
121. Shih, C.-T., Hsu, J.-T., Han, R.-P., Hsieh, B.-T., Chang, S.-J., Wu, J.: A novel method of estimating dose responses for polymer gels using texture analysis of scanning electron microscopy images. PLoS One **8**(7), e67281 (2013). https://doi.org/10.1371/journal.pone.0067281
122. Jiang, S., Liu, S., Feng, W.: PVA hydrogel properties for biomedical application. J. Mech. Behav. Biomed. Mater. **4**(7), 1228–1233 (2011). https://doi.org/10.1016/j.jmbbm.2011.04.005
123. Wang, R.-M., Zheng, S.-R.: Polymer Matrix Composites and Technology. Woodhead Publishing (2011)
124. Maitz, M.F.: Applications of synthetic polymers in clinical medicine. Biosurf. Biotribol. **1**, 161–176 (2015). https://doi.org/10.1016/j.bsbt.2015.08.002
125. Fisher, R.F., Hintenlang, D.E.: Super-size me: adipose tissue-equivalent additions for anthropomorphic phantoms. J. Appl. Clin. Med. Phys. **15**(6), 306–312 (2014). https://doi.org/10.1120/jacmp.v15i6.5007
126. Haddad, R., Clarysse, P., Orkisz, M., Croisille, P., Revel, D., Magnin, I.E.: A realistic anthropomorphic dynamic heart phantom. Comput. Cardiol. (2005). https://doi.org/10.1109/cic.2005.1588226
127. Dąbrowska, A.K., Rotaru, G.M., Derler, S., Spano, F., Camenzind, M., Annaheim, S., Rossi, R.M. et al.: Materials used to simulate physical properties of human skin. Skin Res. Technol, **22**(1), 3–14 (2016)
128. Ayers, F., et al.: Fabrication and characterization of silicone-based tissue phantoms with tunable optical properties in the visible and near infrared domain. Proc. SPIE **6870**, 1–9 (2008)

129. Yoda, R.: Elastomers for biomedical applications. J. Biomater. Sci., 37–41 (2012)
130. Manik, S.P., Banerjee, S.: Determination of chemical cross-links in rubbers. Macromol. Mater. Eng. **6**(71), 171–178 (1979)
131. Jiang, H., Campbell, G., Boughner, D., et al.: Design and manufacture of a polyvinyl alcohol (PVA) cryogel tri-leaflet heart valve prosthesis. Med. Eng. Phys. **26**, 269–277 (2004). https://doi.org/10.1016/j.medengphy.2003.10.007
132. Surry, K.J.M., Austin, H.J.B., Fenster, A., Peters, T.M.: Poly(vinyl alcohol) cryogel phantoms for use in ultrasound and MR imaging. Phys. Med. Biol. **49**, 5529–5546 (2004). https://doi.org/10.1088/0031-9155/49/24/009
133. Zhou, X., Kenwright, D.A., Wang, S., Hossack, J.A., Hoskins, P.R.: Fabrication of two flow phantoms for Doppler ultrasound imaging. IEEE Trans. Ultrason. Ferroelectr. Freq. Control **64**(1), 53–65 (2017). https://doi.org/10.1109/tuffc.2016.2634919
134. Culjat, M.O., Goldenberg, D., Tewari, P., Singh, R.S.: A review of tissue substitutes for ultrasound imaging. Ultrasound Med. Biol. **36**, 861–873 (2010). https://doi.org/10.1016/j.ultrasmedbio.2010.02.012
135. Zell, K., Sperl, J.I., Vogel, M.W., et al.: Acoustical properties of selected tissue phantom materials for ultrasound imaging. Phys. Med. Biol. **52**, N475–N484 (2007). https://doi.org/10.1088/0031-9155/52/20/N02
136. Manickam, K., Machireddy, R.R., Seshadri, S.: Study of ultrasound stiffness imaging methods using tissue mimicking phantoms. Ultrasonics **54**, 621–631 (2014). https://doi.org/10.1016/j.ultras.2013.08.018
137. Chen, R., Shih, A.: Multi-modality gellan gum-based tissue-mimicking phantom with targeted mechanical, electrical, and thermal properties. Phys. Med. Biol. **58**, 5511–5525 (2013). https://doi.org/10.1088/0031-9155/58/16/5511
138. Wang, Y., Tai, B.L., Yu, H., Shih, A.J.: Silicone-based tissue-mimicking phantom for needle insertion simulation. J. Med. Devices **8**, 021001 (2014). https://doi.org/10.1115/1.4026508
139. Fallis, A.: Polyurethane as a base for a family of tissue equivalent materials. J. Chem. Inf. Model. **53**, 1689–1699 (2013). https://doi.org/10.1017/CBO9781107415324.004
140. Fromstein, J.D., Woodhouse, K.A.: Elastomeric biodegradable polyurethane blends for soft tissue applications. J. Biomater. Sci. Polym. Ed. **13**, 391–406 (2002). https://doi.org/10.1163/156856202320253929
141. Pogue, B.W., Patterson, M.S.: Review of tissue simulating phantoms for optical spectroscopy, imaging and dosimetry. J. Biomed. Opt. **11**, 041102-1–041102-16 (2006). https://doi.org/10.1117/1.2335429
142. Cafarelli, A., Miloro, P., Verbeni, A., et al.: Speed of sound in rubber-based materials for ultrasonic phantoms. J. Ultrasound **19**, 251–256 (2016). https://doi.org/10.1007/s40477-016-0204-7
143. Martins, P.A.L.S., Jorge, R.M.N., Ferreira, A.J.M.: A comparative study of several material models for prediction of hyperelastic properties: application to silicone-rubber and soft tissues. Strain, 135–147 (2006)
144. Maggi, L.E., Von Krüger, M.A., Pereira, W.C.A., Monteiro, E.E.C.: Development of silicon-based materials for ultrasound biological phantoms. Proc. IEEE Ultrason. Symp., 1962–1965 (2009). https://doi.org/10.1109/ultsym.2009.5441472
145. Jia, C., Kim, K., Kolias, T.J., et al.: 4D elasticity imaging of PVA LV phantom integrated with pulsatile circulation system using 2D phased array. Proc. IEEE Ultrason. Symp., 876–879 (2007). https://doi.org/10.1109/ultsym.2007.224
146. Hebden, J.C., Price, B.D., Gibson, A.P., Royle, G.: A soft deformable tissue-equivalent phantom for diffuse optical tomography. Phys. Med. Biol. **51**, 5581–5590 (2006). https://doi.org/10.1088/0031-9155/51/21/013

147. Lamouche, G., Kennedy, B.F., Kennedy, K.M., et al.: Review of tissue simulating phantoms with controllable optical, mechanical and structural properties for use in optical coherence tomography. Biomed. Opt. Express **3**, 1381–1398 (2012). https://doi.org/10.1364/BOE.3.001381
148. Chan, R., Manzke, R., Dalal, S., et al.: Image-Based Speckle Tracking for Tissue Motion Characterization in a Deformable Cardiovascular Phantom, vol. 6920, pp. 69200U-1–69200U-7 (2008). https://doi.org/10.1117/12.770631

Tissue-Mimicking Materials for Cardiac Imaging Phantom—Section 2: From Fabrication to Optimization

Nurul Shafiqa Mohd Yusof and Dyah Ekashanti Octorina Dewi

Abstract With complex morphological tissues and dynamic functional mechanism in the heart, evaluation and validation of cardiac imaging systems require proper methods to ensure that all parameters are measured in optimum manners. Physical imaging phantom, a quality assurance device in patient dummy representation, has provided various biophysical imaging measurements for verifying the degree of reliability, reproducibility, accuracy, and usefulness in the imaging systems. Its wide uses in image quality, quantitative imaging system, navigation, biophysical measurement, and other imaging applications, has made phantom an important part in imaging system development. Tissue Mimicking Materials (TMMs), materials that build phantom, have their specific characteristics. For cardiac imaging, fabricating TMMs that mimic biological cardiac structures is not an easy task. The nature of the materials and fabrication processes may alter the characteristics of TMMs, causing unmatched properties with biological tissues. Therefore, systematic processes are needed for efficient phantom development. Seven stages in systematic processes have been proposed in this two chapters to guide researchers in generating physical imaging phantoms in step-by-step procedures. The preparation processes in part 1 have focused on the concept, design, simulation, and materials selection stages. The realization processes in part 2 emphasize on how to actualize the physical imaging phantom design and selected materials to TMMs through fabrication, testing and analysis, as well as optimization stages. This part 2 is aimed to review important techniques in TMMs fabrication from conventional to rapid prototyping, testing methods with physical properties analysis, as well as optimization process. As all systematic processes are strongly interconnected and affecting among others, a careful management in phantom development is vital in generating the best phantom performance.

N. S. M. Yusof · D. E. O. Dewi (✉)
IJN-UTM Cardiovascular Engineering Center, Institute of Human Centered Engineering, Universiti Teknologi Malaysia, Johor Bahru, Malaysia
e-mail: dyah@utm.my

School of Biomedical Engineering and Health Sciences, Faculty of Engineering, Universiti Teknologi Malaysia, Johor Bahru, Malaysia

D. E. O. Dewi et al. (eds.), *Cardiovascular Engineering*, Series in BioEngineering,
https://doi.org/10.1007/978-981-10-8405-8_2

Keywords Tissue mimicking materials · Cardiac imaging phantom · Systematic processes · Fabrication · Testing · Analysis · Optimization

1 Introduction

The incidence of Cardiovascular Diseases (CVDs) and their related health disorders has been notified to keep raising these years [1–3]. Strategies from screening to treatment have been included in CVDs management to solve this epidemic problem [4–7]. As one of important parts in the CVDs management, cardiac imaging systems have developed their roles, not only for basic tasks in assisting diagnostic and treatment procedures, but also for visualizing cardiac anatomical and functional components non-invasively in, real time, high image quality, quantitative, and multi-dimensional representations [8–11]. With these gains, wide options for region segmentation, feature extraction and description, tomographic viewing, volumetric reconstruction and rendering, motion compensation, vascularization tracking, and so on, enable for richer anatomical, physiological, and pathophysiological analysis and phenotyping, from systemic structure to molecular levels [10–12]. Thanks to these remarkable advancements, cardiac imaging has transformed into a more accurate, interactive, and comprehensive diagnostic and treatment technology that certifies for better understanding the diseases, determining the best preventive strategies, and personalized medicine [13, 14].

In line with prompt progression of computing and multimedia systems, integration with sensing and robotic technology has triggered various imaging modalities and supporting devices to enhance their functions. Standardization in data communication and control systems, has also allowed for more expansion in generating new functions, such as multimodality image fusion, object navigation, and ubiquitous synchronization. These has also facilitated for the implementation of image-guided cardiac drug delivery, intervention, and surgery applications in the clinical settings [15, 16]. These promising signs have driven imaging research and industry bases to generate imaging hardware and software systems in more patient specific manners.

Along with these imaging technology penetrations, proper validation and evaluation mechanisms are highly compulsory. Due to ethics regulation and high variability issues, the conventional technique for certifying the imaging systems using image data from human subjects or patients is limited, in some extent. With these concerns, the use of imaging models and phantoms in computational and physical means are growing rapidly as favourable validation and evaluation techniques in most biomedical imaging systems [17, 18]. Even the computational model has high performance in simulating and predicting various parameters in imaging systems, its uses are somehow indirect, as this does not straightforwardly mimic the actual acquisition setting like with the real scanned object. Physical imaging phantom, even its properties do not always guarantee for fully matched with those of biological object, this method can naturally and mentally represent the actual acquisition settings [19]. With this physical imaging phantom, various measurements and tests can be managed more measurable, simpler, and safer [20].

Due to actual practicability, physical imaging phantoms are in high demands, in clinical, research, and industrial settings. A number of commercialized phantoms have been used to accompany imaging systems in verifying acquisition accuracy [21–24]. However, although these phantoms seem to be promising for solving imaging validation and evaluation issues, the implementation is not always smooth. Hampering factors in human and machine variability, unmatched characteristics, durability, lifetime, safety, and so on, may occur. Hence, developing a customized phantom that accommodates these specific needs is preferred. Yet, developing phantom requires proper techniques in materials characterization and fabrication.

Specifically in phantoms for cardiac imaging systems, the need to mimic the heart mechanism is essential. Cardiac phantom must have a dynamically beating and pumping mechanism [25, 26]. Complex morphological structures of the heart also necessitate for detailed fabrication techniques [27]. Incorrect understanding and handling on materials characterization and fabrication may result in inefficient methodologies, thus generate higher errors. To address, systematic processes are introduced to guide phantom development. Adapted from prototyping theory in biomedical product design method [28], systematic processes are iteratively applied through seven stages, concept, design, simulation, material selection, fabrication, testing and analysis, and optimization. Preparation processes, in part 1, have focused on the concept, design, simulation, and materials selection. Realization processes in part 2 highlight on fabrication, testing and analysis, as well as optimization.

In this chapter, our review on cardiac imaging phantom development is started with anatomical and physiological description of the heart as the basic knowledge. Fabrication techniques on Tissue Mimicking Materials (TMMs) are defined from conventional to Rapid Prototyping (RP) processes. Some testing methods are also portrayed in the analysis of physical properties. Optimization process, as the last part of the systematic processes, makes adjustments on the phantom from specified attributes.

2 Anatomy and Physiology of the Heart

Heart is one of important organs in body. The heart general shape that resembles to pyramid with a series of borders and surfaces is anatomically enclosed by pericardium. The heart is positioned in the thoracic cavity, in the middle mediastinum, between the lungs and their pleural coverings. The heart wall is made of three layers; epicardium, myocardium, and endocardium. The thickness of the heart wall varies in different parts of the heart. The atria of the heart do not need to pump blood to the whole body, but only to the nearby ventricles. Therefore, they have a very thin myocardium. The ventricles, on the other hand, have a very thick myocardium, as it functions to pump blood to the lung or systemically throughout the entire body. The right side of the heart has less myocardium in its walls than the left side, because the left side has a task to pump blood through the entire body, while the right side only to the lungs [29].

From the anterior section in Fig. 1, the heart contains 4 chambers: right atrium, left atrium, right ventricle, and left ventricle. The atria are smaller than the ventricles with thinner and less muscular walls than the ventricles. The ventricles are the larger and stronger pumping chambers that distribute the blood out of the heart. Compared to the left side of the heart, the right chambers are smaller and have less myocardium.

The structural basis of myocardial viscoelasticity is ascribed to a combination of cardiac cells and extracellular matrix (ECM) proteins. Myocardial tissue is composed of cardiac muscle cells or myocytes. The myocytes are composed of repeating units named sarcomeres, which contain titin, actin, and myosin proteins. By isolating cardiomyocytes from surrounding ECM structures, viscoelastic properties can be measured. Titin functions as a spring that contributes to viscoelastic behavior due to its interactions [30, 31].

Cardiac cells are surrounded and interconnected by ECM proteins to develop structural integrity to the cardiac muscle. Collagen, protein that majorly develops the extracellular matrix of connective tissue, may also provide myocardial viscoelasticity. Its degradation may decrease the elastic stiffness and viscosity of the cardiac muscle [32, 33]. The existence of intramolecular crosslinks of collagen may generate strength in the connective tissues, and this is varying with age, pathology, and so on [34]. Importantly, the majority of myocardial viscoelastic properties are observed from the left ventricle side. Even the left ventricle differs embryologically, geometrically,

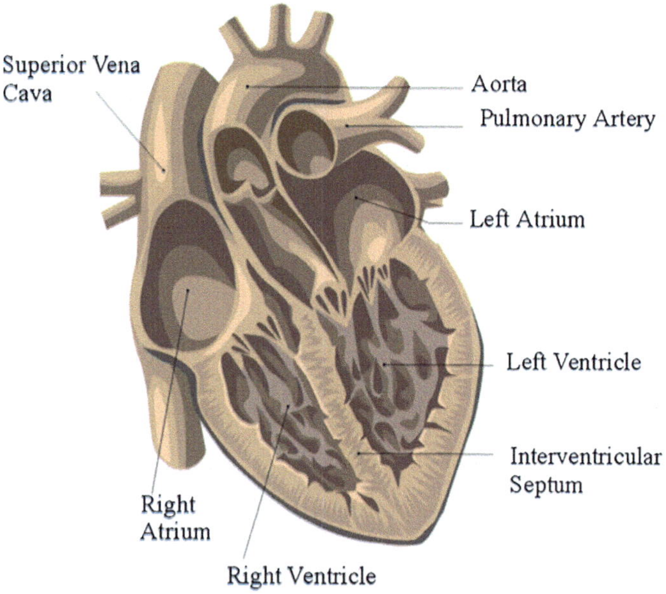

Fig. 1 Anterior sections of the heart in four chamber view [29]

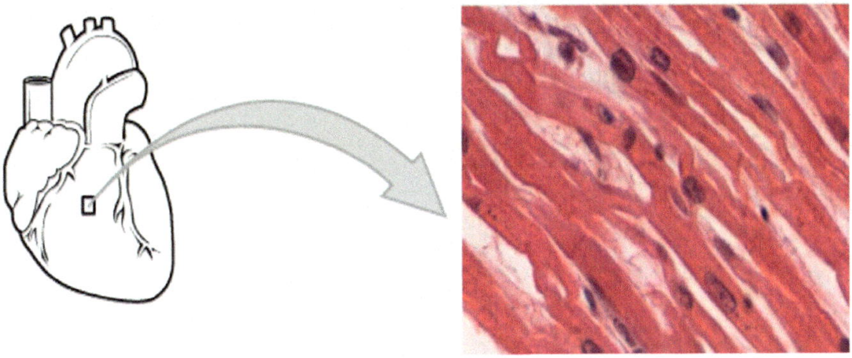

Fig. 2 Schematic structure of cardiac tissue depicted from myocardium region. By using microscopic imaging, detailed structure of the tissue can be observed [36]

and structurally from the right ventricle, however, differences in viscoelastic behavior between these two ventricles are still unclear [35]. Figure 2 shows microscopic image structure of the heart.

3 Schematic Processes in Phantom Development

3.1 Tissue Mimicking Materials (TMMs)

TMMs, the main component in the physical imaging phantom, determine the properties of the phantom. A product prototyping management, named systematic processes, as in Fig. 3, is introduced in the TMMs development to generate the best phantom prototype with high efficiency, accuracy, and fitting with the needs in biomedical imaging application. Preparation processes formulate the phantom development through concept, plan for the structure and function through design, predict the applicability via simulation, and elaborate the TMMs contents in materials selection stages. Realization processes transform the formulae, plans, and materials from the preparation process into a number of steps in fabrication processes. Testing and analysis process generates physical properties that are matched to the biological organ or tissue. Optimization finalizes the process by delivering some necessary adjustments.

4 Fabrication

As illustrated in the systematic processes in Fig. 3, once the concept, design, simulation, and materials composition in the preparation processes of synthesis processes

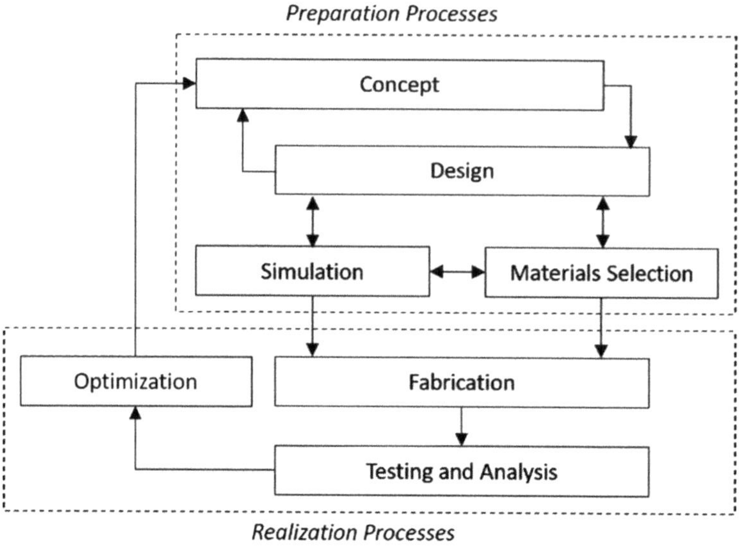

Fig. 3 Systematic processes in physical imaging phantoms development

have been firmly defined, as explained in the previous chapter, the TMMs can continue to proceed with the realization processes that consist of fabrication, testing and analysis, and optimization stages. In the preparation processes as described in the previous chapter, the connectivity among simulation, materials selection, and fabrication processes are made in a triangle closed loop, where simulation and materials selection connection is a reverse process. This reverse process may enable for exchange of similar TMMs parameters that undergo for both sides. In this way, materials properties can be used as similar TMMs parameters that are processed in the simulation to observe whether the materials can fit with the design, and vice versa.

On the other side, the two connectivity branches between simulation and fabrication, and between materials selection and fabrication, are made one way. This can be described that TMMs fabrication process is dependent on the two inputs, the best design and the appropriate materials, as the outputs of these previous two processes. The best design is the result of the simulation process as the translation of the best scenario in phantom development. In this way, the best design from the simulation plays role to assist the fabrication process based on the best materials and parameters on the specified design. However, the best design from the simulation process is the predicted outcome based on the measured parameters only that is the simplified version of the real settings in fabrication. Once the design is brought into the experiment or fabrication, some unpredicted factors may come up complicating the fabrication process. In any case, it is a normal situation.

The appropriate materials are the outputs of the materials selection process from the synthesis mapping of several material candidates. The synthesis mapping in the materials selection justifies the materials based on their properties under certain measurement techniques, for instance imaging and mechanical testing. In the synthesis mapping, the first step of parameter marking strategy, characteristics analysis, needs to generate list of requirements that covers biological, mechanical, chemical, and physical aspects. These aspects can be derived into a number of parameters, such as biocompatibility, reliability, corrosion resistance, controllability, customization among others, and many more [37].

The fabrication process of physical imaging phantom is a part of biomedical device manufacturing methodologies [37, 38] and general manufacturing techniques [39, 40]. In [37], the emphasis is mostly on the implementation of biomechanical manufacturing methodologies for biomedical devices, while [38] is focused more to applying digital design and imaging processes in the development process of biomedical device manufacturing. The fundamental manufacturing as provided by [39, 40] has highlighted about taxonomy of material manufacturing. Still, all have outlined about the current advancements of Rapid Prototyping (RP) through 3D printing that are significantly valuable in the biomedical device fabrication.

TMMs fabrication for physical imaging phantoms is directed to manufacturing the materials in such a way that the imaging properties of the materials are close to the targeted biological tissues. Not only that, the TMMs fabrication needs to accommodate the use of the synthetic tissues in some measurement strategies. Hence, as TMMs can be produced either from biological, synthetic, or combined materials, the fabrication can also involve biological, mechanical, chemical, and physical processes, and their combinations. The choice of fabrication processes highly depends on technical, legal, human, and socio-economic aspects, including their relevancy with the basic list of requirements, cost, time, energy, availability of materials and equipment, human resources and skills, environment, authority regulation, and many more [38]. Therefore, defining the best fabrication approach for TMMs is somehow flexible and conditional as long as it is following the regulated mechanism in biomedical products with its safety procedures, standardized quality and reproducible, and resulting in TMMs with suitable mimicking properties. However, each consideration has its own trade-offs and the end results are typically not far from the prediction and methods.

Generally, the basic approaches in biomedical manufacturing are adapted from general manufacturing technology and biomedical device manufacturing principles, as properly defined in [37, 38]. Based on the way the manufacturing process is controlled, the TMMs fabrication can be categorized into three big groups of approaches, human-based manufacturing (HM), machine-based manufacturing (MM), and their combination.

4.1 Human-Based Manufacturing (HM)

HM is generally defined as an approach to fabricate the product by utilizing conventional manual engineering methods, tools, and procedures. The scope of HM may range from basic conservative to modern manufacturing techniques. The basic conservative techniques mostly utilize manual handling on some manufacturing activities operated by human. The implementation of HM in modern manufacturing technique is more to providing specific mechanical assembly that the machines cannot perform as what the human can do. Typically, this specific mechanical assembly is conducted in the pre- or post-manufacturing processes for ensuring the quality of the product as a part of the preparation step or finalization touch before the product is tested.

The basic conservative technique is considered to be old-fashioned and less efficient in term of time and energy because human has limitations in performing high repetitive manufacturing activities with large amount of product generation goals [41]. Nevertheless, this HM approach is still considered to be important in manufacturing education, research, and industry, especially in the specific process called knowledge capture and extraction of the human expertise in manufacturing a product using his or her senses. Though this is challenging enough to obtain, this knowledge capture and extraction mechanism has been widely studied in the context of obtaining data model of the activity, building a knowledge reuse for the early design phase of product and system, process planning, as well as knowledge management [41–47].

By targeting knowledge capture and extraction in manufacturing, it is beneficial to conduct this manufacturing in high accuracy and reproducibility. Even though the ultimate goal is not the manufactured product itself, this HM approach has been proven to increase the productivity of the workers in performing the manufacture [43–45]. Above all, the knowledge of manual handling technique can be preserved and expanded into computation and automation applications [43–45].

In the implementation of physical phantom fabrication, this HM technique is sometimes still used. As in [48], although the phantom was developed using 3D printing, it still required some conventional manufacturing activities. A smoothing process to the phantom surface was made manually using sandpaper and Dremmel tool after the 3D printed phantom is obtained. A filling process was also done manually by applying adhesive, foam, and sealing materials that were prepared using CAD model. Manual milling technique was also used in the work of [49]. The fabrication of a full-scale sagittal-sliced patient-specific radiotherapy phantom using 3D-printing was also mediated by a manual process in the adjustment of polylactic acid (PLA) dosage and drilling the positioning rod of the phantom [50]. The other work [51] also fabricated the TMMs manually by mixing the raw polydimethylsiloxane (PDMS) and curing agents. Specifically in the cardiac imaging phantom application, the work in [52] used manual immersion technique in generating the silicone-based vessel phantom. These works have certified that HM approach by manual handling needs to be accommodated to guarantee the phantom quality, either a stand-alone process or combined with RP.

4.2 Machine-Based Manufacturing (MM)

Different from HM that employs the human knowledge and capability in handling the manufacturing manually, the MM has focused its manufacturing techniques for avoiding or optimally minimizing the involvement of human handling. As MM approach has fully applied the machining scope, it is important to expand the discussion into the machine-based manufacturing processes.

Based on the geometrical production technique, the MM approach can be divided into two classical types, forming and material removal processes. The forming process develops shape and size by deforming a bulk material into specified shape and size following the design. The bulk materials can be in the form of solid, semi-liquid, or liquid. The material removal process is aimed at developing shape and size by removing parts of the bulk material.

As processing the bulk materials depends on their basic material properties, the material removal processes need to involve chemical, mechanical, and physical processes in generating the model. Nowadays, due to their wider applications [51], material removal processes are growing greatly, generating various novel techniques. Therefore, besides these basic processes, hybrid processes that involve more than one type of processes, such as electro-chemical, electro-mechanical, electro-thermal processes, and so on, have existed recently enhancing the uses of the available material removal processes. However, the applications of these hybrid processes are mostly for metal manufacturing that are rarely used for representing biological tissues.

Instead, the manufacturing techniques can also be grouped according to the properties of the materials that consisting of subtractive, net-shape, and additive processes [37]. The subtractive manufacturing creates the shape by subtracting the material using mechanical, electric, and chemical processes. This is basically almost comparable to that of material removal processes. The net-shape manufacturing is similar to the forming and molding processes that aim at increasing geometrical accuracy and flexibility. The additive manufacturing principally produces 3D model by adding the materials layer by layer [37, 50, 53]. From these types, the other newer methods are expanding. As a part of the additive manufacturing techniques, the so called incress manufacturing technique constructs the 3D model using powder, liquid, semi-liquid materials, and their derivatives [50]. From the way the shape is built, this technique applies direct 3D and 2D layering methods. These methods are then advanced and developed as the basic principle of RP. Thus, the RP technology evolves and improves to become Rapid Manufacturing (RM) to directly produce the actual functional parts and products [37, 38].

4.3 Rapid Prototyping (RP)

Along with the great advancements in manufacturing, the newer generation of manufacturing technology, RP, has fulfilled the need of a low to medium cost manufacturing in a relatively shorter time with high customization, less complications. RP fabricates a scale 3D model of a physical or functional parts, or assembly, by gradually growing the materials to the required shape [37, 54].

As an adaptation from the RP cycle in [37, 38, 55], the RP development processes is performed through three main RP processes, the pre-processing, building, and post-processing, as described in step by step as follows:

1. RP Pre-processing
   This step is started by applying the idea from conception to outlined design and function. Then the idea is transformed from the outlined design and function to a computer-aided design and computer-aided manufacturing (CAD/CAM) software to develop 3D geometric model (3DGM) with predefined geometrical parameters and validated volumetric object. The 3DGM can be obtained by computer drawing or acquired using an imaging device from the actual object. Depending on the function, 3DGM can be developed in an integral design as one single object or comprised of reassembled parts of the model. As early validation, the CAD/CAM system also enables to visualize and simulate the 3DGM of the object before the design is manufactured. Once the 3DGM is obtained, the CAD file is created. A specified STL (Stereolithography) format is made to enable for data transfer from 3DGM to SFF (Solid Freeform Fabrication).
2. RP Building
   This step is fabrication process of the 3DGM using automated RP machinery. The SFF of the 3DGM is driven to perform motion control trajectory of the RP machinery to perform 3D geometric object building in RP layer-by-layer. Each layer is developed as 2D drawing that is scanned line-by-line. In this way, some objects may take long manufacturing time.
3. RP Post-processing
   This final step is for the removal of the 3D geometric object out of the RP machine. Some removal procedures need to be done properly to maintain the lifetime of the machine. An RF machine cleaning operation is usually provided for this matter. In order to support the quality of the 3D geometric object, improvement processes are mostly conducted, such as surface smoothing using sanding and brushing methods.

Compared to the conventional method, RP manufacturing allows for better method in freely generating complex shapes, internal structures, repetitive inner parts, very thin-wall, and cross sectional features in a more convenient way. With this method, any complex design can be fabricated and customized more efficiently.

With computerized manufacturing process, 3D geometric model can be built and simulated using CAD/CAM software. With this process, complex design can be morphologically and functionally simulated, reproduced, and modified much easier with high accuracy and customized volumetric physical visualization that is applicable in any purposes. Yet, some ethical issues related to design copyright and publicity of biological objects may arise and is in need of proper regulation [56].

With the layered manufacturing mechanism, RP constructs the 3D geometric objects from very thin 2D cross sections in stacks iteratively until the 3D physical part is completed. Also, for material variation or other specific purposes, the 3D geometric object materials can be also used from different sources with new polymers or biopolymers in various types [37]. This also makes the RP method more acceptable, flexible, and adjustable in the application settings [57]. However, in some cases, the RP materials are not fitting with the requirements, so that the end results can be easily damaged or biased from the initial design.

Due to its advantages, RP has been brought into wide implementation. In the biomedical applications, RP has been known as modeling and simulation, planning, testing, and tooling media to assist from biological and medical education to research and clinical platforms [58–60]. In CVDs management, RP models have been used as physical model and visualization in teaching and training [61, 62], clinician and patient consultation [63–65], measurement, diagnosis, and treatment [66, 67], pre-operative planning and simulation [68–74], as well as implants [68, 75].

In most clinical applications, relationship of RP, CAD/CAM, and 3DGM with medical imaging modalities are strong, as most raw data source is obtained and modeled from the images acquired by those imaging systems, then used for further purposes [76, 77]. Specifically in radiology environment with extensive imaging modalities utilization, RP has been an important media for visualizing more detailed, tangible, volumetric, and realistic representation of radiological images. Specifically when radiologists need to provide more detailed information about the anatomy, physiology, and pathophysiology in the radiological images [78]. Sometimes, the 2D image information is not sufficient, especially when the image is in the tomographic mode. Therefore, with 3D/4D visualization systems from the 3D geometrical reconstruction as well as perceptible sensing through 3D physical models, a more deeper and interactive information can be obtained in better way, making the communication, knowledge transfer, and learning process easier [79].

On the other way around, the 3D-based imaging modalities, such as CT [80], MRI [81–83], MR Angiography [84], 3DUS [85], NM [86], and so on have also played essential roles in the prototyping and testing stages of various biomedical product developments that involve RP as one of its development processes [37, 38, 58]. Some of the products are medical instruments [87, 88], prosthetic implants [69, 75], drugs [89], phantoms [90–92], and many more. Especially, along with the growing needs of patient-specific management with support from imaging systems, customization of tools, procedures, and dummy patients can be excellently made through the RP technology [65, 93–95].

Thanks to the high performance of the RP technology in generating accurate shape and size for TMMs and physical imaging phantoms development, phantom development studies using RP are mostly found to be applied for imaging quality assurance and optimization [96–99]. For X-rays and CT imaging quality assurance, a low cost 3D printed phantom housing and insert cubes have been built in different materials [97]. In [98], fabrication of Raman multi-layer tissue phantom for tissue spectroscopy analysis has implemented 3D printing for producing castings in complex geometry patterns. However, although it was an accurate and inexpensive phantom, the layering technique complicated the fabrication process. Specifically for molecular imaging applications, the study of [99] used a 3D printer-based functional phantom to observe the acoustic parameters of contrast agents in photoacoustic imaging. This study has shown that an accurate, low cost, and reproducible functional phantom for quality control in imaging system can be easily developed with support of RP technology.

From these observation, it can be inferred that the use of RP is not only providing a reliable and low cost prototyping, but also it has specialties in generating small and complex structures in high accuracy. It also shows the tendencies that the RP-based phantoms are prospectively applied in molecular imaging.

4.4 Chemical, Mechanical, and Physical Processes in TMMs

It is obvious that the RP era has changed the way physical imaging phantoms are fabricated. A number of clinical studies have proven about the essential needs of the RP and its advancements for improving the clinical research and healthcare services [96]. The studies have also confirmed that developing phantoms always imposes a multidisciplinary problem solving. In RP, various materials have been used to model the biological structures. However, up till now, their uses are still limited and difficult to match to those of other materials from the conventional fabrication. Therefore, although RP is low cost, fast, and practical, its implementation is still limited. A combination mechanism is always needed to produce the customized phantom in its best performance. In this way, understanding the properties of the biological tissues as well as synthetic materials is the key in determining the best fabrication process. As mostly chemical, mechanical, and physical properties are also measured in the TMMs, these processes are basically meant to alter the basic characteristics of the base materials to obtain the most fitting characteristics to that of biological tissues [53].

4.5 Chemical Process

By definition, chemical process is a process, including any liquid, solid, or gas, to change the chemical substances, either naturally or by chemical reaction. The chemical reaction involves one or more substances to convert or change the physical characteristics of the reactants to other substances [100, 101]. Basically, the chemical reaction can be divided into five types, synthesis, decomposition, single replacement, double replacement and combustion [101], as described in Table 1.

Chemical reaction is highly utilized in TMMs fabrication. In mixing between base and additive materials, a synthesis formula is needed. This formula highly depends on the type of the involved materials. Hence, it is important to verify the formula during simulation process to define the optimum synthesis formula. Decomposition process in TMMs fabrication is mostly implemented when there is a need to detect any unexpected materials during chemical reaction, such as material toxicity in TMMs that tend to endanger the user or require complicated material treatment [102–104]. Single and double replacement processes may also occur in TMMs fabrication, specifically in the electrochemical phantom for ion displacement measurement. In combustion process, energy is needed for initial input in the process. This combustion process

Table 1 Basic types of chemical reaction for TMMs fabrication [101]

Types of reaction	Definition	Equation
Synthesis (composition)	Two or more elements or compounds combine to make a more complex substance	$A + B \rightarrow AB$
Decomposition	Compounds breakdown into simpler substances	$AB \rightarrow A + B$
Single replacement	Occurs when one element replaces another one in a compound	$AB + C \rightarrow AC + B$
Double replacement	Occurs when different atoms in two different compounds trade places	$AB + CD \rightarrow AC + BD$
Combustion	Occurs when one element or compound reacts with oxygen producing energy	$CH + O_2 \rightarrow CO_2 + H_2O$

includes some inducing factors, such as concentrations of reactants, temperature, and catalysts that may affect reaction rates. Concentration affects the reaction rate when the molecules collide for any reaction. Temperature also influences the reaction rate as it is one of the factors that causes collisions between reactant and sufficient energy. Catalysts also affect reaction rates by providing a faster pathway for reaction can occur. This combustion process can affect physical properties of the TMMs. Thus, selection in types of chemical reaction must be followed by selection in the combustion process [105].

4.6 Mechanical Process

Mechanical process includes any operations using mechanical procedures [106]. Although TMMs fabrication is not complex like those of heavy industrial machining, the basic mechanical process is comparable. One of the applicable mechanical operation in TMMs fabrication is stirring, a mixing process of composite materials in liquid state [107].

Some parameters, such as stirring tooling, temperature, time, and speed, have been known to affect the microstructure and mechanical properties of the composites materials in this process. However, with these parameters, it is possible to easily and simply control the bonding process of the structure. Since the stirring tools may affect the TMMs fabrication result, it is important to ensure that the stirring process is performed optimally and suited with the material. The stirring tooling can be divided into two techniques, manual or automatic stirring, as provided in Table 2. Commonly, manual stirring uses hand pressure to stir the mixture until the material dissolves. It is a fast and simple technique in small production but can be troublesome in accurately repeated production. Alternatively, automatic stirring, either using hand mixer or stand mixer, can provide more constancy in stirring pressure with certain control in some parameters, like temperature, time, and speed, that may affect the

Table 2 Types of stirring technique

Type of stirring	Advantages	Disadvantages
Manual stirring	Well suited to quick tasks, such as materials that starts to cure in 10 s	Weaker capability, non-uniformity
Automatic stirring	Accurately controlled process with cycle time and rotation speed	Difficult stirring, as viscosity reaction increases, risk of harm to the machine also increases

strength of the composited materials. The stirring process at certain temperature can affect the time taken for the materials to turn to solidify in better strength. On the other hand, time and speed parameters can affect the homogeneity of the mixed composite materials that may improve the strength and accuracy properties of the materials [107].

4.7 Physical Process

Physical process involves any process that changes the physical properties of the materials with atoms rearrangement. It includes transition from one form to the other form, such as heating and cooling that causes melting, freezing, sublime states, and changes in physical properties, like strength, durability, crystal form, texture, shape, size, color, volume, and density [108].

Physical process has played significant role in almost all process in the TMM fabrications, especially during mold fabrication process, TMMs mixing process with the additives, accelerators, or fillers, and TMMs curing process in removing air bubbles using vacuum degassing machine. Specifically in the TMMs curing process, the machine triggers expansion process inside of the degassed materials to pull the bubbles to the surface. Once the bubbles burst at the surface, the expansion decreases. The effectiveness of bubble removal depends on the viscosity of the materials. The higher viscosity, the longer and more difficult to degas [109, 110].

This curing process is required for material cross-link when the TMMs are transformed from liquid into solid [111]. Curing parameters, such as time, temperature, and pressure, have shown to affect the curing process. Curing at room temperature usually requires hours and days to harden the TMMs. Heated curing with temperature and pressure level controls has been found to shorten the TMMs hardening time compared to the room temperature condition [111, 112].

From these fabrication processes, TMMs characteristics can be modified to adjust with the required properties to mimic the actual biological tissues. However, defining the best process to alter the material properties needs to be done carefully and accurately, as it may also cause incorrect characteristics in the TMMs. In some ways, some trade-offs in the TMMs fabrication are unavoidable. Researchers need to focus on one particular purpose only to obtain the best performance. For instance, if the focus is to obtain TMMs with excellent anatomical structure, the functional properties need to be compensated, and vice versa.

4.8 Phantom Fabrication

As discusses previously, the fabrication of phantom involves multidisciplinary processes. However, a specified procedure needs to be organized in order to generate a standardized and reproducible phantom fabrication. Therefore, in this review, we propose a schematic procedure, as a part of the systematic processes that can be used for phantom fabrication through TMMs development.

As illustrated in Fig. 4, the generic fabrication process in TMMs for phantom prototyping can be divided into two main processes:

1. *Combined process*
2. *Separated process*

In some extent, the two processes can be executed in combined or separated version, depending on the TMMs design. The combined process is considered to be the simplest fabrication technique, as long as TMMs use the materials that are commonly available. It is mostly performed when the TMMs are developed through RP method using 3D printer or other RP machinery [37–39]. In this version, the 3DGM design

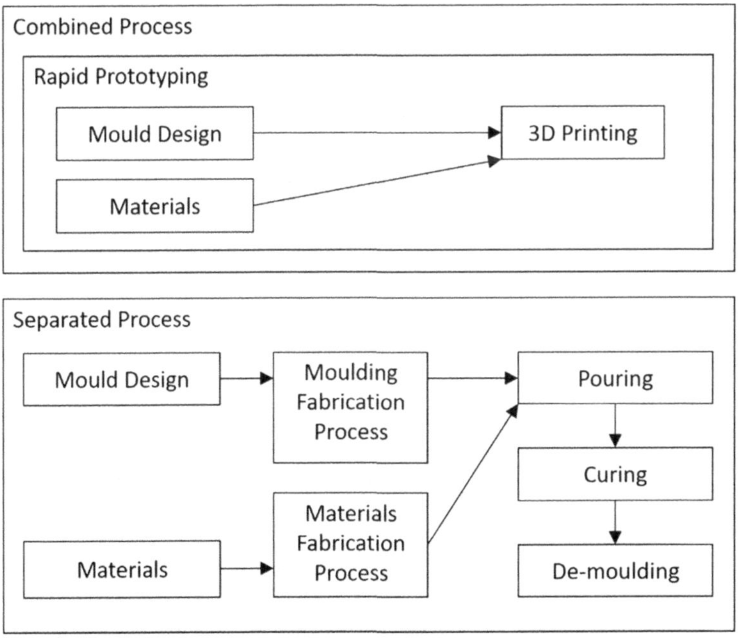

Fig. 4 Generic fabrication process types in TMMs development

of the TMMs must be well prepared in high resolution. In the simplified 3DGM, data pre-processing, such as volume segmentation, feature landmarks, reconstruction, and data formatting, need to be properly performed to obtain an accurate geometrical model of the TMMs. Afterwards, the materials are built through layering technique following the given 3DGM. A number of layering techniques in RP have provided various options with their characteristics [38, 39].

In contrast, the separated process is considered to be more complex. It is done through conventional moulding and materials fabrication processes. The moulding fabrication process is first done to prepare the mould, then the materials fabrication process is performed to prepare the materials to be poured into the mould, followed by curing and de-moulding processes as finalization stage. In this way, involvement of RP is highly possible in the moulding fabrication to simplify the process.

4.9 Moulding Fabrication Process

Moulding fabrication process is the step to develop the outer and inner geometries of the phantom. This can be generally obtained by first designing the moulding system, then fabricating the moulding system. The design of the moulding system is highly determined from the design of the TMMs, either anthropomorphic, functional, or combination of both phantoms. Each design has its own complication depending on the details as required in the phantom. The anthropomorphic phantom fabrication requires a more accurate anatomical structure in the moulding system, while the functional phantom fabrication has simpler moulding process, but it necessitates more complex and accurate functional features that can affect the moulding process. The moulding process for the combined anthropomorphic and functional phantom is generally more complex to fabricate.

Basically, the moulding fabrication process needs to reverse the TMMs design, as the mould plays role as the container of the TMMs during the materials fabrication process [39]. The moulding fabrication process has three main options:

1. 3D printing
2. External forming
3. Model forming

The first option, 3D printing technique requires proper 3DGM of the phantom mould. This technique has high accuracy as it is produced by the considerably accurate machine. In this way, the design of the phantom mould is similar to creating the design of the phantom, but in reverse version. As this reverse phantom design utilizes the 3DGM from technical drawing or 3D volumetric anatomy of CT or MRI that mostly has complicated shapes, it is important to verify the feasibility of using the existing design in the simulation and actual model [113]. This needs to be considered because the 3D mould parts use the reverse design of the phantom which only concentrate on regions that are located along the inner surface of the 3D phantom design. Also, the 3D mould requires openings and closing systems in some parts to ease the pouring and curing mechanisms. Therefore, mould material type is important. Once the mould is fabricated, a careful cutting in the mould is needed to determine the moulding track for opening and closing mechanism.

The second option, external forming technique fabricates the mould using forming system that is triggered externally. This external forming uses flexible thin layer material that is shaped into mould system using some methods, such as thermal treatment, mechanical forming, chemical reaction, and so on. The shaping reference is used to determine the required shape as in the design. However, this external forming technique has limitation if the shape is too complex or small, thus the external trigger is not adequate for making this complex shape.

Almost similar to the external forming, the third option, model forming technique fabricates the mould by using the 3D TMMs model derived from the design phase, then cover the model using flexible mould material, such as silicon rubber, to create the detailed inner form. Curing and de-moulding processes are also implemented once the mould is solidified. The model can be in the form of 3D printed, plastinated, biologically made, and many more [114].

Once the phantom mould design is obtained, the 3D printing is performed on the suitable materials using commercially produced filaments or specifically customized 3D printing materials. However, the choice of materials in the phantom mould system needs to consider about the material interaction between the mould and the phantom materials. Openings and closing of the mould also need to be made.

4.10 Materials Fabrication Process

Materials fabrication process is the stage to develop the TMMs substances before it is formed inside of the phantom mould. This process can be managed as follows:

1. *Biological tissue analysis*: Evaluate the reference from biological tissue and TMMs characterization databanks, if any.
2. *Materials preparation*: Organize the selected components of the materials, consisting of base and additive materials.
3. *Composition formulation*: Vary the estimated composition between base and additive materials into several set-ups for reference matching.
4. *Chemical process*: Materials mixing.
5. *Mechanical process*: Stirring the solution.
6. *Physical process*: Thermal treatment and curing.
7. *Model forming*: Pouring to mould, and de-moulding.

As in the synthesis mapping, the databank for materials characterization can be transformed into reference values for comparison with those of biological tissues and TMMs [115–117]. Data analysis techniques can be used to provide quantitative analysis about how close the characteristics are. Then, the material components and composition between the base and additive materials can be estimated. As the estimation is not proven yet, some set-ups need to be made in the experimental protocol, for instance, taking a range below and above the predicted values in the experimental set-ups for defining the ratio between base and additive materials.

5 Testing and Analysis

Testing and analysis stage is part of TMMs verification to clarify the TMMs performance in mimicking the biological tissues. The testing part is aimed to conduct a set of measurements for TMMs characterization using various techniques. A number of measurement techniques have been widely used to characterize the properties of materials, including TMMs. Compared to testing, analysis part is more intended to evaluate the measurement results and compare between the characteristics of TMMs and those of biological tissues for extracting the information about how far TMMs mimic the biological tissues. As mentioned previously, to obtain a more accurate analysis, it is important to provide databank of materials characterization as the valid reference in analyzing the measurements. In [115], a wide-ranging theory and references in physical property measurements for various biological tissues can also be used as contents in the biological tissue databank. Also, in [108], similar physical property measurements for polymers, as the common material for TMMs, can be used as the databank contents.

As TMMs are mostly used for simulating or modeling human tissues in performing various physical maneuvers, such as incision, needle insertion, biopsy, suturing, surgery, and so on, it is crucial to ensure that the TMMs physical properties have proper matching with those of actual biological tissues. However, depending on the measurement technique, most physical property measurements are performed in single measurement or in one-dimensional (1D) mapping where the single measurement is done periodically relative to time. On the other hand, the nature of TMMs or biological tissues and organs are in 2D layers or 3D volumes, and sometimes in 4D if modified with motion that is relative to time too. Also, the spatial or volumetric analyses in the physical property measurements are quite excessive to measure, as the measurement requires large number of sensors and intricate settings. The option for the 2D, 3D, or 4D measurement is performing image acquisition and analysis for quantifying important imaging properties of the TMMs. With this way, some physical properties that are related to imaging physics and systems can be measured properly, especially when the TMMs are used for verifying the performance of imaging systems in maintaining the image quality, detection, measurement, and analysis of various object structure and geometry. Therefore, in this part, some physical and imaging property measurements and their parameters are elaborated for TMMs testing and analysis.

A large number of measurement techniques have been elaborated in extracting physical and imaging properties of materials. These techniques can be simply categorized into eight big groups, texture, mechanical, thermal, electrical, acoustic, optical, geometrical, to radiation testing. Table 3 shows selected physical and imaging properties for TMMs testing and analysis, specifically for properties that are commonly

Table 3 Physical and imaging properties in TMMs testing and analysis

Testing method	Properties
Mechanical [106, 108, 115–122]	Young's modulus, tensile strength, slip, durability, plasticity, roughness, etc.
Thermal [106, 108, 115, 121–128]	Thermometry, calorimetry, kinetic behavior, thermodynamics, heat transfer, etc.
Electrical [115, 121, 128–134]	Dielectric properties (conductivity, permittivity), resistivity, capacitance, etc.
Acoustic [103, 108, 115, 120, 129, 134–138]	Absorption, speed of sound, impedance, acoustic attenuation, etc.
Texture [139–147]	Intensity, area, surface topology, macrotexture, microtexture, orientation mapping, etc.
Optical [130, 137, 148–152]	Absorption, scattering, color, photosensitivity, transmittance, etc.
Geometrical [138, 153–157]	Shape (ellipticity, rectangularity, triangularity, granularity), size (thickness, area, length, angle)
Radiation [158–161]	Attenuation, mass density, mass attenuation coefficient, mass energy-absorption coefficient, etc.

used in the development of physical cardiac imaging phantom. For imaging properties, the measurement also highly depend on the physics of imaging modality. Therefore, if similar material is scanned under two different imaging modalities, and measured the same imaging properties, it is common that the imaging properties of both imaging modalities have discrepancies, in some extent.

6 Optimization

As it is defined, optimization is the process to obtain the optimum or best outcomes. In the testing and analysis process, the characteristics of the TMMS obtained from the measurement clarify how far the TMMs match to the biological tissues to be mimicked. Optimization uses the information from this degree of closeness as feedbacks for improving the TMMs performance.

In the TMMs systematic processes, a number of optimization techniques can be adapted for obtaining the most mimicking materials. As the design and fabrication processes are the most influencing process in the systematic processes to obtain the most optimum physical imaging phantom, the optimization can be in the range of design optimization and/or fabrication optimization, depending on which part of the process that needs to be optimized.

Optimization techniques in design processes are aimed at finding the best design parameters that satisfy the TMMs requirements. In this way, some design optimization techniques can be applied. One of them is called Design of Experiments (DoE) or

Quality by Design (QbD). This DoE is basically aimed at determining combination of parameters under hypothesized factors to predict the best condition. By varying all combinations of factors in certain levels, the factor of combination interactions is obtained. From this, the understanding of parameter interaction and interdependence is built, providing values for considering the best design [162]. This DoE is also implemented in the simulation process of the systematic processes for TMMs. On the other hand, in the context of optimization technique for fabrication processes, some optimization techniques can be performed, such as Topology Optimization (TO) and Design for Manufacturability (DfM). The TO is based on Finite Element Method to optimize the parameters in the fabrication process within a given design space [163]. The DfM highlights about how to make a design process so that the fabrication cost can be made in its most efficient way [164, 165]. Above all, many other optimization techniques can always be chosen to support in producing the best TMMs. Surely, the usability of these technique is all depending on many factors, mainly the suitability with the problem, capacity, and availability.

7 Summary

The great importance of physical imaging phantoms in verifying the accuracy of imaging systems has brought big concerns in the phantom development processes. The TMMs realization processes that emphasize on actualizing the phantom preparation through fabrication, testing and analysis, and optimization have certify that those lengthy processes need to be managed properly so that the best phantom can be achieved. Furthermore, the influence RP in the TMMs and phantom fabrications has accelerated any fabrication processes with faster, simpler, more accurate, and better representation of the phantom prototype. The testing and analysis process is also important in providing a comprehensive TMMs characterization for comparing with the biological tissues. Lastly, optimization is also one last process to ensure that the phantom is in its best performance.

Acknowledgements The authors are grateful for funding supports by Universiti Teknologi Malaysia and Ministry of Higher Education Malaysia under FRGS Grant R.J130000.7845.4F764 and GUP Tier 1 Grant Q.J130000.2545.20H36.

References

1. Writing Group Members, Benjamin, E.J., Blaha, M.J., et al.: Heart disease and stroke statistics—2017 update: a report from the American Heart Association. Circulation **135**(10), e146–e603 (2017). https://doi.org/10.1161/cir.0000000000000485
2. Misra, A., Tandon, N., Ebrahim, S., Sattar, N., Alam, D., Shrivastava, U., Narayan, K.M., Jafar, T.H.: Diabetes, cardiovascular disease, and chronic kidney disease in South Asia: current status

and future directions. BMJ (Clin. Res. Ed.) **357**, j1420 (2017). ISSN 0959-8138. https://doi.org/10.1136/bmj.j1420

3. Atlas Writing Group, Timmis, A., Townsend, N., Gale, C., Grobbee, R., Maniadakis, N., Flather, M., Wilkins, E., Wright, L., Vos, R., Bax, J., Blum, M., Pinto, F., Vardas, P., European Society of Cardiology: Cardiovascular disease statistics 2017. Eur. Heart J. **39**(7), 508–579 (2018). https://doi.org/10.1093/eurheartj/ehx628
4. Vogelmeier, C.F., Criner, G.J., Martinez, F.J., Anzueto, A., Barnes, P.J., Bourbeau, J., Celli, B.R., Chen, R., Decramer, M., Fabbri, L.M., Frith, P., Halpin, D.M., López Varela, M.V., Nishimura, M., Roche, N., Rodriguez-Roisin, R., Sin, D.D., Singh, D., Stockley, R., Vestbo, J., Wedzicha, J.A., Agustí, A.: Global strategy for the diagnosis, management, and prevention of chronic obstructive lung disease 2017 report. GOLD executive summary. Am. J. Respir. Crit. Care Med. **195**(5), 557–582 (2017). https://doi.org/10.1164/rccm.201701-0218pp
5. Duffy, J.Y., et al.: Cardiovascular disease screening. Semin. Perinatol. **39**(4), 264–267 (2015)
6. Turner, L.R., et al.: Cardiovascular disease screening in general practice: general practitioner recording of common risk factors. Prev. Med. **99**, 282–285 (2017)
7. Noc, M., et al.: Invasive coronary treatment strategies for out-of-hospital cardiac arrest: a consensus statement from the European Association for Percutaneous Cardiovascular Interventions (EAPCI)/Stent for Life (SFL) groups. EuroIntervention **10**, 31–37 (2014)
8. Lu, K.J., Chen, J.X., Profitis, K., Kearney, L.G., DeSilva, D., Smith, G., Ord, M., Harberts, S., Calafiore, P., Jones, E., Srivastava, P.M.: Right ventricular global longitudinal strain is an independent predictor of right ventricular function: a multimodality study of cardiac magnetic resonance imaging, real time three-dimensional echocardiography and speckle tracking echocardiography. Echocardiography **32**(6), 966–974 (2015). https://doi.org/10.1111/echo.12783. Epub 7 Oct 2014
9. Goebel, J., Nensa, F., Bomas, B., et al.: Real-time SPARSE-SENSE cardiac cine MR imaging: optimization of image reconstruction and sequence validation. Eur. Radiol. **26**, 4482 (2016). https://doi.org/10.1007/s00330-016-4301-y
10. Reichek, N.: Imaging cardiac morphology in hypertrophic cardiomyopathy: recent advances. Curr. Opin. Cardiol. **30**(5), 461–467 (2015). https://doi.org/10.1097/HCO.0000000000000209
11. Zhou, R., Huang, W., Yang, Y., Chen, X., Weller, D.S., Kramer, C.M., Kozerke, S., Salerno, M.: Simple motion correction strategy reduces respiratory-induced motion artifacts for k-t accelerated and compressed-sensing cardiovascular magnetic resonance perfusion imaging. J. Cardiovasc. Magn. Reson. **20**(1), 6 (2018). https://doi.org/10.1186/s12968-018-0427-1
12. Liang, X., Garnavi, R., Wail, S., Liang, S., Prasanna, P.: Automatic segmentation of the left ventricle into 17 anatomical regions in cardiac MR imaging. Conf. Proc. IEEE Eng. Med. Biol. Soc. **2015**, 6531–6535 (2015). https://doi.org/10.1109/embc.2015.7319889
13. Vasconcellos, H.D., Moreira, H.T., Ciuffo, L., Nwabuo, C.C., Yared, G.S., Ambale-Venkatesh, B., Armstrong, A.C., Kishi, S., Reis, J.P., Liu, K., Lloyd-Jones, D.M., Colangelo, L.A., Schreiner, P.J., Sidney, S., Gidding, S.S., Lima, J.A.C.: Cumulative blood pressure from early adulthood to middle age is associated with left atrial remodelling and subclinical dysfunction assessed by three-dimensional echocardiography: a prospective post hoc analysis from the coronary artery risk development in young adults study. Eur. Heart J. Cardiovasc. Imaging (2018). https://doi.org/10.1093/ehjci/jey086
14. Dilsizian, S.E., Siegel, E.L.: Artificial intelligence in medicine and cardiac imaging: harnessing big data and advanced computing to provide personalized medical diagnosis and treatment. Curr. Cardiol. Rep. **16**, 441 (2014). https://doi.org/10.1007/s11886-013-0441-8
15. Behar, J.M., Mountney, P., Toth, D., Reiml, S., Panayiotou, M., Brost, A., Fahn, B., Karim, R., Claridge, S., Jackson, T., Sieniewicz, B., Patel, N., O'Neill, M., Razavi, R., Rhode, K., Rinaldi, C.A.: Real-time X-MRI-guided left ventricular lead implantation for targeted delivery of cardiac resynchronization therapy. JACC Clin. Electrophysiol. **3**(8), 803–814 (2017). https://doi.org/10.1016/j.jacep.2017.01.018. Epub 26 Apr 2017
16. Xu, R., Athavale, P., Nachman, A., Wright, G.A.: Multiscale registration of real-time and prior MRI data for image-guided cardiac interventions. IEEE Trans. Biomed. Eng. **61**(10), 2621–2632 (2014). https://doi.org/10.1109/TBME.2014.2324998. Epub 16 May 2014

17. Wu, X., von Birgelen, C., Li, Z., Zhang, S., Huang, J., Liang, F., Li, Y., Wijns, W., Tu, S.: Assessment of superficial coronary vessel wall deformation and stress: validation of in silico models and human coronary arteries in vivo. Int. J. Cardiovasc. Imaging **34**(6), 849–861 (2018). https://doi.org/10.1007/s10554-018-1311-7. Epub 3 Feb 2018
18. Gharaie, S.H., Mosadegh, B., Morsi, Y.: In vitro validation of a numerical simulation of leaflet kinematics in a polymeric aortic valve under physiological conditions. Cardiovasc. Eng. Technol. **9**(1), 42–52 (2018). https://doi.org/10.1007/s13239-018-0340-7. Epub 10 Jan 2018
19. Allard, L., Soulez, G., Chayer, B., Qin, Z., Roy, D., Cloutier, G.: A multimodality vascular imaging phantom of an abdominal aortic aneurysm with a visible thrombus. Med. Phys. **40**(6), 063701 (2013). https://doi.org/10.1118/1.4803497
20. Lebacq, A.L., Bruggeman, M., Vanhavere, F.: Efficiency calibration of a whole-body-counting measurement setup using a modular physical phantom. Radiat. Prot. Dosimetry **144**(1–4), 411–414 (2011). https://doi.org/10.1093/rpd/ncq575. Epub 6 Jan 2011
21. Fernandez, M.C., Venencia, C., Garrigó, E., Caussa, L.: SU-E-T-454: dosimetric comparison between pencil beam and Monte Carlo algorithms for SBRT lung treatment using IPlan V4.1 TPS and CIRS thorax phantom. Med. Phys. **39**(6Part17), 3809 (2012). https://doi.org/10.1118/1.4735543
22. Pallotta, S., Calusi, S., Foggi, L., Lisci, R., Masi, L., Marrazzo, L., ... & Simontacchi, G.: ADAM: a breathing phantom for lung SBRT quality assurance. Physica. Medica. **49**, 147–155 (2018).
23. Kitagawa, K., George, R.T., Arbab-Zadeh, A., Lima, J.A.C., Lardo, A.C.: Characterization and correction of beam-hardening artifacts during dynamic volume CT assessment of myocardial perfusion. Radiology **256**(1) (2010)
24. Snyder, K.C., Xhaferllari, I., Huang, Y., Siddiqui, M.S., Chetty, I.J., Wen, N.: Evaluation and verification of the QFix Encompass™ couch insert for intracranial stereotactic radiosurgery. J. Appl. Clin. Med. Phys. **19**(4), 222–229 (2018). https://doi.org/10.1002/acm2.12387. Epub 15 Jun 2018
25. Boltz, T., Pavlicek, W., Paden, R., Renno, M., Jensen, A., Akay, M.: An anthropomorphic beating heart phantom for cardiac X-ray CT imaging evaluation. J. Appl. Clin. Med. Phys. (2010). https://doi.org/10.1120/jacmp.v11i1.3129
26. Matusiak, K., Wasilewska-Radwanska, M., Stepien, A.: Dynamic heart phantom for the quality control of SPECT equipment. Phys. Med. **24**(2), 112–116 (2008). https://doi.org/10.1016/j.ejmp.2008.01.015. Epub 10 Mar 2008
27. Ho, C.K., Chee, A.J., Yiu, B.Y., Tsang, A.C., Chow, K.W., Yu, A.C.: Wall-less flow phantoms with tortuous vascular geometries: design principles and a patient-specific model fabrication example. IEEE Trans. Ultrason. Ferroelectr. Freq. Control **64**(1), 25–38 (2017). https://doi.org/10.1109/TUFFC.2016.2636129. Epub 6 Dec 2016
28. Verkerke, G.J., van der Houwen, E.B.: Design of biomedical products. In: Rakhorst, G., Ploeg, R. (eds.) Biomaterials in Modern Medicine: The Groningen Perspective, pp. 23–38. Biomechanical Engineering, World Scientific Publishing (2008)
29. Gray, H., Pick, T.P., Howden, R.: Anatomy, Descriptive and Surgical (1901)
30. Humphrey, J.D., McCulloch, A.D.: The cardiovascular system—anatomy, physiology and cell biology. In: Holzapfel, G.A., Ogden, R.W. (eds.) Biomechanics of Soft Tissue in Cardiovascular Systems. International Centre for Mechanical Sciences (Courses and Lectures), vol. 441. Springer, Vienna (2003)
31. Moorman, A.F.M., Lamers, W.H.: Molecular anatomy of the developing heart. Trends Cardiovasc. Med. **4**(6), 257–264 (1994)
32. Jugdutt, B.I., et al.: Remodeling of the myocardium and potential targets in the collagen degradation and synthesis pathways. Curr. Drug Targets Cardiovasc. Haematol. Disord. (2003)
33. Burlew, B.S., Weber, K.T.: Connective tissue and the heart: functional significance and regulatory mechanisms. Cardiol. Clin. **18**(3), 435–442 (2000)
34. Bailey, A.J., Shimokomaki, M.S.: Age related changes in the reducible cross-links of collagen. FEBS Lett. **16**(2), 86–88 (1971)

35. Mangual, J.O., Jung, B., Ritter, J.A., Kheradvar, A.: Modeling radial viscoelastic behavior of left ventricle based on MRI tissue phase mapping. Ann. Biomed. Eng. **38**(10), 3102–3111 (2010). https://doi.org/10.1007/s10439-010-0079-2. Epub 27 May 2010
36. Cardiac Muscle Tissue | Anatomy and Physiology, Chapter 10, Muscle Tissue. https://opentextbc.ca/anatomyandphysiology/chapter/10-7-cardiac-muscle-tissue/. Accessed 16 Dec 2017
37. Özel, T., Bártolo, P.J., Ceretti, E., Gay, J.D.C., Rodriguez, C.A., Da Silva, J.V.L. (eds.): Biomedical Devices: Design, Prototyping, and Manufacturing. Wiley (2016). ISBN: 978-1-118-47892-9
38. Lantada, A.D. (ed.): Handbook on Advanced Design and Manufacturing Technologies for Biomedical Devices. Springer Science + Business Media, LLC (2013)
39. Kruth, J.P.: Material incress manufacturing by rapid prototyping techniques. CIRP Ann. **40**(2), 603–614 (1991)
40. Snoeys, R., Staelens, F., Dekeyser, W.: Current trends in non-conventional material removal processes. CIRP Ann. **35**(2), 467–480 (1986)
41. Mourtzis, D., Doukas, M.: Knowledge capturing and reuse to support manufacturing of customised products: a case study from the mould making industry. 24th CIRP design conference. Procedia CIRP **21**, 123–128 (2014)
42. Han, W., Liu, X., Radcliffe, J.H., Ghariban, M., Wei, J., Chung, K.C., Beling, P.A.: Analysis of manual manufacturing processes using motion sensing technologies. In: 2014 Systems and Information Engineering Design Symposium (SIEDS), pp. 244–249. https://doi.org/10.1109/sieds.2014.6829876
43. Feng, S.C., Bernstein, W.Z., Hedberg Jr., T., Feeney, A.B.: Towards knowledge management for smart manufacturing. ASME J. Comput. Inf. Sci. Eng. **17**(3) (2017). Paper No: JCISE-16-2051. https://doi.org/10.1115/1.4037178
44. Kimura, F., Ariyoshi, H., Ishikawa, H., Naruko, Y., Yamato, H.: Capturing expert knowledge for supporting design and manufacturing of injection moulds. CIRP Ann. Manuf. Technol. **53**(1), 147–150 (2004)
45. Nakao, M., Tsuchiya, K., Iino, K.: Design knowledge extraction from scenario-based databases using associative search engine for FR-induced decisions. CIRP Ann. **57**(1), 163–166 (2008)
46. Roy, R., Hinduja, S., Teti, R.: Recent advances in engineering design optimisation: challenges and future trends. CIRP Ann. **57**(2), 697–715 (2008)
47. Zheng, L.Y., Dong, H.F., Vichare, P., Nassehi, A., Newman, S.T.: Systematic modeling and reusing of process knowledge for rapid process configuration. Rob. Comput. Integr. Manuf. **24**(6), 763–772 (2008)
48. Wood, S., Krishnamurthy, N., Santini, T., Raval, S.B., Farhat, N., et al.: Correction: design and fabrication of a realistic anthropomorphic heterogeneous head phantom for MR purposes. PLoS One **13**(2), e0192794 (2018). https://doi.org/10.1371/journal.pone.0192794
49. Shi, J., Stenzel, R., Wenger, T., Lueth, T.C.: Accuracy study of a new assistance system under the application of Navigated Control® for manual milling on a head phantom. In: 2010 Annual International Conference of the IEEE Engineering in Medicine and Biology. https://doi.org/10.1109/iembs.2010.5627578
50. Craft, D.F., Howell, R.M.: Preparation and fabrication of a full-scale, sagittal-sliced, 3D-printed, patient-specific radiotherapy phantom. J. Appl. Clin. Med. Phys. **18**(5), 285–292 (2017)
51. Saager, R.B., Quach, A., Rowland, R.A., Baldado, M.L., Durkin, A.J.: Low-cost tissue simulating phantoms with adjustable wavelength-dependent scattering properties in the visible and infrared ranges. J. Biomed. Opt. **21**(6), 067001 (2016). https://doi.org/10.1117/1.JBO.21.6.067001
52. Arcaute, K., Wicker, R.B.: Patient-specific compliant vessel manufacturing using dip-spin coating of rapid prototyped molds. J. Manuf. Sci. Eng. **130**(5), 051008 (2008). https://doi.org/10.1115/1.2898839

53. Wong, K.V., Hernandez, A.: A review of additive manufacturing. ISRN Mech. Eng., 10, Article ID 208760 (2012). https://doi.org/10.5402/2012/208760
54. Lantada, A.D., Morgado, P.L.: Rapid prototyping for biomedical engineering: current capabilities and challenges. Annu. Rev. Biomed. Eng. **14**, 73–96 (2012)
55. Cooper, K.: Rapid Prototyping Technology: Selection and Application (Mechanical Engineering), 1st edn. Marcel Dekker Inc., CRC Press. ISBN-13: 978-0824702618, ISBN-10: 0824702611
56. Kleijnen, J.: Ethical issues in engineering models: an operations researcher's reflections. Sci. Eng. Ethics **17**(3), 539–552 (2011). https://doi.org/10.1007/s11948-010-9215-5
57. Haverman, T.M., Karagozoglu, K.H., Prins, H.J., Schulten, E.A., Forouzanfar, T.: Rapid prototyping: a very promising method. Ned. Tijdschr. Tandheelkd. **120**(3), 136–141 (2013)
58. Petzold, R., Zeilhofer, H.-F., Kalender, W.A.: Rapid prototyping technology in medicine—basics and applications. Comput. Med. Imaging Graph. **23**(5), 277–284 (1999)
59. Krauel, L., Fenollosa, F., Riaza, L., Pérez, M., Tarrado, X., Morales, A., Gomà, J., Mora, J.: Use of 3D prototypes for complex surgical oncologic cases. World J. Surg. **40**(4), 889–894 (2016). https://doi.org/10.1007/s00268-015-3295-y
60. Esses, S.J., Berman, P., Bloom, A.I., Sosna, J.: Clinical applications of physical 3D models derived from MDCT data and created by rapid prototyping. AJR Am. J. Roentgenol. **196**(6), W683–W688 (2011). https://doi.org/10.2214/ajr.10.5681
61. Torres, K., Staśkiewicz, G., Śnieżyński, M., Drop, A., Maciejewski, R.: Application of rapid prototyping techniques for modelling of anatomical structures in medical training and education. Folia Morphol. (Warsz.) **70**(1), 1–4 (2011)
62. Hawkins, J., Madden, K., Fletcher, A., Midgley, L., Grant, A., Cox, G., Moore, L., Campbell, R., Murphy, S., Bonell, C., White, J.: Development of a framework for the co-production and prototyping of public health interventions. BMC Public Health **17**(1), 689 (2017). https://doi.org/10.1186/s12889-017-4695-8
63. Biglino, G., Koniordou, D., Gasparini, M., et al.: Piloting the use of patient-specific cardiac models as a novel tool to facilitate communication during clinical consultations. Pediatr. Cardiol. **38**, 813 (2017). https://doi.org/10.1007/s00246-017-1586-9
64. Biglino, G., Capelli, C., Wray, J., Schievano, S., Leaver, L.K., Khambadkone, S., Giardini, A., Derrick, G., Jones, A., Taylor, A.M.: 3D-manufactured patient-specific models of congenital heart defects for communication in clinical practice: feasibility and acceptability. BMJ Open. **5**(4), e007165 (2015). https://doi.org/10.1136/bmjopen-2014-007165
65. Kim, M.S., Hansgen, A.R., Carroll, J.D.: Use of rapid prototyping in the care of patients with structural heart disease. Trends Cardiovasc. Med. **18**(6), 210–216 (2008). https://doi.org/10.1016/j.tcm.2008.11.001
66. Vukicevic, M., Mosadegh, B., Min, J.K., Little, S.H.: Cardiac 3D printing and its future directions. JACC Cardiovasc. Imaging **10**(2), 171–184 (2017). https://doi.org/10.1016/j.jcmg.2016.12.001
67. Kim, M.S., Hansgen, A.R., Wink, O., Quaife, R.A., Carroll, J.D.: Rapid prototyping: a new tool in understanding and treating structural heart disease. Circulation **117**(18), 2388–2394 (2008). https://doi.org/10.1161/circulationaha.107.740977
68. Singare, S., Lian, Q., Wang, W.P., et al.: Rapid prototyping assisted surgery planning and custom implant design. Rapid Prototyp. J. **15**(1), 19–23 (2009)
69. Heuts, S., Daemen, J.H.T., Streukens, S.A.F., Olsthoorn, J.R., Vainer, J., Cheriex, E.C., Maessen, J.G., Nia, P.S.: Planning of transapical beating heart mitral valve repair for safe adaptation in clinical practice. Innovations (Phila) **13**(3), 200–206 (2018). https://doi.org/10.1097/imi.0000000000000506.preoperative
70. Mottl-Link, S., Hübler, M., Kühne, T., Rietdorf, U., Krueger, J.J., Schnackenburg, B., De Simone, R., Berger, F., Juraszek, A., Meinzer, H.P., Karck, M., Hetzer, R., Wolf, I.: Physical models aiding in complex congenital heart surgery. Ann. Thorac. Surg. **86**(1), 273–277 (2008). https://doi.org/10.1016/j.athoracsur.2007.06.001
71. Bhatla, P., Tretter, J.T., Chikkabyrappa, S., Chakravarti, S., Mosca, R.S.: Surgical planning for a complex double-outlet right ventricle using 3D printing. Echocardiography **34**(5), 802–804 (2017). https://doi.org/10.1111/echo.13512. Epub 19 Mar 2017

72. Zhao, L., Zhou, S., Fan, T., Li, B., Liang, W., Dong, H.: Three-dimensional printing enhances preparation for repair of double outlet right ventricular surgery. J. Card. Surg. **33**(1), 24–27 (2018). https://doi.org/10.1111/jocs.13523
73. Pellegrino, P.L., Fassini, G., Di Biase, M., Tondo, C.: Left atrial appendage closure guided by 3D printed cardiac reconstruction: emerging directions and future trends. J. Cardiovasc. Electrophysiol. **27**(6), 768–771 (2016). https://doi.org/10.1111/jce.12960. Epub 5 Apr 2016
74. Jacobs, S., Grunert, R., Mohr, F.W., Falk, V.: 3D-imaging of cardiac structures using 3D heart models for planning in heart surgery: a preliminary study. Interact. Cardiovasc. Thorac. Surg. **7**(1), 6–9 (2008). https://doi.org/10.1510/icvts.2007.156588
75. Díaz Lantada, A., Valle-Fernández, R.D., Morgado, P.L., Muñoz-García, J., Muñoz Sanz, J.L., Munoz-Guijosa, J.M., Otero, J.E.: Development of personalized annuloplasty rings: combination of CT images and CAD-CAM tools. Ann. Biomed. Eng. **38**(2), 280–290 (2010). https://doi.org/10.1007/s10439-009-9805-z. Epub 14 Oct 2009
76. Rengier, F., Mehndiratta, A., von Tengg-Kobligk, H., et al.: 3D printing based on imaging data: review of medical applications. Int. J. CARS **5**, 335 (2010). https://doi.org/10.1007/s11548-010-0476-x
77. Filippou, V., Tsoumpas, C.: Recent advances on the development of phantoms using 3D printing for imaging with CT, MRI, PET, SPECT and ultrasound. Med. Phys. (2018). https://doi.org/10.1002/mp.13058. Epub ahead of print
78. Mitsouras, D., Liacouras, P., Imanzadeh, A., Giannopoulos, A.A., Cai, T., Kumamaru, K.K., George, E., Wake, N., Caterson, E.J., Pomahac, B., Ho, V.B., Grant, G.T., Rybicki, F.J.: Medical 3D printing for the radiologist. Radiographics **35**(7), 1965–1988 (2015). https://doi.org/10.1148/rg.2015140320
79. Ngan, E.M., Rebeyka, I.M., Ross, D.B., Hirji, M., Wolfaardt, J.F., Seelaus, R., Grosvenor, A., Noga, M.L.: The rapid prototyping of anatomic models in pulmonary atresia. J. Thorac. Cardiovasc. Surg. **132**(2), 264–269 (2006)
80. Bibb, R., Winder, J.: A review of the issues surrounding three-dimensional computed tomography for medical modelling using rapid prototyping techniques. Radiography **16**(1), 78–83 (2010)
81. Canstein, C., Cachot, P., Faust, A., Stalder, A.F., Bock, J., Frydrychowicz, A., Küffer, J., Hennig, J., Markl, M.: 3D MR flow analysis in realistic rapid-prototyping model systems of the thoracic aorta: comparison with in vivo data and computational fluid dynamics in identical vessel geometries. Magn. Reson. Med. **59**(3), 535–546 (2008)
82. Giannopoulos, A.A., Mitsouras, D., Yoo, S.-J., Liu, P.P., Chatzizisis, Y.S., Rybicki, F.J.: Applications of 3D printing in cardiovascular diseases. Nat. Rev. Cardiol. (2016)
83. Bartel, T., Rivard, A., Jimenez, A., Mestres, C.A., Müller, S.: Medical three-dimensional printing opens up new opportunities in cardiology and cardiac surgery. Eur. Heart J. **39**(15), 1246–1254 (2018). https://doi.org/10.1093/eurheartj/ehx016
84. Markl, M., Schumacher, R., Küffer, J., et al.: Rapid vessel prototyping: vascular modeling using 3t magnetic resonance angiography and rapid prototyping technology. MAGMA **18**, 288 (2005). https://doi.org/10.1007/s10334-005-0019-6
85. Farooqi, K.M., Sengupta, P.P.: Echocardiography and three-dimensional printing: sound ideas to touch a heart. J. Am. Soc. Echocardiogr. **28**(4), 398–403 (2015)
86. Miller, B.W., Moore, J.W., Barrett, H.H., Fryé, T., Adler, S., Sery, J., Furenlid, L.R.: 3D printing in X-ray and gamma-ray imaging: a novel method for fabricating high-density imaging apertures. Nucl. Instrum. Methods Phys. Res. A **659**(1), 262–268 (2011)
87. Rankin, T.M., Giovinco, N.A., Cucher, D.J., Watts, G., Hurwitz, B., Armstrong, D.G.: Three-dimensional printing surgical instruments: are we there yet? J. Surg. Res. **189**(2), 193–197 (2014)
88. Taylor, R.H., Menciassi, A., Fichtinger, G., Fiorini, P., Dario, P.: Medical robotics and computer-integrated surgery. In: Siciliano, B., Khatib, O. (eds.) Springer Handbook of Robotics. Springer, Cham (2016)
89. Kappler, M.A.: Software for rapid prototyping in the pharmaceutical and biotechnology industries. Curr. Opin. Drug Discov. Dev. **11**(3), 389–392 (2008)

90. Wang, K., Ho, C.-C., Zhang, C., Wang, B.: A review on the 3D printing of functional structures for medical phantoms and regenerated tissue and organ applications. Engineering **3**(5), 653–662 (2017)
91. Bieniosek, M.F., Lee, B.J., Levin, C.S.: Technical note: characterization of custom 3D printed multimodality imaging phantoms. Med. Phys. **42**(10), 5913–5918 (2015). https://doi.org/10.1118/1.4930803
92. Kurenov, S.N., Ionita, C., Sammons, D., Demmy, T.L.: Three-dimensional printing to facilitate anatomic study, device development, simulation, and planning in thoracic surgery. J. Thorac. Cardiovasc. Surg. **149**(4), 973–979.e1 (2015)
93. Meess, K.M., Izzo, R.L., Dryjski, M.L., Curl, R.E., Harris, L.M., Springer, M., Siddiqui, A.H., Rudin, S., Ionita, C.N.: 3D printed abdominal aortic aneurysm phantom for image guided surgical planning with a patient specific fenestrated endovascular graft system. In: Proceedings, Medical Imaging 2017: Imaging Informatics for Healthcare, Research, and Applications, vol. 10138, p. 101380P (2017) https://doi.org/10.1117/12.2253902
94. Laing, J., Moore, J.T., Vassallo, R., Bainbridge, D., Drangova, M., Peters, T.M.: Patient-specific cardiac phantom for clinical training and preprocedure surgical planning. J. Med. Imaging **5**(2), 021222 (2018). https://doi.org/10.1117/1.JMI.5.2.021222
95. Kamomae, T., Shimizu, H., Nakaya, T., Okudaira, K., Aoyama, T., Oguchi, H., Komori, M., Kawamura, M., Ohtakara, K., Monzen, H., Itoh, Y., Naganawa, S.: Three-dimensional printer-generated patient-specific phantom for artificial in vivo dosimetry in radiotherapy quality assurance. Phys. Med. **44**, 205–211 (2017)
96. Leng, S., et al.: Anatomic modeling using 3D printing: quality assurance and optimization. 3D Print. Med., 3–6 (2017)
97. Groenewald, A., Groenewald, W.A.: Development of a universal medical X-ray imaging phantom prototype. J. Appl. Clin. Med. Phys. **17**(6) (2016)
98. Esmonde-White, F.W.L., Esmonde-White, K.A., Kole, M.R., Goldstein, S.A., Roessler, B.J., Morris, M.D.: Biomedical tissue phantoms with controlled geometric and optical properties for Raman spectroscopy and tomography. Analyst **136**(21), 4437–4446 (2011)
99. Arconada-Alvarez, S.J., Lemaster, J.E., Wang, J., Jokerst, J.V.: The development and characterization of a novel yet simple 3D printed tool to facilitate phantom imaging of photoacoustic contrast agents. Photoacoustics **5**, 17–24 (2017). https://doi.org/10.1016/j.pacs.2017.02.001
100. Hougen, O.A., Watson, K.M., Ragatz, R.A.: Chemical Process Principles. Wiley, New York (1952)
101. Brown, T.E., LeMay, H.E.H., Bursten, B.E., Murphy, C., Woodward, P.: Chemistry: The Central Science. Prentice Hall, Upper Saddle River (2008)
102. Dabbagh, A., Abdullah, B.J.J., Ramasindarum, C., Abu Kasim, N.H.: Tissue-mimicking gel phantoms for thermal therapy studies. Ultrason. Imaging **36**(4), 291–316 (2014)
103. Zell, K., Sperl, J.I., Vogel, M.W., Niessner, R., Haisch, C.: Acoustical properties of selected tissue phantom materials for ultrasound imaging. Phys. Med. Biol. **52**, N475–N484 (2007). https://doi.org/10.1088/0031-9155/52/20/N02
104. Hellerbach, A., Schuster, V., Jansen, A., Sommer, J.: MRI phantoms—are there alternatives to agar? **8**(8), e70343 (2013). https://doi.org/10.1371/journal.pone.0070343
105. Krevelen, D.W.: Some basic aspects of flame resistance of polymeric materials. Polymer **16**(8), 615–620 (1975)
106. Jones, R.M.: Mechanics of Composite Materials. Taylor & Francis (1999)
107. Dwivedi, S.P., Gupta, B., Chaudhary, D.: The effect of process parameters on mechanical stir casting process. NIET J. Eng. Technol. (2014)
108. Mark, J.: Physical Properties of Polymers Handbook, 2nd edn. Springer Science + Business Media, LLC (2007)
109. Gogos, G.: Bubble removal in rotational molding. Polym. Eng. Sci. **44**, 388–394 (2004). https://doi.org/10.1002/pen.20035
110. Kontopoulou, M., Vlachopoulos, J.: Bubble dissolution in molten polymers and its role in rotational molding. Polym. Eng. Sci. **39**, 1189–1198 (1999). https://doi.org/10.1002/pen.11505

111. Nielsen, L.E.: Cross-linking–effect on physical properties of polymers. J. Macromol. Sci. Part C Polym. Rev. **3**(1), 69–103 (1969)
112. Chen, X., Dam, M.A., Ono, K., Mal, A., Shen, H., Nutt, S.R., Sheran, K., Wudl, F.: A thermally re-mendable cross-linked polymeric material. Science **295**(5560), 1698–1702 (2002). https://doi.org/10.1126/science.1065879
113. Hoi, Y., Woodward, S.H., Kim, M., Taulbee, D.B., Meng, H.: Validation of CFD simulations of cerebral aneurysms with implication of geometric variations. J. Biomech. Eng. **128**(6), 844–851 (2006). https://doi.org/10.1115/1.2354209
114. Latorre, R., Bainbridge, D., Tavernor, A., Lopez Albors, O.: Plastination in anatomy learning: an experience at Cambridge University. JVME **43**(3), 8 (2016). AAVMC. https://doi.org/10.3138/jvme.0715-113r1
115. Duck, F.A.: Physical Properties of Tissue: A Comprehensive Reference Book. Academic, Harcourt Brace Jovanovich, London (1990)
116. Li, C., Huang, Z., Wang, R.K.: Elastic properties of soft tissue-mimicking phantoms assessed by combined use of laser ultrasonics and low coherence interferometry. Opt. Express **19**, 10153–10163 (2011)
117. Fromageau, J., Gennisson, J.-L., Schmitt, C., Maurice, R.L., Mongrain, R., Cloutier, G.: Estimation of polyvinyl alcohol cryogel mechanical properties with four ultrasound elastography methods and comparison with gold standard testings. IEEE Trans. Ultrason. Ferroelectr. Freq. Control **54**(3) (2007)
118. Hollister, S.J., Maddox, R.D., Taboas, J.M.: Optimal design and fabrication of scaffolds to mimic tissue properties and satisfy biological constraints. Biomaterials **23**(20), 4095–4103 (2002)
119. Lamouche, G., Kennedy, B.F., Kennedy, K.M., Bisaillon, C.-E., Curatolo, A., Campbell, G., Pazos, V., Sampson, D.D.: Review of tissue simulating phantoms with controllable optical, mechanical and structural properties for use in optical coherence tomography. Biomed. Opt. Express **3**(6), 1381–1398 (2012). https://doi.org/10.1364/BOE.3.001381
120. Bisaillon, C.-E., Lamouche, G., Maciejko, R., Dufour, M., Monchalin, J.-P.: Deformable and durable phantoms with controlled density of scatterers. Phys. Med. Biol. **53**(13) (2008)
121. Chen, R., Shih, A.: Multi-modality gellan gum-based tissue-mimicking phantom with targeted mechanical, electrical, and thermal properties. Phys. Med. Biol. **58**, 5511–5525 (2013). https://doi.org/10.1088/0031-9155/58/16/5511
122. Shukla, R., Ghosh, S.K., Chakrabarti, D., Chatterjee, S.: Microstructure, texture, property relationship in thermo-mechanically processed ultra-low carbon microalloyed steel for pipeline application. Mater. Sci. Eng. A **587**(10), 201–208 (2013)
123. Graedel, N.N., Polimeni, J.R., Guerin, B., Gagoski, B., Wald, L.L.: An anatomically realistic temperature phantom for radiofrequency heating measurements. Magn. Reson. Med. **73**, 442–450 (2015)
124. McDannold, N., Hynynen, K.: Quality assurance and system stability of a clinical MRI-guided focused ultrasound system: four-year experience. Med. Phys. **33**, 4307 (2006). https://doi.org/10.1118/1.2352853
125. Wu, T., Felmlee, J.P.: A quality control program for MR-guided focused ultrasound ablation therapy. J. Appl. Clin. Med. Phys. **3**(2), 162–167 (2002)
126. Holt, R.G., Roy, R.A.: Measurements of bubble-enhanced heating from focused, MHz-frequency ultrasound in a tissue-mimicking material. Ultrasound Med. Biol. **27**(10), 1399–1412 (2001)
127. Yuan, Y., Wyatt, C., Maccarini, P., Stauffer, P., Craciunescu, O., MacFall, J., Dewhirst, M., Das, S.K.: A heterogeneous human tissue mimicking phantom for RF heating and MRI thermal monitoring verification. Phys. Med. Biol. **57**(7) (2012)
128. Choi, M.J., Guntur, S.R., Lee, K.I., Paeng, D.G., Coleman, A.: A tissue mimicking polyacrylamide hydrogel phantom for visualizing thermal lesions generated by high intensity focused ultrasound. Ultrasound Med. Biol. **39**(3), 439–448 (2013)
129. Chin, R.B., Madsen, E.L., Zagzebski, J.A., Jadvar, H., Wu, X.-K., Frank, G.R.: A reusable perfusion supporting tissue-mimicking material for ultrasound hyperthermia phantoms. Med. Phys. **17**(3), 380–390 (1990)

130. Gabriel, C., Gabriel, S., Corthout, E.: The dielectric properties of biological tissues: I. Literature survey. Phys. Med. Biol. **41**(11) (1996)
131. Gabriel, S., Lau, R.W., Gabriel, C.: The dielectric properties of biological tissues: II. Measurements in the frequency range 10 Hz to 20 GHz. Phys. Med. Biol. **41**(11) (1996)
132. Gabriel, S., Lau, R.W., Gabriel, C.: The dielectric properties of biological tissues: III. Parametric models for the dielectric spectrum of tissues. Phys. Med. Biol. **41**(11) (1996)
133. Joines, W.T., Zhang, Y., Li, C., Jirtle, R.L.: The measured electrical properties of normal and malignant human tissues from 50 to 900 MHz. Med. Phys. **21**(4), 547–550 (1994)
134. Belmont, B., Dodde, R.E., Shih, A.J.: Impedance of tissue-mimicking phantom material under compression. J. Electr. Bioimpedance **4** (2013). http://dx.doi.org/10.5617/jeb.443
135. Roberjot, V., Bridal, S.L., Laugier, P., Berger, G.: Absolute backscatter coefficient over a wide range of frequencies in a tissue-mimicking phantom containing two populations of scatterers. IEEE Trans. Ultrason. Ferroelectr. Freq. Control **43**(5) (1996)
136. Asami, R., Ikeda, T., Azuma, T., Umemura, S., Kawabata, K.-I.: Acoustic signal characterization of phase change nanodroplets in tissue-mimicking phantom gels. Jpn. J. Appl. Phys. **49**(7S) (2010)
137. Spirou, G.M., Oraevsky, A.A., Vitkin, I.A., Whelan, W.M.: Optical and acoustic properties at 1064 nm of polyvinyl chloride-plastisol for use as a tissue phantom in biomedical optoacoustics. Phys. Med. Biol. **50**, N141 (2005)
138. Abbott, J.G., Thurstone, F.L.: Acoustic speckle: theory and experimental analysis. Ultrason. Imaging **1**(4), 303–324 (1979)
139. Haralick, R.M.: Statistical and structural approaches to texture. Proc. IEEE **67**(5), 786–804 (1979)
140. Szczesniak, A.S.: Texture is a sensory property. Food Qual. Prefer. **13**(4), 215–225 (2002)
141. Garland, M., Heckbert, P.S.: Simplifying surfaces with color and texture using quadric error metrics. In: Proceedings Visualization '98 (Cat. No. 98CB36276). https://doi.org/10.1109/visual.1998.745312
142. Bunge, H.J.: Texture Analysis in Materials Science: Mathematical Methods. Butterworths (1982)
143. Randle, V., Engler, O.: Introduction to Texture Analysis Macrotexture, Microtexture and Orientation Mapping, 1st edn. CRC Press (2014)
144. Sharan, L., Liu, C., Rosenholtz, R., et al.: Recognizing materials using perceptually inspired features. Int. J. Comput. Vis. **103**, 348 (2013). https://doi.org/10.1007/s11263-013-0609-0
145. Weszka, J.S., Rosenfeld, A.: An application of texture analysis to materials inspection. Pattern Recogn. **8**(4), 195–200 (1976)
146. Haindl, M., Filip, J.: Visual Texture: Accurate Material Appearance Measurement, Representation and Modeling. Springer-Verlag, London (2013)
147. González, E., Bianconi, F., Álvarez, M.X., Saetta, S.A.: Automatic characterization of the visual appearance of industrial materials through colour and texture analysis: an overview of methods and applications. Adv. Opt. Technol. **2013**, 11, Article ID 503541 (2013). https://doi.org/10.1155/2013/503541
148. Park, J., Bae, Y., Bae, Y., Kang, H., Lee, K.-J., Jung, B.: Fabrication of double layer optical tissue phantom by spin coating method: mimicking epidermal and dermal layer. In: Proceedings of the SPIE 8583, Design and Performance Validation of Phantoms Used in Conjunction with Optical Measurement of Tissue V, p. 85830G, 28 Feb 2013. https://doi.org/10.1117/12.2003603; https://doi.org/10.1117/12.2003603
149. Kennedy, B.F., McLaughlin, R.A., Kennedy, K.M., Chin, L., Curatolo, A., Tien, A., Latham, B., Saunders, C.M., Sampson, D.D.: Optical coherence micro-elastography: mechanical-contrast imaging of tissue microstructure. Biomed. Opt. Express **5**, 2113–2124 (2014)
150. Gunadi, S., Leung, T.S.: Regional sensitivity comparison between optical and acousto-optic sensing. In: Proceedings of the SPIE 7899, Photons Plus Ultrasound: Imaging and Sensing 2011, p. 78992Q, 28 Feb 2011. https://doi.org/10.1117/12.874033; https://doi.org/10.1117/12.874033

151. Wabnitz, H., Taubert, D.R., Funane, T., Kiguchi, M., Eda, H., Pifferi, A., Torricelli, A., Macdonald, R.: Characterization of homogeneous tissue phantoms for performance tests in diffuse optics. In: Proceedings of the SPIE 9700, Design and Quality for Biomedical Technologies IX, p. 970004, 18 Mar 2016. https://doi.org/10.1117/12.2209038; https://doi.org/10.1117/12.2209038
152. Jacques, S.L.: Optical properties of biological tissues: a review. Phys. Med. Biol. **58**, 5007 (2013)
153. Rosin, P.L.: Measuring shape: ellipticity, rectangularity, and triangularity. Mach. Vis. Appl. (2000). https://doi.org/10.1007/s00138-002-0118-6
154. Jähne, B.: Shape analysis. In: Digital Image Processing. Springer, Berlin, Heidelberg (1997)
155. Zhang, M., Golland, P.: Statistical shape analysis: from landmarks to diffeomorphisms. Med. Image Anal. **33**, 155–158 (2016). https://doi.org/10.1016/j.media.2016.06.025
156. Montalvo, L.S.S., Pulido, L.F., Flores, P.M.Q., Cuamatzi, P.H.: Analysis of material shape for recycling through digital image processing (ANFOR). In: 2014 9th International Conference on Software Engineering and Applications (ICSOFT-EA), 29–31 Aug 2014
157. Jones, L.A., Higgins, G.C., Stultz, K.F.: Photographic granularity and graininess. VIII.* A method of measuring granularity in terms of the scanning area giving a threshold luminance gradient. J. Opt. Soc. Am. **45**(2), 107–112 (1955)
158. Jones, A.K., Hintenlang, D.E., Bolch, W.E.: Tissue-equivalent materials for construction of tomographic dosimetry phantoms in pediatric radiology. Med. Phys. **30**(8), 2072–2081 (2003)
159. Cubeddu, R., Pifferi, A., Taroni, P., Torricelli, A., Valentini, G.: A solid tissue phantom for photon migration studies. Phys. Med. Biol. **42**(10)
160. Aranda-Lara, L., et al.: Biological tissue modeling with agar gel phantom for radiation dosimetry of 99mTc. Open J. Radiol. **4**(1) (2014)
161. Perisinakis, K., Damilakis, J., Theocharopoulos, N., Manios, E., Vardas, P., Gourtsoyiannis, N.: Accurate assessment of patient effective radiation dose and associated detriment risk from radiofrequency catheter ablation procedures. Circulation **104**(1) (2001)
162. Fisher, R.A.: The arrangement of field experiments. J. Min. Agric. Gr. Brit. **33**, 503–513 (1926)
163. Huang, X., Xie, M.: Evolutionary Topology Optimization of Continuum Structures: Methods and Applications. Wiley (2010)
164. Mansfield, S.M., Han, G., Liebmann, L.W.: Through-process modeling for design-for-manufacturability applications. J. Micro Nanolithogr. MEMS MOEMS **6**(3), 031007 (2007). https://doi.org/10.1117/1.2774987
165. King, P.H., Fries, R.C., Johnson, A.T.: Design of Biomedical Devices and Systems, 3rd edn. CRC Press (2015)

Computational Human Models in Cardiovascular Imaging: From Design to Generations

Nurulazirah Md Salih and Dyah Ekashanti Octorina Dewi

Abstract Computational human models have evolved into important tools in simulating various sets of cardiovascular circumstances to support the understanding, diagnosis, and treatment of diseases. The involvement of cardiac imaging in cardiac models have shown great improvement in representing the anatomy and physiology of the heart, not only with better functions, but also equipped with more realistic geometrical and structural features. In consort with the advancements of computing and imaging methods, different types of cardiac models have also been built in large range of applications and purposes. The implementations of most cardiac models are mainly referred to the general computational human model types, from Stylized, Voxelized, to Boundary representation (BREP) models, from the whole heart to partial cardiac regions, and from single to multiple cardiac applications. These wide range of cardiac models are interesting to analyze as a part of learning process, model improvements, and application adaptation. This chapter reviews the development of generations in computational human models for cardiovascular applications. The objective of this review is to map the state-of-the-art in the development of computational human models in cardiovascular applications. The review highlights on the uses of different types of computational human model approaches, the evolution of cardiac models in various anthropometry, as well as the advancements of computational models as whole and partial regions of the heart. Some technical hints are also elaborated to provide brief understanding on the development process of cardiac models.

Keywords Computational human model · Cardiovascular imaging · Development · Simulation · Generation · Stylized · Voxelized · BREP

N. M. Salih · D. E. O. Dewi (✉)
IJN-UTM Cardiovascular Engineering Center, Institute of Human Centered Engineering, Johor Bahru, Malaysia
e-mail: dyah@utm.my

School of Biomedical Engineering and Health Sciences, Faculty of Engineering, Universiti Teknologi Malaysia, Johor Bahru, Malaysia

D. E. O. Dewi et al. (eds.), *Cardiovascular Engineering*, Series in BioEngineering,
https://doi.org/10.1007/978-981-10-8405-8_3

1 Introduction

Computational human models and physical phantoms have become common tools in simulation, evaluation, and validation of various biomedical applications [1, 2]. In conjunction with higher necessities in revealing the anatomical structures, models with spatiotemporal description in higher-dimensional representation are seen to be more comprehensive in describing the human tissue, organ, or body [2]. Electrophysiology and biomechanics have also been used in simulating the physiological mechanisms of the body [3, 4]. Thus, the direction of the computational human models has expanded largely into more complex modelling techniques [5]. Also, in line with the trend of personalized and preventive medicine, advanced models with patient-specific characteristics are required in supporting the understanding, diagnosis, treatment of diseases [6]. Not only that, clinical trials of medical intervention and drug deliveries also require proper modelling techniques to establish the behaviour, predict the impacts, and aid the developments [7].

Unlike the physical phantoms that represent the actual matters of human directly, the computational human models, or digital phantoms, represent the human by using biomedical data modelling and simulations in digital manners [1, 2, 8]. With this way, computational human models have more freedom from physical phantom problems, for instance inadequate fabrication techniques or unmatched material properties with synthetic chemical materials. Also, these human models can be developed with more variations and features by adding some functions or adjusting a number of components and parameters. Therefore, even physical tactile representation is still on going, its implementations have much wider scopes.

As the main source for obtaining anatomical and physiological representation of the human tissue, organ, and body, biomedical imaging systems play important roles in gathering actual medical images from real human, either in individual, group, or population scale [2, 9]. The involvement of imaging algorithms, such as image processing, computer vision, artificial intelligence, and visualization, in the development of the human models has added the values with more realistic geometrical features and better functions [10, 11]. On the other way around, these human models can also be applied in biomedical imaging applications to simulate certain processes, such as verification of imaging quality, radiotherapy simulation, planning and learning in image-based intervention and surgery, and so on [1, 2, 12].

In the case of cardiovascular diseases (CVDs), cardiac models are crucial in supporting clinicians in defining the best strategies in diagnostic and treatment management. The well-established techniques based on electrophysiology and biomechanics have shown their significances in simulating the biophysical signals, heartbeats, and blood flows that describe various cardiac pathophysiology states [4, 13–16]. Also, the substantial innovations in biomedical imaging technologies have evolved the cardiac models into cardiac imaging models with combined geometrical, structural, and functional representation of the heart in a more realistic way. Different types of cardiac imaging models have been introduced in large range of applications and purposes, from simple quadratic equations in Stylized models to advanced Boundary representation (BREP) models, from the whole heart to partial cardiac regions, and from single to multiple cardiac applications [2, 13–18].

This review study highlights on development of cardiovascular computational human models, simulation methods for developing computational human models, generations and types of computational models in cardiac modelling, evolution of cardiac models, cardiac models in whole and partial versions, as well as their benefits and challenges. The objective of this review is *to map the state-of-the-art in the development of computational human models in cardiovascular applications*. Some technical hints are also elaborated to give an understanding on important tools in the development process of cardiac models.

2 Computational Human Models Developments

A computational model of human anatomy is generally defined as a virtual representation of the external shape and structures of human body with internal organs and tissues boundaries. This model simulates and studies the characteristics and mechanisms of human body by applying fundamental techniques in mathematics, physics, imaging, and computing [1, 2, 8]. Similar to physical phantoms, developing a computational human model for biomedical applications necessitates an understanding that the model is a biomedical product that requires systematic standardized processes. However, as it is also in the means of computational method, it has to be seen as a software product that entails a set of software design too. Therefore, computational human models must be created and developed through combination of generic biomedical product design as taken from biomedical prototyping theory and model-driven software development as adapted from Model Driven Architecture (MDA) theory [19, 20].

In the context of seeing computational human models as biomedical product as in [19], the development processes have to consider about complex series of high standard requirements that match between biomedical and engineering solutions before the models are designed. The requirements are basically about functional analysis, matrix check list, and regulatory guidelines [21]. When these development processes are met with the software engineering design part as in [20], the model-driven software development plays role to translate the requirements into a set of meta-models that describe the processes in the development of computational human model software. Thus, adapted from biomedical product development in [19] and software product development in [20], the generic development processes of computational human models can be divided into six stages; concept, functional design, meta-modelling, code implementation, testing and analysis, and optimization, as described in Fig. 1.

Specifically in computational human models development for cardiovascular imaging applications, this model development scheme is certainly beneficial for building a well-structured process and identifying all detailed elements that build the cardiovascular imaging model. In this way, biomedical model standards must also be

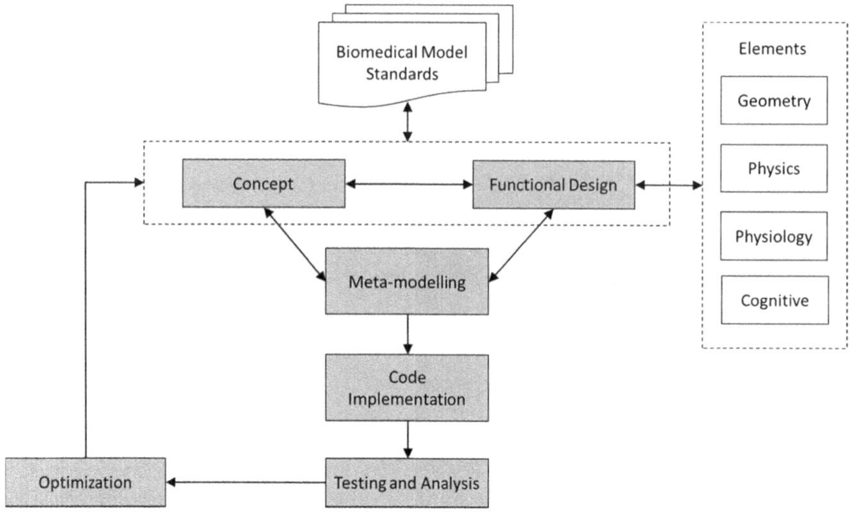

Fig. 1 Six stages, from concept to optimization, with four elements, in computational human model development referring to biomedical model standards for biomedical applications

adjusted with standards in cardiovascular imaging that involve all formal regulatory aspects from cardiology, radiology, and other related fields.

2.1 Concept

Concept stage, as the initial process in developing computational human model, is the process where the model requirements from the clinical side are placed, described, elaborated, analysed, and integrated with the engineering side. In this process, all components, including problems, goals, functions, and parameters are systematically organized to generate a model concept. Each component is then referred to the biomedical model standards to comply with the biomedical product requirements. The components are then delivered into the next stage to develop the functional design. In cardiovascular imaging models, the concept is related to solving problems in the clinical practices or scientific research in cardiovascular imaging that range from prevention, diagnostic, treatment monitoring, surgery and intervention, and many more.

2.2 Functional Design

In functional design stage, the big picture of the design and all elements that build the computational human model are created. The functions of each element are also defined in more quantitative or with descriptive manners with measurable mechanisms and parameters. Once the functions are created, the networks and interconnectivities of all elements are generated by denoting to the characteristics and mechanisms of human body that are modelled.

Based on their functional design, computational human models can be constructed by four major levelled elements which are built systematically depending on the required details in the design [9]:

1. Geometrical element.
2. Physical element.
3. Physiological element.
4. Cognitive element.

Geometrical Element

As elaborated in [9], the basic level of the model design, the geometrical element, adopts anatomical structures of the human tissue, vessels, organ, or body to be modelled into a mathematical representation in digital form. In this way, this element is responsible in digitally capturing, geometrically characterizing, and designing the boundary, region, shape, and size of the objects. This first element of this functional design basically facilitates the generation of the anthropometrical characteristics in anthropomorphic anatomical model [1, 2, 9, 22, 23].

In cardiovascular imaging application, geometrical element is represented from the boundary, region, shape, and size of anatomical structure of the heart and its vascular system. It is simply defined as the heart organ with four chambers, valves, and their structures, and blood vessels structures with their interconnections. The other element that needs to be taken into account is imaging modality. In this way, it needs to be understood that imaging physics determine the characteristics of heart anatomical features. Even capturing the same object, different imaging modality may generate different anatomical characteristics. Hence, it is essential to specify the imaging modality and which part of anatomical structures to be modelled.

Physical Element

The second level of the model design, the physical element, addresses the physical and biomechanical modelling of the object [9]. This element involves description of various physical properties of the structures in the human tissue, vessels, organ, or body. Finite element analysis is mostly chosen for assessing structure-properties relationships through physical and mechanical characteristics of the living materials [24]. The geometrical element and this physical element can be combined together, resulting in the combined anatomical and physical representation of the structure for specifying certain physiological functions.

As a very complex circulation system that dynamically moves supplying the blood into the whole human body, physical elements only cannot fully describe the actual mechanism of the heart and its vascular systems. As in [9], the model design is supposed to cover more various parameters, such as muscular activity, electrochemical activation, external temperature, state of health, blood pressure, and flow rate, and so on. Thorough understanding on the clinical problem that needs to be solved may describe the specific parameters to be modelled. Again, it also needs to refer to the physical properties that imaging modality can acquire.

Physiological Element

The third level of the model design, the physiological element, constructs the functions and mechanisms of the physiological systems in the model design, such as cardiovascular, respiratory, digestive, endocrine, musculoskeletal, and nervous systems. In this level, biological, chemical, and physical functions and mechanisms are involved within a unified living system [9, 25]. Besides normal physiological states, pathological metabolisms of diseases or disorders can also be parts of this physiological system model design [26]. This third element build an integrated system by merging several second element in physical functions and first elements in anatomical geometries and interconnecting all elements in certain degree of complexities for defining involved processes in the body.

In describing physiological functions and mechanisms in cardiovascular imaging models, larger scale of elements are involved, interconnected, and working together [9]. To abridge, it is better to limit the modelling scope by focusing on certain physiological state, such as blood flow; myocardium structure; electrical conduction system, cardiac cycle and cardiac output, and systemic blood circulation system. As most models are derived from biophysical representation of the human heart and circulation, medical imaging may generate anatomical and physiological information that reveal various mechanical and metabolic processes in cell, tissue, organ, and body levels. This principle also applies in modelling pathological metabolism of the heart. In this regard, characteristics between normal and abnormal state of the heart are required.

Cognitive Element

The fourth level of the model design, the cognitive element, develops the advanced functions of the model with close interactions among systems in the body, and possibly relates to the behaviour phenomena of the model [9]. This highest element is purposely generated to support the comprehension and response of complex mechanism in human cognition and intelligence [27]. With its complexity, developing a model in this level necessitates readily available elements in the lower levels and description of interaction as the stimuli.

Even not directly focused for cardiovascular applications, cognitive element may also be applied to describe the cognitive impacts related to cardiovascular complication as in [28]. However, not all models aim to represent the object up to this high level. Hence, the two previous stages, concept and functional design, are the determining stages in which level the model is focused on. Reconfirmation with biomedical model standards is also needed to conform as the biomedical product. Once all detailed functional design elements are described, the next stages are more about the software design stages.

2.3 Meta-modelling

Meta-modelling stage is process of expressing models by generating meta-models through statements. This stage translates the concept and functional design as a model that is expressed in a well-defined notation to system understanding and consists of interconnected and integrated representations of meta-models. These meta-models impose a series of transformations among models that are organized into an architectural framework of layers and transformation. The meta-models can be in the forms of mathematical relations with inputs, abstraction, properties, parameters, and outputs, representing the mechanism or behaviour that are arranged in such a way that it conforms the used coding platform [20].

Meta-modelling scheme is chosen in the computational human models is due to the capability of this scheme in generating the abstraction based on models by representing the elements in a simpler way. With this way, the characteristics, properties, and functions of the system can be analysed and interconnected easier to be used to generate the meta-models. Then, the meta-models are transformed into code implementation based on the architecture in the software system. According to our knowledge, there should be no difference in meta-modelling, coding, testing, and optimization between cardiovascular application and other applications.

2.4 Code Implementation

Code implementation stage is the process to execute mathematical plans in meta-models by coding or writing algorithms as program, software, or system with specific standardized language syntax. As programming languages are too many, it is beneficial to define an apt programming language and platform for an optimum model. High level programming languages have been used for prototyping as they are relatively easier and faster to code compared to lower level languages, especially when sets of libraries are available. Yet, the trade-offs are in memory allocation and computation time. In some points, combination of several programming languages and platforms may provide better performance. The use of parallel computing may also increase the operational performance of the model [29].

2.5 Testing and Analysis

Testing and analysis stage is aimed at verifying the code performance in computational model. This mechanism is vital to value the correctness and reliability of the codes. This stage applies proper sets of experimental design and performance measurement from simple to complex cases, from dummy to actual datasets, and from small to large number of datasets. Quantitative and qualitative analyses are made to rate the performance. The outcomes of this stage are generally the degree of confidence of the model performance.

2.6 Optimization

Optimization stage is the end stage that functions as feedback loop to the concept stage for iterative process. This stage receives the input from testing and analysis when the degree of confidence of the model performance is in need of improvements. Optimization techniques in computation can also become options for improving the model performance.

3 Simulation Methods in Computational Human Models

Playing role as the main engine in the development of computational human models, the rapid growing field of simulation methods have largely contributed in the fundamental development of the models, triggered faster establishment of newer generations in computational human models [1, 2, 22]. Although no formal rule mentioning about classification of simulation methods based on function, by adapting to the functional design classification from Ref. [9], it is somehow beneficial to refer to the functional design in the computational human model development before selecting the simulation method. This is because each simulation method has its specific mathematical characteristics that suit with certain model, type of user, technical tools, biomedical application, and all that. Thus, to ease, we classify the simulation methods into geometry, physical, physiological, and cognitive aspects, following the levels in the functional design. However, choosing the best simulation method is always the authority of the researcher. Also, combination of more than one simulation methods is quite common to obtain an integrated model.

3.1 Geometrical Simulation

Geometrical-based simulation methods are seen as the most deep-rooted methods, because geometrical characteristics in boundary, region, shape, and size are the initial anthropometrical features obtained for the computational human models. Large number of methods in modelling and rendering have been proposed as mixture among mathematics, computer graphics, and geometric modelling. Based on concept modelling, geometrical-based simulation methods can be divided into NURBS (Non-Uniform Rational Basis Spline), Polygons, and Subdivision surfaces [30, 31].

NURBS, as rooted from B-spline model, has shown to be major methods for representing curve properties in anthropomorphic computational human models for biomedical applications [9]. For its forte in interpolating and approximating smooth curves and surfaces with high degree of accuracy and controllable rounding-off error through weighting factors, NURBS is highly efficient in representing deformable geometry of the model [32, 33]. In imaging-based models with multidimensional representations, NURBS, B-spline models, and their derivative variations have been used in anatomical and motion representation, image reconstruction, surgical simulations, and many more, including cardiovascular imaging models [1, 2, 8, 9, 32–35].

Besides B-spline models, Polygons and Subdivision surfaces have also sounded in multi-dimensional geometrical-based modelling and rendering. Polygons are developed based on approximation of surfaces using Polygon Mesh as polyhedral object built and shaped from arrangements of elements consisting of individual vertices, edges, and faces that are represented in specific data structure for geometry processing [36]. By using variety of mesh operations to build the surface and volume, multi-dimensional polygonal representation can be flexibly developed for various modelling applications. In computational human models, any modification of human body and organ can be made by altering the control points of polygon mesh wireframes, then represented into voxels [37].

Subdivision surfaces, as generalization of spline surfaces, have better approximation with smoother surface by iteratively subdivide the polygonal face into smaller faces, then new vertices and new faces are created generating refined scheme. Although smoother, this method suffers from sampling artifacts and loss of information when it is over-subdivided. Therefore, through a coarse control mesh, Subdivision surfaces are capable of processing surfaces in arbitrary topology with better flexibility and efficiency [36].

3.2 Physical Simulation

In physical-based simulation methods, physical processes, such as mechanical, electrical, fluid, magnetic, and thermal systems, are used to generate physical representations of human tissue, organ, and body. Systematic modelling techniques are implemented to model physical phenomenon for simulating the physical characteristics and supporting diagnosis and treatment. With these computational simulations, gaps due to complicated experimental observation of some biomechanical characteristics in spatiotemporal scales can be easily solved in more efficient and interactive manners [9, 38].

With its capability of describing wide range of mechanisms from macroscopic to microscopic scales, physical-based simulation methods using Finite Difference Methods (FDM) and Finite Element Methods (FEM) principles have been widely applied in modelling applications in biomedical and non-biomedical fields, such as modelling using computational fluid dynamics for blood flow analysis in human arteries [39, 40], ergonomics and human factor analysis [41, 42], heat transfer modelling in human body [43, 44], and telecommunication health impacts [45].

The other biomechanical modelling scheme, for instance radiation transfer model, has also brought huge benefits in biomedical fields. One of the renowned methods that have been extensively used in radiation dose estimation and radiation therapy management is Monte Carlo simulation [1, 2, 8, 22, 46–48]. With its strength in inherent random process, Monte Carlo reduces computational complexity in simple, flexible, and scalable way, making it efficient for difficult sampling, estimation, and optimization problems [46–48]. In modelling radiation transport for radiation dose analysis, Monte Carlo also has well-established modelling through multifactorial aspects in anatomy, tissue and organ radio-sensitivity, computational efficiency, and geometrical compatibility [2, 22]. Cardiovascular imaging has gained beneficial impacts from Monte Carlo simulation. In radiographic imaging modalities, this simulation is capable of determining optimum radiation exposures from X-ray and Computed Tomography (CT) Angiography during catheterization and surgical monitoring while maintaining image quality during intervention [49].

Mass-spring model is also one other important method that is grown for various types of mechanical-based computational human models and biomedical simulators [9]. With spring connected nodes mesh to simulate the deformation, this physical model has been known for its simplicity of implementation and lower computational complexity in calculating the complex deformation, specifically in soft tissues [9, 50]. The works in [51–53] have focused on modelling the soft tissue deformation in the heart tissues.

3.3 *Physiological Simulation*

Computational representation in higher level of physiological and pathological states with functional and metabolic mechanisms has demanded for more complex modelling methods. Biomedical feature levels are not only characterized from human body or organ, but also certain deeper degree of details up to microscopic scales of tissue, cellular, or molecular structures [9]. With its complex components, quantitative experimental approaches are not sufficient to describe the insights into functional and metabolic mechanisms. Integration of quantitative experimental and computational modelling is considered to be the best way to unravel details of functional and metabolic information in molecular levels [54, 55].

Referring to [55], techniques in computational biology or bioinformatics can be categorized into two, knowledge-based discovery and simulation-based analysis. Knowledge-based discovery extracts knowledge in hidden patterns from large volume of experimental data based on heuristics on statistical discriminators using tools from statistics, machine learning, and pattern recognition. Data mining has been employed to search for patterns of interests for prediction and description [54–56]. On the other hand, simulation-based analysis uses in silico experiments to predict the systems dynamics in the in vitro and in vivo studies [54, 55, 57, 58].

Along with the growth of physiological simulation, initiatives of roadmap on physiological simulation come up [59–61]. Discussions on roadmap for visions and strategies in physiological simulation have been organized in term of organization, development, adoption, dissemination, and translation of virtual human concept and its extensions. This planning has also been realized into project consortiums for engaging scientific and industrial networks, common standards, open source software, accessible data and model repositories, and training and activities. Moreover, its implementations in cardiovascular applications have also shown been made, especially in biomarker discovery and Physiome [62, 63].

3.4 Cognitive Simulation

In human behaviours, neuroscience, and social interaction applications, cognitive simulations have been widely employed for numerous analyses and predictions. Computational cognitive modelling examines and extracts cognitive functionalities through computational models by descripting cognitions in computer programs [64]. As mentioned in [64, 65], models in cognitive science consist of computational, mathematical, and verbal-conceptual models. The computational models uses algorithmic descriptions in clarifying cognitive processes. In mathematical models, relationship among variables are defined mathematically. Verbal-conceptual models elaborates all involved entities, relations, and processes intuitively by using less formal natural languages.

4 Basic Generations in Computational Human Models

Back to the history, revolutions in computational human models have been on track since 1960s [1, 2, 8, 22, 46]. Along with the current swift uprising in computing, computational human models have moved forward from generation to generation. Faster computing transformations with multi-scale modelling have enabled to create human models at multiple levels in shorter time. Supported by sophisticated data acquisition and analysis, newer generations of models are equipped with more realistic and comprehensive features. With the superiorities in capturing many human characteristics and operating in various scenarios, computational human models have key features that attract more users from time to time [66].

Utilization of computational models in biology, chemistry, physics, medicine, social sciences, and other fields has shown to transform the conventional way of working into predictive, well-planned, and analytical structures, thus generate more productivity and benefits [1, 2, 8]. In biomedical applications, the establishment of computational human models has brought larger impacts where the analysis of big volume of biomedical data can be made easier, faster, more accurate, and efficient. Also, any difficult and high risk actions, for instance in radiation safety or medical interventions and surgeries, can be demonstrated without actually performing in real life or sacrificing living creatures, and even in repeated versions with many scenarios.

Therefore, using simulations, realization of any strategies can be predicted in advance and best strategies can be decided more efficiently [9].

Nowadays, penetrations of high-performance computation, multimedia, and virtual reality technologies have prompted bigger demands on innovative ideas for computational human models to answer higher needs in biomedical applications [18–22]. By referring to the directions into personalized and preventive medicine [67], the key specifications of today's computational human models are concentrated to personalization of the model with tailored level of details [68], sources of data and modelling techniques [69], interactivity with users and systems [70], types of biomedical applications and involvement of pathological cases that are related to the purpose of the computational human models.

Extensive observations and analyses in [1–3, 8–10, 22, 66, 68] have brought significant impacts in disseminating the understanding on the importance of computational human models. Besides, combination of mathematical basics, computational methods, and fundamental anatomy and physiology that build the models have become essential knowledge for developing the models in wider scales. The reviews in [1, 2, 8, 22, 46] have also highlighted about generations of computational human models, including their components and characteristics. Realization in general radiology and radiotherapy has also been largely reported.

Based on how computation generates and represents the human details, [1, 2, 8, 22, 46] grouped the computational human model evolution into three basic generations:

1. First Generation—Stylized models.
2. Second Generation—Voxelized models.
3. Third Generation—BREP (Boundary Representation) models.

In brief, the Stylized models apply quadratic equations in developing the model. These simple models are then improved with more personalization and anatomical organs. However, the drawback is about incompleteness in the internal organ details. The Voxelized models enhance the simulation accuracy of the previous generation by using voxel-based tomographic image data from three-dimensional (3D) imaging. Statistical shape modelling to produce morphological characteristics of the model is also used in this generation. Yet, issues in capturing and handling large image data are still challenging. Using boundary representation and computerized operations, BREP models are introduced to generate better anatomical geometry and structures with dynamic deformation and adjustment options [1, 2, 8].

4.1 First Generation—Stylized Models

The general stylized computational models are the simplest human models which apply mathematical or analytical formulae in quadratic equations to represent human body [1, 2, 8]. By using simple mathematical forms, biophysical interaction in the body could be computationally simulated and analysed. Biophysical concepts were assumed through geometrically shaped analytical representation (dimension, size, and position) with adjustable physical parameters. As part of optimization, shape parameter descriptions through surface equations were used to modify the personalization, anatomical organs, functions, and applications [22, 46].

Anthropometrically, the stylized models evolution according to timeline and research organization can be grouped into five, ICRU, ORNL and MIRD, GSF, NASA, and MCAT [1]. The first type of stylized model, ICRU sphere model with 'effective radius' was the initial stylized model led by ICRU (International Commission on Radiation Units and Measurements) as basic radiation phantom [1, 2]. Though most ICRU models were more on radiation physics and not into anthropomorphic models, the ICRU models have become the stepping point for next anthropomorphic models [1, 2, 71, 72]. The second type of stylized model, ORNL, is said as the first anthropomorphic human model generation released by Oak Ridge National Laboratory (ORNL) [73]. This model was built with rough organ approximation, homogenous tissue distribution, and lack of anatomical localization [1, 2, 22]. To enhance, MIRD-5 (Medical Internal Radiation Dosimetry) with adult personalization, heterogeneous organ and tissue distributions, anatomical positioning, and geometrical parameterization from human population was created as Reference Man [1, 2, 74]. Since then, by customizing Akin principles, anthropomorphic models with various type of human in different gender, age, size, ethnic group, health condition, and applications could be generated. In the third type, GSF (*Gesellschaft für Strahlenforschung*) extended the second type concept into a pair of gender-specific adult phantoms with additional calculation for external radiation dose [1, 75]. NASA (National Aeronautics and Space Administration) also took part in aeronautic human model by issuing the fourth type model as CAM/CAF (Computational Anatomical Male/Female). Built using CSG (Constructive Solid Geometry) modelling and CAMERA program, CAM/CAF was reported to have highly detailed anatomical model with thousand unique geometric surfaces and solid regions of internal body [1, 76]. The newest version of the stylized model, Mathematical Cardiac Torso (MCAT), was meant for cardiac nuclear medicine [1, 77]. MCAT and its derivations are still growing and extended into dynamic models [2, 78].

In the context of concept and functional design, the mathematical basis in the stylized models brought significant benefits in calculation-based models with anatomical variability and temporal changes in efficient memory allocation [2, 22, 46, 79]. Likewise, with delineation-based surface equations, creation of geometrical element of the anatomical organs, from simplified single spherical shape to integration of complex anthropometrical shapes in size, localization, time, tissue distribution, and basic physical parameters, could be analytically produced for various types of personalization and scenario [2, 22]. Instead, with limitations in unrealistic irregular geometry, undetailed internal structures of the organ, and merely static modelling, most traditional stylized models are not sustained nowadays. However, revolution in the MCAT type from static to (four-dimensional/4D) dynamic representation that incorporates time-dependent geometric modeling has brought new directions for the stylized model. Also, combination of the dynamic representation with biomechanical principles as in [80] has provided important physical element of the model for representing motion parameters. With this method, physiological elements in motion can be elaborated in more realistic, accurate, precise, flexible, expandable, and suitable for clinical applications, ranging from cardiovascular, pulmonology, radiology, radiation therapy, epidemiology, and so on [1, 2, 46, 78–80].

4.2 Second Generation—Voxelized Models

Voxelized computational models were conceptualized based on utilization of voxels in building the model, representing its internal structures, and linking them to various biophysical parameters. These models expand in consort with the growth of tomographic imaging modalities, image reconstruction, and high-dimensional image computations, generating more realistic anthropometrical, anatomical, and physiological structures [1, 2, 8]. Based on concept and functional design, voxels are reconstructed through volumetric acquisition process in generating 3D volumetric images. Voxels, cuboidal volume elements which are naturally the smallest part of 3D volume or reconstructed from stacks of two-dimensional (2D) slice in pixels, are the main components in the model [81, 82]. Arrangements of voxels into one group or more represent anatomical structures. The contents of each voxel signify physiological parameters through attenuation map. Through this voxelization concept, more realistic anthropomorphic human models based on volumetric structures of human body, organs, or tissues from tomographic imaging modalities can be generated by arranging forms of shapes, surface, volume, internal anatomy, landmark positioning, and their compositions [46, 79–82]. The design processes of the voxelized models can be divided into four basic steps, as adapted from Refs. [1, 46];

- Image acquisition for producing high resolution volumetric image.
- Anatomical representation through object classification and segmentation for identifying biophysical markers.
- Tissue specification for characterizing the composition of the object.
- Geometric registration for matching image slice and visualized 3D model.

Also, as tomographic imaging modalities are supported with DICOM (Digital Imaging and Communications in Medicine) standards, utilization of image data has been designed for wide range of research and clinical works. One of the protocols has also declared about the use of voxelized models in radiation therapy [83, 84]. This is also to provide the needs of patient models as part of translational research in clinical practice as in silico test with more realistic tomographic representation.

However, the published voxelized models are still limited for exposure dose and safety applications with some clinical cases as reviewed by [1, 2, 8, 46, 85]. Though, the voxelized models have grown into two directions, based on imaging modality and human group variants. For the voxelized model groups based on imaging modality, even the earliest voxelized models were initiated using 2D X-ray radiography for dentomaxillofacial radiation dose [86, 87] as reported in [1, 85], the voxelized models rely on 3D imaging modalities to capture the 3D image volumes. Later, the models went into 3D voxelized forms from CT and MRI volumes with various qualities and functionalities. For the voxelized model groups based on human group variants, population-based human models are created by altering size, shape, and other parameters of the models according to anatomical and physiological characteristics. Mostly, this grouping refers to epidemiological studies which classify the groups based on gender, age, and ethnic group, then correlate to certain health and disease profiles. The roles of this grouping have been addressed in International Commission on Radiological Protection (ICRP) and World Health Organization (WHO) recommendations [88–90].

In CT, the models represent organ structures from grey level values and mainly for radiation exposure quantification. As reported in [1, 85], CT-based individual adult human models range from VoxelMan with MANTISSUE3-6 and VOXTISS8 with MRI [82, 91], Golem [92], ADELAIDE [93–95], and FRANK, DONNA, HELGA, LAURA, GODWIN, KLARA, and IRENE from the GSF family [46, 96, 97] focusing for partial or whole body, from cadavers, patients, to healthy volunteers. As sizes, shapes, and other parameters are modifiable, the other types of CT-based human models from different ages, physical health, and ethnic groups are also growing in pairs, or family sets. As part of its family set, GSF also developed the BABY and CHILD models [46, 96–100]. Specifically for paediatric groups, [101] produced on head phantoms in different child age and anthropometry. As complementary, pregnant models have also been created in [101–103]. In adult pairs, MAX (Male Adult voXel) and FAX [104, 105], and OTOKO and ONAGO [106, 107] have become most discussed couple models for estimating radiation exposure. In newer generations, voxelized phantom has expanded as simulator for imaging functionality, as in [108] with VOXSI (VOXelized CT SImulator), [109, 110] with VIPRIS, and [111, 112] with MOBY for pre-clinical animal models.

In MRI, the models mostly simulate electromagnetic field effect to the human body. Although MRI has excellent anatomical structures with rich of information, whole body scanning time is too long. Also, motion artefacts and poor bone imaging still exist [85]. Similar to CT, MRI-based models have been developed through variations of parameters. NORMAN and NAOMI with their improvements [113–116], Nagaoka Man and Nagaoka Woman [117], TARO and HANAKO with pregnant HANAKO [118, 119], KORMAN and KORWOMAN [120, 121], and CNMAN [122] are the couple model examples that have been widely reviewed. Also, KATJA, pregnant model made from MRI is one of GSF family members [123]. In the family set, the IT'IS group has also developed voxelized human models representing various health conditions and treatment strategies [124–127]. Besides CT and MRI, some voxelized models have also been created from Nuclear Medicine, either sole, such as SPECT (Single-photon emission computed tomography) and Positron Emission Tomography (PET), or hybrid modalities, like SPECT/CT and PET/CT [79, 85, 128, 129]. Furthermore, multimodality imaging expansion has triggered the growth of new voxelized models based on hybrid modalities, like SILVY [130, 131] and MEET Man [132], and hybrid computation [133–136]. Along with time, the development of voxelized models has evolved into high resolution models with more complex mechanism and purposes.

Nevertheless, besides its great advantages for diagnostically realistic details of anatomy and physiology, voxelized models have also been common for many medical physics applications [79, 80]. With these strengths, the voxelized computational models have been endorsed as ICRP standards for radiation protections. In this way, radiation transport processes are mostly simulated using Monte Carlo techniques for radiation dosimetry [1, 2, 46, 85–87]. On the other hand, besides their advantages, several studies reported about some drawbacks in the implementation of voxelized models for radiation dosimetry simulations [1, 2, 46]. Discretization and approximation processes in generating geometric voxels have been found to create inherent errors that degrade the model quality and cause loss of structural information [46, 79]. These errors are basically easy to solve by increasing the resolution into finer sampling scales and performing some smoothing techniques if necessary. However, these efforts may also trigger other computational problems. As stated in [46, 79, 85–90], higher resolution models have demanded for bigger computational memory and processing time that consequently necessitate for higher hardware and software specification. The other challenges in the development of voxelized models are about constraints in whole-body imaging, insufficiency in image region segmentation and outlining, and incompleteness of physical information [46, 88]. Issues related to extensive efforts in model preparation, population variability, anatomical, physiological, and pathological complexities have also triggered for finding better solutions. In tackling these drawbacks, advanced restoration techniques from interpolation, smoothing, enhancement, to parallel processing have been used.

4.3 BREP Models

BREP (Boundary Representation) models, the third generation phantoms, are created based on advanced primitives and contained deformable topological and geometrical features, producing models with high adaptability and flexibility. Derived from 3D and 4D tomographic medical image data, BREP models are capable of representing realistic characteristics of various human organs and bodies, making these models highly applicable to generate uniquely complex organ structures and signify different populations [1, 8, 22, 131].

The BREP models are developed through computerized operations and modelling. As in [8, 22, 137], the models are composed from topological and geometrical elements of its boundary. The topological elements are generated from the boundary of a shape as the relationships among faces, edges, vertices, and their orientations, including their adjacencies and incidences. The geometrical elements are produced based on the surfaces, curves and points creating shapes, sizes, and volumes. A face is part of surfaces, an edge is segment of curves, and vertex becomes points at the edge boundary. These basic elements can also be extended into advanced elements, such as shell, loop, and loop-edge links. A shell is a set of connected faces. A loop is the connected edges that describes the closed face boundaries. A circuit of edges bounding a face makes the loop. A loop-edge links creates edge circuits.

These advanced elements enable for creating holes in faces and voids in volumes [137]. These elements are used for computerized operations, for instance extrusion, chamfer, blending, drafting, shelling, tweaking and other operations. By arranging these elements into certain construction, complex anatomical and motion features of BREP models can be built.

In BREP development, surface contours extract image data and integrate with organ details to build whole body model. Organ surface geometries, from shape and location of organ boundaries in a uniform density, are defined in the forms of [1, 8]:

- Non-uniform Rational B-splines (NURBS).
- Polygon mesh.
- Mixture of both data structures by voxels.

NURBS and Polygonal meshes can be considered as interchangeable BREP data structures. The contours convert the voxels into NURBS or Polygon mesh surfaces producing smooth and anatomically realistic representation. NURBS is defined by underlying formulas. With its computational efficiency and geometrical flexibility, NURBS geometries are easy to adopt for real time and non-real time modes. However, complex topologies of organs are probably losing their details, especially for the hard-corner shapes. Therefore, NURBS are not much used nowadays and replaced by Polygonal mesh due to the fact that shapes are represented in arbitrary topology with high flexibility, expandability, and anatomical accuracy. As Polygonal mesh is developed through vertices and their relationships, it can be extended to smoother surfaces with excellent details of anatomical geometries. However, it is memory consuming [8, 137, 138]. For this purpose, specific strategy has to be made to define specific applications and user preference. Along with the smaller size of computer systems and faster performance, BREP models eventually lead to the evolution of stylized anthropomorphic models.

The anatomical NURBS-based models were pioneered by Segars with NURBS-based cardiac-torso (NCAT) model using Visible Human CT [139]. From this concept, the model was expanded for more advanced cardiac imaging models with dynamic beating features [140–142]. Along with these works, other applications have also been released, from NCAT-based respiratory models [143–146], models for imaging simulations [1, 147–150], population models with wide ranges of age, gender, physical circumstances, and clinical applications [151–156], to animal models with MOBY and Digimouse that are useful for pre-clinical research [1, 157–159].

For Polygonal mesh models, the works of Xu in [1, 131, 160] on BREP-based models for Rensselaer Polytechnic Institute (RPI) to develop family and pregnancy models using ICRP-89 parameters have brought significant contribution to the radiation dose calculation. Casolla's group has also produced polygonal mesh-based FASH and MASH models consisting of 3D human organs, tissues, skeletons and bodies based on anatomical atlases in standing posture [161–163]. In the other population group, Virtual Korean Man (VKH-Man) has also been developed by Korean as the reference Korean male (PSRK-Man) for Geant4 Monte Carlo simulation [164–166].

5 Computational Human Models of Cardiac Organ

Due to high needs of performing radiation planning, most computational models were rooted from radiation dosimetry applications. However, their vast growth and emerging benefits in clinical settings has endorsed the larger uses of computational models in more new applications. One of the crucial applications is for modelling cardiovascular diseases. Although development of computational models for cardiac organ is principally similar and using common techniques, certain specific issues still come up in digital modelling of cardiac organs and tissues, especially for its consistent dynamic heart beating and respiratory mechanisms, complex arrangements in myocardial tissues, blood flows, characteristics of electrical, chemical, and mechanical properties, interconnection with other organs, as well as cardiac imaging performances for diagnostic and treatment monitoring in single and multimodality forms.

Cardiac imaging modalities have contributed largely as the main sources in capturing anatomical structures for the computational models. Along with rapid advancements in cardiac imaging technologies, the acquired image information has also advanced, not only anatomical information, but also physiological and pathophysiological information. Additionally, thanks to image reconstruction technologies, the image is not only in 2D modes, but has expanded into 3D/4D, and multi-dimensional options. Especially, integration of one imaging modality with other modalities or diagnostic and therapeutic tools has enquired computational models for cardiac organs more parameters to take into account. With these comprehensive aspects, cardiac imaging and computational models have significantly played role to contribute in many cardiovascular research, especially in the development of quantitative methods for patient-specific cardiovascular diagnosis and treatment. Following the development phase of computational human modelling technology, computational models of cardiac organ have undergone improvements in term of quality, functionality, and applicability. In this part, the discussion can be divided into two parts, evolution of cardiac models and anthropometry of cardiovascular organ models.

5.1 *Evolution of Cardiac Models*

The evolution of 3D cardiac model was started with simple models which represented cardiac anatomy using geometrical shapes. The first generation was the stylized models based on quadratic equation. Most cardiac models in this generation only consisted of left ventricle (LV) which was represented by a pair of concentric ellipsoids shortened at the base in order to approximately follow the shape of the LV [167, 168]. Along the way, this cardiac model approach was still in use for certain applications even the model was not critically having anatomical realism [169, 170]. The early basic stylized computational model for cardiac region was developed by Tsui's group and named as Mathematical Cardiac Torso (MCAT) [171]. The MCAT model was developed as improvement for the original MIRD-5 stylized model by implementing mathematical formulae to create realistic model with size and shape of cardiothoracic organs and structures for SPECT and PET [172]. This model consisted of ellipsoids which were configured to mimic the changes in chamber volume, myocardium thickness, and cardiac rotation [173] which happened during cardiac cycle. The ellipsoids were used to define the outer and inner boundaries of ventricles and atria regions. The advantage of MCAT model is that it can be simply tailored for variety of patient anatomies simulation. Yet, it has drawbacks that it is anatomically less realistic compared to the voxel models which are developed from high-resolution CT or MRI.

Improving the crude stylized cardiac model, several voxel torso models, including cardiac regions, were reported by Refs. [40, 82, 174]. Voxel torso model, ADELAIDE, developed by Caon [40] from a set of CT images of a teenage girl. The interesting aspect of this model is that it consisted of non-adult voxel data, and the result showed most reliable CT dose estimation for this age group. The other voxel model developed from detailed CT study of a male by Zubal [82] was actually a head-torso model called Voxel Man. The original model was developed for all organs from the head to hip and used for Nuclear Medicine imaging optimization. Important thoracic structures and organs, especially the heart, were segmented and assigned to a specific numbering in order to identify each voxel that belonged to a specific organ. Pediatric voxel models developed by Lee [174] represented several age and gender groups. These models were based of organs segmentation from chest-abdomen-pelvis and head CT scans. The CT data were collected from normal and near normal anatomy of the patient CT data. Although these voxel models are not specifically for cardiac region, the cardiac organ has been defined. However, these models lack of detailed discussion on the cardiovascular organs.

A specifically dedicated voxel cardiac model was presented by Werner based on the images of Visible Man dataset [175]. The images were processed through a set of digital image processing, reconstructed into 3D voxel model, and simulated for several cardiovascular pathologies. Another interesting voxel model of the heart was developed by Deng [176], reconstructed from CT heart images of donated heart specimen, segmented into 3D heart anatomy using ScanIP software, and consisted of anatomically detailed mathematical representation. This model was one of the most detailed human heart voxel models in the context of experimentally fibres orientation measurement.

The BREP computational cardiothoracic models have been intensively developed by Segars' group through a series of heart models, NCAT, 4D NCAT, beating heart model, and 4D extended cardiac-torso (XCAT) models [139–150]. The high resolution 3D/4D CT imaging was implemented for optimizing the image quality and radiation dose measurements. The models were developed using Visible Human anatomy from CT data of both male and female anatomies. Specific models based on gated MRI data were also used for mimicking cardiac and respiratory motions. The development of these models was made using SURFdriver reconstruction and Rhinoceros 3D NURBS software applications. The developed models looked extremely realistic as it contained detailed anatomical structures. In these periods, Segars' group has performed simulations of PET, SPECT, and CT. the other BREP computational models were the female torso models for a study on morphological effects of counting efficiency curves in the in vivo lung monitoring of several workers by IRSN [177]. NURBS and Mesh surfaces were used to produce the torso model based on reference data from the ICRP adult female reference model. Different chest size, breast tissue composition, and internal organ volumes were shaped from basic torso model using Rhinoceros 3D modelling software. Simulations analysis using the models had produced comparable result to the previous simulation using voxel models. Potse's model did a study on the action potential propagation using a BREP model of the whole human heart [178]. The geometrical element of this model was developed based on CT data at 1 mm resolution. Detailed technical description of the model included the 3D finite difference grid with the resolution of 200 μm and large size of model consisting around 45 million polygon mesh nodes. Fibre orientations for the model were calculated based on mathematical functions to mimic tissue structures.

5.2 *Anthropometry of Cardiovascular Organ Models*

For its measurement of physical properties of the human, anthropometry plays an important role in the design process of equipment or facilities where statistical data about the distribution of body dimensions in the population are used for optimization [179]. Therefore, most standardized population for computational models were developed based on anthropometry characteristics from specific Reference Man data, a representation of average of anatomical, physiological, and metabolic features from a specific ethnic population, as defined by the ICRP. However, the currently available data is merely from Caucasian population of Western Europe and North America. Even some Asian ethnic groups have generated Reference Man, these have not represented the general Asian people.

In 1998, International Atomic Energy Agency (IAEA) was working on a research to compile the anatomical and physiological characteristics for establishing the Asian Reference Man (ARM) [180]. In this work, the IAEA group collected the anthropomorphic data, including organ mass, physiological measurements, and food intake from six groups of different ages in nine Asian countries. The result shows that the Asian population tends to have less organ masses compared to those of ICRP values and this is the starting point of the effort in establishing Asian reference values. Extensive studies in [1, 2, 9, 181] have also revealed various models with their anthropometric measurements and ethnics of origin, including Asia. As excerpted from Ref. [1], Table 1 summarizes basic anthropometry of average heart and lung organ in density measurement for Reference Man data. For detailed measurements of the cardiothoracic organ mass and volume data of heart and heart wall, Table 2 recaps cardiac anthropometry data from Caucasian and ARM [1, 181]. For the Stylized and Voxelized types, Table 3 portrays anthropometry measurements in organ mass and volume from Ref. [1]. For the BREP models, Table 4 describes the anthropometries of voxel and polygon sizes [1, 181].

Table 1 Basic anthropometry of average Reference Man density values of cardiothoracic organ for heart, lung, bronchial tree, and heart wall regions [1]

Cardiac organ	Density (g/cm^3)
Heart	1.06
Lung	0.296
Bronchial tree	1.1
Heart wall	1.03

Table 2 Anthropometry data of cardiac organ mass and volume from Caucasian and Asian Reference Man [1, 182]

Model	Organ mass (x)		Volume (x)	
	Heart	Heart wall	Heart	Heart wall
ICRP reference man	500.0	330.0	–	–
ICRP reference woman	350.0	240.0	–	–
ARM	400.0	380.0	337.4	368.9
ARMF	300.0	320.0	283.0	310.7
Japanese male	362.0	–	341.5	–
Japanese female	306.0	–	288.7	–
Korean male	348.8	380.0	329.1	368.9
Korean female	301.6	287.0	284.5	278.6

6 Computational Models of Partial Cardiac Region

Besides the whole organ models, several studies have also worked on computational models in partial cardiac region. The studies can be divided into ventricular, atrial, and blood vessels models. Partial cardiac region models are usually developed for biological and physiological simulations which can assist in defining and analysing experimental dataset, that in the future can contribute to the detection, diagnosis, and treatment of complex cardiac diseases.

6.1 Computational Models of Ventricular Region

In exploring the voxelized ventricle models, Panfilov's group had developed the model based on the geometrical data collected from histological heart data of a normal subject which has 500 μm image resolutions [182]. The model mapped the fibre orientation of canine heart and described the anisotropic heart structure. The model was mainly used to analyse the effect of heterogeneous action potential duration (ADP) restitution on ventricular fibrillation organization.

In the BREP models, Fleureau developed a realistic B-spline model of LV region based on Multi-slice CT (MSCT) data [183]. The geometrical surface of the model

Table 3 Anthropometry data of cardiac organ mass and volume from Stylized and Voxelized cardiac computational models [1]

Model	Ethnics	Organ mass (x)		Volume (x)	
		Heart	Heart wall	Heart	Heart wall
ICRP male	Caucasian	510.0	329.9	–	–
ICRP female	Caucasian	370.0	250.5	–	–
RPI-AM	Caucasian	–	330.0	–	–
RPI-AF	Caucasian	–	250.0	–	–
MIRD	Caucasian	–	316.0	–	–
VIP-man	Caucasian	–	398.7	–	–
Zubal phantom	Caucasian	628.59	–	598.66	–
ADELAIDE	Australian	705.0	–	–	–
MCAT	Caucasian	–	–	640.0	–
Otoko	Japan	476.0	–	449.1	–
Onago	Japan	476.0	–	449.1	–
JM	Japan	–	528.0	–	512.6
JF	Japan	–	280.0	–	271.8
HDRK-man	Korea	379.3	383.4	357.8	372.2
PSRK-man	Korea	–	380.0	–	368.9
KMIRD	Korea	388.0	260.6	366.0	253.0
KORMAN	Korea	800.9	–	–	–
Taiwanese male	Taiwan	413.4	278.1	390	270
Taiwanese female	Taiwan	329.7	222.5	311	216
CNMAN	China	503.9	–	475.4	–
VCH	China	384.7	238.0	362.9	231.1

Table 4 Voxel size anthropometry for voxel computational models and polygon size anthropometry for BREP computational models [1, 181]

Model	Voxel size (mm^3)	Polygon size (cm^2)
Zubal phantom	64.00	–
ADELAIDE	64.01	
Otoko and Onago	9.60	–
HDRK-man	8.18	–
PSRK-man	–	0.55
KORMAN	40.00	–
CNMAN	0.026	–
VCH	0.002	–

was derived from segmented cardiac volumes. Region growing slice segmentation was performed along the ventricular axis for extracting endocardium and implementing mesh construction. Anatomically, the model was divided into 18 segments with consistent mechanical and electrical properties for standard cardiac function analysis. Recently, Pravdin's group also developed a NURBS analytical model of LV from sonography data [184]. The model was developed using two and four chamber data views of one specific patient's heart.

A cubic Hermite Spline model of the human ventricles was made by Krishnamurthy to perform cardiac biomechanics simulations [185]. The Hermite Spline meshes have the capability to represent complex geometries with fewer elements and smaller computational time. Moreover, the cubic Hermite Spline meshes are very suitable for patient-specific cardiac models. Due to fewer degrees of freedom in Hermite Spline meshes, they are useful for creating template meshes for average cardiac dimensions of a population.

6.2 Computational Models of Atrial Region

Along with the impressive progress of ventricular model, several attempts to develop atria model have also been made. Recently, Loewe's group developed voxel models of atrial with fibre orientation for electrophysiological simulations using segmented MRI data [186]. The model was based on the influence of the Earliest Activated Site (EAS) and intactness of the posterior inter-atrial connections approach to investigate the P-wave terminal force (PTF). Then, Chang's group used the 3D digital model of the anatomically realistic human atria to investigate the effects of increased calcium-driven alternans (CDA) on arrhythmia susceptibility. The atria voxel model was constructed using the Visible Human female dataset [187].

Also, Harrild and Henriquez developed a human atria model to study the normal conduction within atrial structures. The model was created in 3D using the commercially available surface meshes (Viewpoint Digital) and consisting of the atria major muscle bundles [188]. However, as the model geometry was built using only surface information, uniform muscle thicknesses were configured for the left and right atria with 3 and 2 mm respectively. In the model, one muscle bundles were represented as anisotropic structures and incorporated with faster conduction along the muscle bundle axes. Then, the other muscle bundle of the atrial tissue was defined to be isotropic.

The first structural MRI-based cardiac model was developed by Jacquemet's group by constructing a monolayer model of human atria to analyse electrogram morphologies during atrial fibrillation (AF) [189]. The fibre orientations mapping onto the model's surface was made manually without considering the atria wall thickness. The other work from Reumann's group made the most detail structural atria model and studied the ablation strategies in AF prevention [190]. The model was generated from cryosection images of the Visible Female dataset. Fibre orientations in the model were synthesized through rule-based methods. Besides Reumann's group, the other study by Plank's group also used similar atrial model by improving the fibre orientations [191].

6.3 Computational Model of Blood Vessels in Cardiovascular Flow

Current technology improvement in Computational Fluid Dynamics (CFD) and image-based modelling has provided assistance to coronary flow and pressure fields calculations based on anatomical data [192]. Computation of coronary flow can be performed by using constructed anatomic model of the coronary arteries. The coronary anatomy model is capable of deriving the boundary conditions which represent the cardiac functions and numerical solution of fluid dynamics. One example of coronary anatomy models in [192] used CT angiography (CTA) data to develop an image-based coronary lumen anatomical model. Image segmentation based on boundary limits derived by CTA resolution was used to extract the luminal surface properties of the main vessels and branches as well as the topology information of the coronary artery tree, delineation of coronary plaques in each vessel, and luminal boundary. Another coronary anatomy model was developed by Biglino's group [193] as a patient specific aortic arch model and implemented for simulation of insertion in a CMR-compatible mock circulatory system. The 3D whole heart dataset from CMR examination was used to model the aortic root, aortic arch, head-and-neck vessels, and descending aorta using Mimics software. Computational meshes of the model were created using tetrahedral elements with ICEM software.

7 Technical Hints

Technical hints may help in the development of computational cardiac models. As the cardiac models are highly dependent on imaging modalities, image analysis algorithms and software, and modelling software, it is important to generate a mapping of the type of cardiac models which are related to the imaging modalities, segmentation algorithms, as well as the essential tools for developing the computational human models. Table 5 summarizes the use of imaging modalities and segmentation algorithms for developing whole cardiac organ and partial cardiac region models based on the abovementioned review. Then, Table 6 also provides a list of computer software in different steps of cardiac model development based on the reviewed studies.

Table 5 Summary of imaging modalities and segmentation algorithms used in developing several cardiac models

Cardiac models	Imaging modalities	Segmentation algorithms
Whole cardiac organ model	CT SPECT MRI	Region growing [175] Thresholding [176]
Ventricular region model	MRI CT Ultrasound	Region growing [183]
Atrial region model	MRI	–
Blood vessels model	CT MRI	Boundary based [193]

Table 6 List of computer software in cardiac models development

Process	Software
Organs segmentation	MATLAB [194] ScanIP [176] Mimics [193]
Voxel reconstruction	OsiriX MD [195] ScanIP [176] SURFdriver [139]
Polygon surface rendering/mesh	MeVisLab [196] 3D blender [185] Rhinoceros [139, 146] TrueGrid [188]

8 Benefits and Challenges

Cardiac computational models have shown to bring significant benefits to the area of cardiac imaging and cardiology. In modern cardiac research, computer modelling has capability to assist for better understanding of normal heart and diseases cardiac function and solve difficulties or impossible actions through simulation of failure conditions before performing the actual works. Furthermore, patient-specific cardiac models are nowadays used to integrate the anatomical and functional details of a specific patient through various approaches in a very comprehensive manner. With this way, it has the ability to open up a new path in cardiac therapeutic planning, management, and follow up.

Even though cardiac modelling is in a positive progress, it still has constraint due to several challenges. Noticeable challenge is the high complexity and inter- and intra-subject variability of cardiac anatomy. Due to this complexity, it is challenging for researcher to fully understand the relationships among cardiac components, multiple influences, and broader context of time and space. In the development of cardiac model, sufficient data is compulsory to represent the actual complexity of the cardiac characteristics and model the biological activities. The challenge can be in understanding the relationships of multiple variables operating at different scales and represent them computationally and into models. Besides, processing big population data from multiple patients is another challenge in the cardiac model development. Therefore, at this stage, only small number of cardiac features can be fully implemented and personalised.

On the other hand, development of a complex computational cardiac model requires suitable conceptual basis and implementation. Technically, the current mathematical and image processing tools for cardiac modelling are quite limited for many institutions. The challenges may evolve from data handling techniques, to the simulation algorithms. It is still in question on how to represent high complexity of cardiac characteristics in the most effective understood ways. One of the most critical image processing problems in the development of patient specific cardiac models is cardiac images segmentation, reconstruction, and registration. In the context of voxelized cardiac model, 3D reconstruction by alignment of adjacent 2D slices is also still a problem for generating an accurate modelling. In the computer programming point of view, the challenge can be to provide a flexible and affordable programming language to incorporate all interesting biological elements and integrate models with different level of concepts and different mathematical patterns. Other main challenges in development of cardiac model are the high computational expense and the time required for model construction.

Acknowledgements The authors are grateful for funding supports by Universiti Teknologi Malaysia and Ministry of Higher Education Malaysia under FRGS Grant. J130000.7845.4F764 and GUP Tier 1 Grant Q.J130000.2545.20H36

References

1. Xu, X.G., Eckerman, K.F. (eds.): Handbook of Anatomical Models for Radiation Dosimetry. CRC press (2009)
2. Zaidi, H., Tsui, B.M.: Review of computational anthropomorphic anatomical and physiological models. Proc. IEEE **97**(12), 1938–1953 (2009)
3. Hunter, P., Nielsen, P.: A strategy for integrative computational physiology. Physiology **20**(5), 316–325 (2005)
4. Trayanova, N.A.: Whole-heart modeling: applications to cardiac electrophysiology and electromechanics. Circ. Res. **108**(1), 113–128 (2011)
5. Lopez-Perez, A., Sebastian, R., Ferrero, J.M.: Three-dimensional cardiac computational modelling: methods, features and applications. Biomed. Eng. Online **14**(1), 35 (2015)
6. Antiga, L., Piccinelli, M., Botti, L., Ene-Iordache, B., Remuzzi, A., Steinman, D.A.: An image-based modeling framework for patient-specific computational hemodynamics. Med. Biol. Eng. Comput. **46**(11), 1097 (2008)
7. Clark, D.E., Pickett, S.D.: Computational methods for the prediction of 'drug-likeness'. Drug Discov. Today **5**(2), 49–58 (2000)
8. DeWerd, L.A.: The Phantoms of Medical and Health Physics. M. Kissick (ed.). Springer, New York (2014)

9. Ayache, N., Lions, J.L. (eds.): Computational Models for the Human Body, vol. 12. Gulf Professional Publishing (2004)
10. Metaxas, D.N.: Physics-Based Deformable Models: Applications to Computer Vision, Graphics and Medical Imaging, vol. 389. Springer, New York (2012)
11. Henriet, J., Leni, P.E., Laurent, R., Salomon, M.: Case-based reasoning adaptation of numerical representations of human organs by interpolation. Expert Syst. Appl. **41**(2), 260–266 (2014)
12. Lee, L.C., Genet, M., Dang, A.B., Ge, L., Guccione, J.M., Ratcliffe, M.B.: Applications of computational modeling in cardiac surgery. J. Card. Surg. Incl. Mech. Biol. Support Heart Lungs **29**(3), 293–302 (2014)
13. Sermesant, M., Delingette, H., Ayache, N.: An electromechanical model of the heart for image analysis and simulation. IEEE Trans. Med. Imag. **25**(5), 612–625 (2006)
14. Kim, H.J., Vignon-Clementel, I.E., Coogan, J.S., Figueroa, C.A., Jansen, K.E., Taylor, C.A.: Patient-specific modeling of blood flow and pressure in human coronary arteries. Ann. Biomed. Eng. **38**(10), 3195–3209 (2010)
15. Clayton, R.H., Bernus, O., Cherry, E.M., Dierckx, H., Fenton, F.H., Mirabella, L., Panfilov, A.V., Sachse, F.B., Seemann, G., Zhang, H.: Models of cardiac tissue electrophysiology: progress, challenges and open questions. Prog. Biophys. Mol. Biol. **104**(1–3), 22–48 (2011)
16. Wong, J., Kuhl, E.: Generating fibre orientation maps in human heart models using Poisson interpolation. Comput. Methods Biomech. Biomed. Eng. **17**(11), 1217–1226 (2014)
17. Dössel, O., Krueger, M.W., Weber, F.M., Wilhelms, M., Seemann, G.: Computational modeling of the human atrial anatomy and electrophysiology. Med. Biol. Eng. Comput. **50**(8), 773–799 (2012)
18. Deng, D., Jiao, P., Ye, X., Xia, L.: An image-based model of the whole human heart with detailed anatomical structure and fiber orientation. Comput. Math. Methods Med. **2012**, 1–16 (2012)
19. Verkerke, G.J., Houwen, E.V.D.: Design of biomedical products. In: Biomaterials in Modern Medicine: The Groningen Perspective, pp. 23–38 (2008)
20. Völter, M., Stahl, T., Bettin, J., Haase, A., Helsen, S.: Model-Driven Software Development: Technology, Engineering, Management. Wiley, London (2013)
21. Shefelbine, S., Clarkson, P. J., Farmer, R.: Good design practice for medical devices and equipment-requirements capture (2002)
22. Xu, X.G.: An exponential growth of computational phantom research in radiation protection, imaging, and radiotherapy: a review of the fifty-year history. Phys. Med. Biol. **59**(18), R233 (2014)
23. Grenander, U., Miller, M.I.: Computational anatomy: an emerging discipline. Q. Appl. Math. **56**(4), 617–694 (1998)
24. Dingreville, R., Karnesky, R.A., Puel, G., Schmitt, J.H.: Review of the synergies between computational modeling and experimental characterization of materials across length scales. J. Mater. Sci. **51**(3), 1178–1203 (2016)
25. Ottesen, J.T., Olufsen, M.S., Larsen, J.K.: Applied Mathematical Models in Human Physiology. Society for Industrial and Applied Mathematics (2004)
26. Jerby, L., Shlomi, T., Ruppin, E.: Computational reconstruction of tissue-specific metabolic models: application to human liver metabolism. Mol. Syst. Biol. **6**(1), 401 (2010)
27. O'reilly, R.C.: Biologically based computational models of high-level cognition. Science **314**(5796), 91–94 (2006)
28. Frances, A., Sandra, O., Lucy, U.: Vascular cognitive impairment, a cardiovascular complication. World J. Psychiatry **6**(2), 199 (2016)
29. Brodland, G.W.: How computational models can help unlock biological systems. In: Seminars in Cell & Developmental Biology, vol. 47, pp. 62–73. Academic Press (2015, December)
30. Wronecki, J.: Concept modeling with NURBS, polygon and subdivision surfaces. In: Proceedings of the 2006 American Society for Engineering Education Annual Conference & Exposition (2006)

31. Wilhelms, J., Van Gelder, A.: Anatomically based modeling. In: Proceedings of the 24th annual conference on Computer Graphics and Interactive Techniques, pp. 173–180. ACM Press/Addison-Wesley Publishing Co. (1997, August)
32. Yang, Y.J., Cao, S., Yong, J.H., Zhang, H., Paul, J.C., Sun, J.G., Gu, H.J.: Approximate computation of curves on B-spline surfaces. Comput.-Aided Des. **40**(2), 223–234 (2008)
33. Gibson, S.F., Mirtich, B.: A survey of deformable modeling in computer graphics. Technical Report, Mitsubishi Electric Research Laboratories (1997)
34. Prautzsch, H., Boehm, W., Paluszny, M.: Bézier and B-spline techniques. Springer, New York (2013)
35. Yoo, D.J.: Three-dimensional surface reconstruction of human bone using a B-spline based interpolation approach. Comput.-Aided Des. **43**(8), 934–947 (2011)
36. Botsch, M., Pauly, M., Kobbelt, L., Alliez, P., Lévy, B., Bischoff, S., Röossl, C.: Geometric modeling based on polygonal meshes (2007)
37. Cassola, V. F., de Melo Lima, V. J., Kramer, R., Khoury, H. J.: FASH and MASH: female and male adult human phantoms based on polygon mesh surfaces: I. Development of the anatomy. Phys. Med. Biol. **55**(1), 133 (2009)
38. Wellstead, P.E.: Introduction to Physical System Modelling, pp. 17–32. Academic Press, London (1979)
39. Sherwin, S.J., Formaggia, L., Peiro, J., Franke, V.: Computational modelling of 1D blood flow with variable mechanical properties and its application to the simulation of wave propagation in the human arterial system. Int. J. Numer. Methods Fluids **43**(6–7), 673–700 (2003)
40. Steinman, D.A.: Image-based computational fluid dynamics modeling in realistic arterial geometries. Ann. Biomed. Eng. **30**(4), 483–497 (2002)
41. Sørensen, D.N., Voigt, L.K.: Modelling flow and heat transfer around a seated human body by computational fluid dynamics. Build. Environ. **38**(6), 753–762 (2003)
42. Duffy, V.G.: Handbook of Digital Human Modeling: Research for Applied Ergonomics and Human Factors Engineering. CRC press (2016)
43. Fiala, D., Havenith, G.: Modelling human heat transfer and temperature regulation. In: The Mechanobiology and Mechanophysiology of Military-Related Injuries, pp. 265–302. Springer, Cham (2015)
44. Bhatti, M.M., Zeeshan, A., Ellahi, R.: Heat transfer analysis on peristaltically induced motion of particle-fluid suspension with variable viscosity: clot blood model. Comput. Methods Prog. Biomed. **137**, 115–124 (2016)
45. Wessapan, T., Srisawatdhisukul, S., Rattanadecho, P.: Specific absorption rate and temperature distributions in human head subjected to mobile phone radiation at different frequencies. Int. J. Heat Mass Transfer **55**(1–3), 347–359 (2012)
46. Zaidi, H., Xu, X.G.: Computational anthropomorphic models of the human anatomy: the path to realistic Monte Carlo modeling in radiological sciences. Annu. Rev. Biomed. Eng. **9**, 471–500 (2007)
47. Zaidi, H.: Relevance of accurate Monte Carlo modeling in nuclear medical imaging. Med. Phys. **26**(4), 574–608 (1999)
48. Kroese, D.P., Brereton, T., Taimre, T., Botev, Z.I.: Why the Monte Carlo method is so important today. Wiley Interdiscip. Rev. Comput. Stat. **6**(6), 386–392 (2014)
49. Schultz, F.W., Geleijns, J., Spoelstra, F.M., Zoetelief, J.: Monte Carlo calculations for assessment of radiation dose to patients with congenital heart defects and to staff during cardiac catheterizations. Br. J. Radiol. **76**(909), 638–647 (2003)
50. Mollemans, W., Schutyser, F., Van Cleynenbreugel, J., Suetens, P.: Tetrahedral mass spring model for fast soft tissue deformation. In: Surgery Simulation and Soft Tissue Modeling, pp. 145–154. Springer, Berlin, Heidelberg (2003)
51. Mohr, M.B., Blümcke, L.G., Seemann, G., Sachse, F.B., Dössel, O.: Volume modeling of myocardial deformation with a spring mass system. In: Surgery Simulation and Soft Tissue Modeling, pp. 332–339. Springer, Berlin, Heidelberg (2003)
52. Hammer, P.E., Sacks, M.S., Pedro, J., Howe, R.D.: Mass-spring model for simulation of heart valve tissue mechanical behavior. Ann. Biomed. Eng. **39**(6), 1668–1679 (2011)

53. Votta, E., Le, T.B., Stevanella, M., Fusini, L., Caiani, E.G., Redaelli, A., Sotiropoulos, F.: Toward patient-specific simulations of cardiac valves: state-of-the-art and future directions. J. Biomech. **46**(2), 217–228 (2013)
54. Tavassoly, I., Goldfarb, J., Iyengar, R.: Systems biology primer: the basic methods and approaches. Essays Biochem. **62**(4), 487–500 (2018)
55. Kitano, H.: Computational systems biology. Nature **420**(6912), 206 (2002)
56. Brusic, V., Zeleznikow, J.: Knowledge discovery and data mining in biological databases. Knowl. Eng. Rev. **14**(3), 257–277 (1999)
57. Toni, T., Stumpf, M.P.: Simulation-based model selection for dynamical systems in systems and population biology. Bioinformatics **26**(1), 104–110 (2009)
58. Borenstein, E.: Computational systems biology and in silico modeling of the human microbiome. Brief. Bioinform. **13**(6), 769–780 (2012)
59. Hunter, P., Chapman, T., Coveney, P.V., De Bono, B., Diaz, V., Fenner, J., Frangi, A.F., Harris, P., Hose, R., Kohl, P., Lawford, P.: A vision and strategy for the virtual physiological human: 2012 update. Interface Focus **3**(2), 20130004 (2013)
60. Viceconti, M., Hunter, P.: The virtual physiological human: ten years after. Ann. Rev. Biomed. Eng. **18**, 103–123 (2016)
61. Hoekstra, A.G., van Bavel, E., Siebes, M., Gijsen, F., Geris, L.: Virtual physiological human 2016: translating the virtual physiological human to the clinic (2017)
62. Azuaje, F., Devaux, Y., Wagner, D.: Computational biology for cardiovascular biomarker discovery. Briefings Bioinform. **10**(4), 367–377 (2009)
63. Smith, N.P., Crampin, E.J., Niederer, S.A., Bassingthwaighte, J.B., Beard, D.A.: Computational biology of cardiac myocytes: proposed standards for the physiome. J. Exp. Biol. **210**(9), 1576–1583 (2007)
64. Sun, R.: Introduction to computational cognitive modeling. In: Cambridge Handbook of Computational Psychology, pp. 3–19 (2008)
65. Bechtel, W., Graham, G., Balota, D.A. (eds.): A Companion to Cognitive Science, pp. 1–104. Blackwell, Oxford (1998)
66. Winslow, R.L., Trayanova, N., Geman, D., Miller, M.I.: Computational medicine: translating models to clinical care. Sci. Transl. Med. **4**(158), 158rv11 (2012)
67. Hamburg, M.A., Collins, F.S.: The path to personalized medicine. N. Engl. J. Med. **363**(4), 301–304 (2010)
68. Neufeld, E., Lloyd, B., Kainz, W., Kuster, N.: Functionalized anatomical models for computational life sciences. Front. Physiol. **9**, 1594 (2018)
69. Schank, R.C., Abelson, R.P.: Scripts, Plans, Goals, and Understanding: An Inquiry into Human Knowledge Structures. Psychology Press (2013)
70. Taylor, R.H., Menciassi, A., Fichtinger, G., Fiorini, P., Dario, P.: Medical robotics and computer-integrated surgery. In: Springer Handbook of Robotics, pp. 1657–1684. Springer, Cham (2016)
71. Valentin, J.: Basic anatomical and physiological data for use in radiological protection: reference values: ICRP Publication 89. Ann. ICRP **32**(3–4), 1–277 (2002)
72. Williams, G., Swanson, W.P., Kragh, P., Drexler, G.: Calculation and analysis of photon dose equivalent distributions in the ICRU sphere (No. GSF-S–958). Gesellschaft fuer Strahlen-und Umweltforschung mbH Muenchen (1983)
73. Fisher, H.L.J. Snyder, W.S.: Variation of dose delivered by 137Cs as a function of body size from infancy to adulthood. ORNL-4007. Oak Ridge National Laboratory, Oak Ridge, TN, pp. 221 (1966)
74. Snyder, W.S., Ford, M.R., Warner, G.G.: MIRD Pamphlet No. 5, Revised Estimates of specific absorbed fractions for monoenergetic photon sources uniformly distributed in various organs of a heterogeneous phantom. Society of Nuclear Medicine, New York (1978)
75. Kramer, R., Zankl, M., Williams, G., Drexler, G.: The Calculation of Dose from External Photon Exposures Using Reference Human Phantoms and Monte Carlo Methods: Part I. The Male (ADAM) and Female (EVA) Adult Mathematical Phantoms GSF-Report S-885. Institut fuer Strahlenschutz, GSF-Forschungszentrum fuer Umwelt und Gesundheit, Neuherberg (1982)

76. Billings, M.P., Yucker, W.R.: The computerized anatomical man CAM model, NASA CR-134043. Government Printing Office, Washington, DC (1973)
77. Tsui, B.M.W., Terry, J.A., Gullberg, G.T.: Evaluation of cardiac cone-beam single-photon emission computed-tomography using observer performance experiments and receiver operating characteristic analysis. Invest. Radiol. **28**, 1101 (1993)
78. Segars, W.P., Tsui, B.M.: MCAT to XCAT: the evolution of 4-D computerized phantoms for imaging research: computer models that take account of body movements promise to provide evaluation and improvement of medical imaging devices and technology. Proc. IEEE (Institute of Electrical and Electronics Engineers) **97**(12), 1954 (2009)
79. Peter, J., Tornai, M.P., Jaszczak, R.J.: Analytical versus voxelized phantom representation for Monte Carlo simulation in radiological imaging. IEEE Trans. Med. Imag. **19**(5), 556–564 (2000)
80. Segars, W.P., Lalush, D.S., Tsui, B.M.: Modeling respiratory mechanics in the MCAT and spline-based MCAT phantoms. IEEE Trans. Nucl. Sci. **48**(1), 89–97 (2001)
81. Kaufman, A.: Voxels as a computational representation of geometry. The computational representation of geometry. SIGGRAPH **94**, 45 (1994)
82. Zubal, I.G., Harrell, C.R., Smith, E.O., Rattner, Z., Gindi, G., Hoffer, P.B.: Computerized three-dimensional segmented human anatomy. Med. Phys. **21**(2), 299–302 (1994)
83. Law, M.Y., Liu, B.: DICOM-RT and its utilization in radiation therapy. Radiographics **29**(3), 655–667 (2009)
84. Pinter, C., Lasso, A., Wang, A., Jaffray, D., Fichtinger, G.: SlicerRT: radiation therapy research toolkit for 3D Slicer. Med. Phys. **39**(10), 6332–6338 (2012)
85. Caon, M.: Voxel-based computational models of real human anatomy: a review. Radiat. Environ. Biophys. **42**(4), 229–235 (2004)
86. Gibbs, S., Pujol, J.: A Monte Carlo method for patient dosimetry from diagnostic x-ray. Dentomaxillofac Radiol. **11**, 25 (1982)
87. Gibbs, S.J., Pujol Jr., A., Chen, T.S., Carlton, J.C., Dosmann, M.A., Malcolm, A.W., James Jr., A.E.: Radiation doses to sensitive organs from intraoral dental radiography. Dentomaxillofacial Radiol. **16**(2), 67–77 (1987)
88. Valentin, J.: Basic anatomical and physiological data for use in radiological protection: reference values: ICRP Publication 89. Ann. ICRP **32**(3–4), 1–277 (2002)
89. Valentin, J.: The 2007 recommendations of the international commission on radiological protection, pp. 1–333. Elsevier, Oxford (2007)
90. World Health Organization. Extremely low frequency fields (2007)
91. Sjogreen, K.: The Zubal Phantom Data, Voxel-Based Anthropomorphic Phantoms. http://noodle.med.yale.edu/phantom (1998)
92. Zankl, M., Wittmann, A.: The adult male voxel model "Golem" segmented from whole-body CT patient data. Radiat. Environ. Biophys. **40**, 153–162 (2001)
93. Caon, M., Bibbo, G., Pattison, J.: An EGS4-ready tomographic computational model of a 14-year-old female torso for calculating organ doses from CT examinations. Phys. Med. Biol. **44**, 2213–2225 (1999)
94. Caon, M., Bibbo, G., Pattison, J.: Monte Carlo calculated effective dose to teenage girls from computed tomography examinations. Radiat. Prot. Dosim. **90**(4), 445–448 (2000)
95. Caon, M., Sedlář, J., Bajger, M., Lee, G.: Computer-assisted segmentation of CT images by statistical region merging for the production of voxel models of anatomy for CT dosimetry. Australas. Phys. Eng. Sci. Med. **37**(2), 393–403 (2014)
96. Petoussi-Henss, N., Zankl, M., Fill, U., Regulla, D.: The GSF family of voxel phantoms. Phys. Med. Biol. **47**, 89–106 (2002)
97. Fill, U.A., Zankl, M., Petoussi-Henss, N., Siebert, M., Regulla, D.: Adult female voxel models of different stature and photon conversion coefficients for radiation protection. Health Phys. **86**(3), 253–272 (2004)
98. Petoussi, N., et al.: Organ doses for fetuses, babies, children and adults from environmental gamma-rays. Radiat. Prot. Dosim. **37**, 31 (1991)

99. Veit, R., Zankl, M., Petoussi, N., Mannweiler, E., Williams, G., Drexler, G.: Tomographic anthropomorphic models, Part i: construction technique and description of models of an 8 week old baby and a 7 year old child. GSF-Report **3**, 89 (1989)
100. Zankl, M., Panzer, W., Drexler, G.: Tomographic anthropomorphic models: part II: organ doses from computed tomographic examination in paediatric radiology. GSF-Bericht No. 30/93 (1993)
101. Stratis, A., Touyz, N., Zhang, G., Jacobs, R., Bogaerts, R., Bosmans, H., DIMITRA project partners.: Development of a paediatric head voxel model database for dosimetric applications. Br. J. Radiol. **90**(1078), 20170051 (2017)
102. Shi, C., Xu, X.G.: Development of a 30-week-pregnant female tomographic model from computed tomography (CT) images for Monte Carlo organ dose calculations. Med. Phys. **31**(9), 2491–2497 (2004)
103. Loftis, K., Halsey, M., Anthony, E., Duma, S.M., Stitzel, J.: Pregnant female anthropometry from ct scans for finite element model development. Biomed. Sci. Instrum. **44**, 355–360 (2008)
104. Kramer, R., Vieira, J.W., Khoury, H.J., Lima, F.R.A., Fuelle, D.: All about MAX: a male adult voxel phantom for Monte Carlo calculations in radiation protection dosimetry. Phys. Med. Biol. **48**(10), 1239 (2003)
105. Kramer, R., Khoury, H.J., Vieira, J.W., Loureiro, E.C.M., Lima, V.J.M., Lima, F.R.A., Hoff, G.: All about FAX: a female adult voXel phantom for Monte Carlo calculation in radiation protection dosimetry. Phys. Med. Biol. **49**(23), 5203 (2004)
106. Sato, K., Noguchi, H., Emoto, Y., Koga, S., Saito, K.: Japanese adult male voxel phantom constructed on the basis of CT images. Radiat. Prot. Dosim. **123**(3), 337–344 (2006)
107. Sato, K., Noguchi, H., Emoto Koga, Y., Saito, K.: Construction of a Japanese adult female voxel phantom for internal dosimetry. Radiat. Environ. Biophys. (2007)
108. van der Heyden, B., Schyns, L.E., Podesta, M., Vaniqui, A., Almeida, I.P., Landry, G., Verhaegen, F.: VOXSI: a voxelized single-and dual-energy CT scenario generator for quantitative imaging. Phys. Imag. Radiat. Oncol. **6**, 47–52 (2018)
109. Winslow, M., Xu, X.G., Yazici, B.: Development of a simulator for radiographic image optimization. Comput. Methods Programs Biomed. **78**(3), 179–190 (2005)
110. Son, I.Y., Winslow, M., Yazici, B., Xu, X.G.: X-ray imaging optimization using virtual phantoms and computerized observer modelling. Phys. Med. Biol. **51**(17), 4289 (2006)
111. Larsson, E., Strand, S.E., Ljungberg, M., Jönsson, B.A.: Mouse S-factors based on Monte Carlo simulations in the anatomical realistic Moby phantom for internal dosimetry. Cancer Biother. Radiopharm. **22**(3), 438–442 (2007)
112. Kostou, T., Papadimitroulas, P., Loudos, G., Kagadis, G.C.: A preclinical simulated dataset of S-values and investigation of the impact of rescaled organ masses using the MOBY phantom. Phys. Med. Biol. **61**(6), 2333 (2016)
113. Dimbylow, P.J.: FDTD calculations of the whole-body averaged SAR in an anatomically realistic voxel model of the human body from 1 MHz to 1 GHz. Phys. Med. Biol. **42**, 479 (1997)
114. Dimbylow, P.: Development of the female voxel phantom, NAOMI, and its application to calculations of induced current densities and electric fields from applied low frequency magnetic and electric fields. Phys. Med. Biol. **50**, 1047 (2005)
115. Dimbylow, P., Bolch, W.: Whole-body-averaged SAR from 50 MHz to 4 GHz in the University of Florida child voxel phantoms. Phys. Med. Biol. **52**(22), 6639 (2007)
116. Ferrari, P., Gualdrini, G.: An improved MCNP version of the NORMAN voxel phantom for dosimetry studies. Phys. Med. Biol. **50**(18), 4299 (2005)
117. Nagaoka, T., Watanabe, S., Sakurai, K., Kuneida, E., Watanabe, S., Taki, M., Yamanka, Y.: Development of realistic high resolution whole-body voxel models of Japanese adult male and female of average height and weight, and application of models to radio-frequency electromagnetic-field dosimetry. Phys. Med. Biol. **49**, 1–15 (2004)
118. Lee, C., Nagaoka, T., Lee, J.K.: Implementation of Japanese male and female tomographic phantoms to multi-particle Monte Carlo code for ionizing radiation dosimetry. J. Nucl. Sci. Technol. **43**, 937 (2006)

119. Nagaoka, T., et al.: An anatomically realistic whole-body pregnant-woman model and specific absorption rates for pregnant-woman exposure to electromagnetic plane waves from 10 MHz to 2 GHz. Phys. Med. Biol. **52**, 6731 (2007)
120. Lee, C., Lee, J., Lee, C.: Korean adult male voxel model KORMAN segmented from magnetic resonance images. Med. Phys. **31**, 1017 (2004)
121. Park, S.H., et al.: In vivo organ mass of Korean adults obtained from whole-body magnetic resonance data. Radiat. Prot. Dosim. **118**, 275 (2006)
122. Li, J.L., et al.: Organ dose conversion coefficients for external photon irradiation using the Chinese voxel phantom (CVP). Radiat. Prot. Dosim. (2009)
123. Becker, J., Zankl, M., Fill, U., Hoeschen, C.: Katja—the 24th week of virtual pregnancy for dosimetric calculations. Pol. J. Med. Phys. Eng. **14**(1), 13–20 (2008)
124. Gosselin, M.C., Neufeld, E., Moser, H., Huber, E., Farcito, S., Gerber, L., Jedensjö, M., Hilber, I., Di Gennaro, F., Lloyd, B., Cherubini, E.: Development of a new generation of high-resolution anatomical models for medical device evaluation: the virtual population 3.0. Phys. Med. Biol. **59**(18), 5287 (2014)
125. Rispoli, J.V., Wright, S.M., Malloy, C.R., McDougall, M.P.: Automated modification and fusion of voxel models to construct body phantoms with heterogeneous breast tissue: application to MRI simulations. J. Biomed. Graph. Comput. **7**(1), 1 (2017)
126. Lucano, E., Liberti, M., Lloyd, T., Apollonio, F., Wedan, S., Kainz, W., Angelone, L.M.: A numerical investigation on the effect of RF coil feed variability on global and local electromagnetic field exposure in human body models at 64 MHz. Magn. Reson. Med. **79**(2), 1135–1144 (2018)
127. Li, C., Chen, Z., Yang, L., Lv, B., Liu, J., Varsier, N., Hadjem, A., Wiart, J., Xie, Y., Ma, L., Wu, T.: Generation of infant anatomical models for evaluating electromagnetic field exposures. Bioelectromagnetics **36**(1), 10–26 (2015)
128. Sakellios, N., Rubio, J.L., Karakatsanis, N., Kontaxakis, G., Loudos, G., Santos, A., Nikita, K., Majewski, S.: GATE simulations for small animal SPECT/PET using voxelized phantoms and rotating-head detectors. In: 2006 IEEE Nuclear Science Symposium Conference Record, vol. 4, pp. 2000–2003. IEEE (2006, October)
129. Jackson, P.A., Beauregard, J.M., Hofman, M.S., Kron, T., Hogg, A., Hicks, R.J.: An automated voxelized dosimetry tool for radionuclide therapy based on serial quantitative SPECT/CT imaging. Med. Phys. **40**(11), 112503 (2013)
130. Cech, R., Leitgeb, N., Pediaditis, M.: Fetal exposure to low frequency electric and magnetic fields. Phys. Med. Biol. **52**(4), 879 (2007)
131. Xu, X.G., Taranenko, V., Zhang, J., Shi, C.: A boundary-representation method for designing whole-body radiation dosimetry models: pregnant females at the ends of three gestational periods—RPI-P3,-P6 and-P9. Phys. Med. Biol. **52**(23), 7023 (2007)
132. Sachse, F.B., Werner, C., Müller, M., Meyer-Waarden, K.: MEET Man-Models for Simulation of Electromagnetic, Elastomechanic and Thermic Behavior of Man. Erstellung und technische Parameter. Institut für Biomedizinische Technik, Technische Universität Karlsruhe, Karlsruhe (1997)
133. Bibin, L., Anquez, J., Angelini, E., Bloch, I.: Hybrid 3D pregnant woman and fetus modeling from medical imaging for dosimetry studies. Int. J. Comput. Assist. Radiol. Surg. **5**(1), 49 (2010)
134. Hurtado, J.L., Lee, C., Lodwick, D., Goede, T., Williams, J.L., Bolch, W.E.: Hybrid computational phantoms representing the reference adult male and adult female: construction and applications for retrospective dosimetry. Health Phys. **102**(3) (2012)
135. Rauwendaal, R.: Hybrid computational voxelization using the graphics pipeline (2012)
136. Janßen, C.F., Koliha, N., Rung, T.: A fast and rigorously parallel surface voxelization technique for GPU-accelerated CFD simulations. Commun. Comput. Phys. **17**(5), 1246–1270 (2015)
137. Nyirenda, P.J., Bronsvoort, W.F.: Numeric and curve parameters for freeform surface feature models. Comput.-Aided Des. **40**(8), 839–851 (2008)
138. Kobbelt, L.P., Bischoff, S., Botsch, M., Kähler, K., Rössl, C., Schneider, R., Vorsatz, J.: Geometric Modeling Based on Polygonal Meshes, vol. 1. Max-Planck-Institut für Informatik (2000)

139. Segars, W.P.: Development and application of the new dynamic Nurbs-based Cardiac-Torso (NCAT) phantom (2002)
140. Segars, W.P., Tsui, B.M.: Study of the efficacy of respiratory gating in myocardial SPECT using the new 4-D NCAT phantom. IEEE Trans. Nucl. Sci. **49**(3), 675–679 (2002)
141. Veress, A.I., Segars, W.P., Weiss, J.A., Tsui, B.M., Gullberg, G.T.: Normal and pathological NCAT image and phantom data based on physiologically realistic left ventricle finite-element models. IEEE Trans. Med. Imag. **25**(12), 1604–1616 (2006)
142. Segars, W.P., Taguchi, K., Fung, G.S.K., Fishman, E.K., Tsui, B.M.W.: Effect of heart rate on CT angiography using the enhanced cardiac model of the 4D NCAT. In: Medical Imaging 2006: Physics of Medical Imaging, vol. 6142, p. 61420I. International Society for Optics and Photonics (2006, March)
143. Garrity, J.M., Segars, W.P., Knisley, S.B., Tsui, B.M.: Development of a dynamic model for the lung lobes and airway tree in the NCAT phantom. IEEE Trans. Nucl. Sci. **50**(3), 378–383 (2003)
144. Segars, W.P., Mori, S., Chen, G.T.Y., Tsui, B.M.W.: Modeling respiratory motion variations in the 4D NCAT phantom. In: 2007 IEEE Nuclear Science Symposium Conference Record, vol. 4, pp. 2677–2679. IEEE (2007, October)
145. McGurk, R., Seco, J., Riboldi, M., Wolfgang, J., Segars, P., Paganetti, H.: Extension of the NCAT phantom for the investigation of intra-fraction respiratory motion in IMRT using 4D Monte Carlo. Phys. Med. Biol. **55**(5), 1475 (2010)
146. Zhang, J., Xu, X. G., Shi, C., Fuss, M.: Development of a geometry-based respiratory motion–simulating patient model for radiation treatment dosimetry. J. Appl. Clin. Med. Phys. **9**(1), 16–28 (2008)
147. Segars, W.P., Tsui, B.M., Frey, E.C., Fishman, E.K.: Extension of the 4D NCAT phantom to dynamic x-ray CT simulation. In: 2003 IEEE Nuclear Science Symposium. Conference Record (IEEE Cat. No. 03CH37515), vol. 5, pp. 3195–3199. IEEE (2003, October)
148. Segars, W.P., Mahesh, M., Beck, T., Frey, E.C., Tsui, B.M.W.: Validation of the 4D NCAT simulation tools for use in high-resolution x-ray CT research. In: Medical Imaging 2005: Physics of Medical Imaging, vol. 5745, pp. 828–835. International Society for Optics and Photonics (2005, April)
149. Segars, W.P., Tsui, B.M., Da Silva, A.J., Shao, L.: CT-PET image fusion using the 4D NCAT phantom with the purpose of attenuation correction. In: 2002 IEEE Nuclear Science Symposium Conference Record, vol. 3, pp. 1775–1779. IEEE (2002, November)
150. Segars, W.P., Tsui, B.M.: MCAT to XCAT: The evolution of 4-D computerized phantoms for imaging research. Proc. IEEE **97**(12), 1954–1968 (2009)
151. Segars, W.P., Bond, J., Frush, J., Hon, S., Eckersley, C., Williams, C.H., Feng, J., Frush, D., Tward, D.J., Ratnanather, J.T., Miller, M.I., Frush, D.: Population of anatomically variable 4D XCAT adult phantoms for imaging research and optimization. Med. Phys. **40**(4), 043701 (2013)
152. Norris, H., Zhang, Y., Bond, J., Sturgeon, G.M., Minhas, A., Tward, D.J., Ratnanather, J.T., Miller, M.I., Frush, D., Samei, E., Segars, W.P.: A set of 4D pediatric XCAT reference phantoms for multimodality research. Med. Phys. **41**(3) (2014)
153. Geyer, A.M., O'Reilly, S., Lee, C., Long, D.J., Bolch, W.E.: The UF/NCI family of hybrid computational phantoms representing the current US population of male and female children, adolescents, and adults—application to CT dosimetry. Phys. Med. Biol. **59**(18), 5225 (2014)
154. Segars, W., Tsui, B.: 4D MOBY and NCAT phantoms for medical imaging simulation of mice and men. J. Nucl. Med. **48**(Supplement 2), 203P–203P (2007)
155. Keenan, M.A., Stabin, M.G., Segars, W.P., Fernald, M.J.: RADAR realistic animal model series for dose assessment. J. Nucl. Med. **51**(3), 471–476 (2010)
156. Stabin, M., Xu, X., Emmons, M., Segars, W., Shi, C., Fernald, M.: RADAR reference adult, pediatric, and pregnant female phantom series for internal and external dosimetry. J. Nucl. Med. **53**(11), 1807–1813 (2012)
157. Kostou, T., Papadimitroulas, P., Loudos, G., Kagadis, G.C.: A preclinical simulated dataset of S-values and investigation of the impact of rescaled organ masses using the MOBY phantom. Phys. Med. Biol. **61**(6), 2333 (2016)

158. Dogdas, B., Stout, D., Chatziioannou, A.F., Leahy, R.M.: Digimouse: a 3D whole body mouse atlas from CT and cryosection data. Phys. Med. Biol. **52**(3), 577 (2007)
159. Xie, T., Zaidi, H.: Development of computational small animal models and their applications in preclinical imaging and therapy research. Med. Phys. **43**(1), 111–131 (2016)
160. Zhang, J., Na, Y., Caracappa, P., Xu, X.: RPI-AM and RPI-AF, a pair of mesh-based, size-adjustable adult male and female computational phantoms using ICRP-89 parameters and their calculations for organ doses from monoenergetic photon beams. Phys. Med. Biol. **54**(19), 5885–5908 (2009)
161. Cassola, V.F., de Melo Lima, V.J., Kramer, R., Khoury, H.J.: FASH and MASH: female and male adult human phantoms based on polygon mesh surfaces: I. Development of the anatomy. Phys. Med. Biol **55**(1), 133 (2009)
162. Kramer, R., Cassola, V.F., Khoury, H.J., Vieira, J.W., de Melo Lima, V.J., Brown, K.R.: FASH and MASH: female and male adult human phantoms based on polygon mesh surfaces: II. Dosimetric calculations. Phys. Med. Biol. **55**(1), 163 (2009)
163. Cassola, V.F., Milian, F.M., Kramer, R., de Oliveira Lira, C.A.B., Khoury, H.J.: Standing adult human phantoms based on 10th, 50th and 90th mass and height percentiles of male and female caucasian populations. Phys. Med. Biol. **56**(13), 3749 (2011)
164. Kim, C.H., Jeong, J.H., Bolch, W.E., Cho, K.W., Hwang, S.B.: A polygon-surface reference Korean male phantom (PSRK-Man) and its direct implementation in Geant4 Monte Carlo simulation. Phys. Med. Biol. **56**(10), 3137 (2011)
165. Kim, C.H., Yeom, Y.S., Nguyen, T.T., Wang, Z.J., Kim, H.S., Han, M.C., Lee, J.K., Zankl, M., Petoussi-Henss, N., Bolch, W.E., Lee, C.: The reference phantoms: voxel vs polygon. Ann. ICRP **45**(1_suppl), 188–201 (2016)
166. Park, J.S., Jung, Y.W., Lee, J.W., Shin, D.S., Chung, M.S., Riemer, M., Handels, H.: Generating useful images for medical applications from the visible Korean human. Comput. Methods Programs Biomed. **92**(3), 257–266 (2008)
167. Ghista, D., Sandler, H.: An analytic elastic-viscoelastic model for the shape and the forces in the left ventricle. J. Biomech. **2**, 35–47 (1969)
168. van den Broek, J., van den Broek, M.: Application of an ellipsoidal heart model in studying left ventricular contractions. J. Biomech. **13**(6), 493–503 (1980)
169. Kerckhoffs, R.C.P., Bovendeerd, P.H.M., Kotte, J.C.S., Prinzen, F.W., Smits, K., Arts, T.: Homogeneity of cardiac contraction despite physiological asynchrony of depolarization: a model study. Ann. Biomed. Eng. **31**, 536–547 (2003)
170. Sermesant, M., Moireau, P., Camara, O., Sainte-Marie, J., Andriantsimiavona, R., Cimrman, R., et al.: Cardiac function estimation from MRI using a heart model and data assimilation: advances and difficulties. Med. Image Anal. **10**, 642–656 (2006)
171. Tsui, B., Terry, J., Gullberg, G.: Evaluation of cardiac cone-beam single photon emission computed tomography using observer performance experiments and receiver operating characteristic analysis. Invest. Radiol. **28**(12), 1101–1112 (1993)
172. Pretorius, P.H., Xia, W., King, M.A., Tsui, B.M.: Evaluation of right and left ventricular volume and ejection fraction using a mathematical cardiac torso phantom. J. Nucl. Med. **38**(10), 1528 (1997)
173. Gibb, M., Bishop, M., Burton, R., Kohl, P., Grau, V., Plank, G., Rodriguez, B.: The role of blood vessels in rabbit propagation dynamics and cardiac arrhythmias. In: International Conference on Functional Imaging and Modeling of the Heart, pp. 268–276. Springer, Berlin, Heidelberg (2009)
174. Lee, C., Williams, J., Lee, C., Bolch, W.: The UF series of tomographic computational phantoms of pediatric patients. Med. Phys. **32**(12), 3537–3548 (2005)
175. Werner, C.D., Sachse, F.B., Dössel, O.: Electrical excitation propagation in the human heart. Int. J. Bioelectromagn. **2**(2) (2000)
176. Deng, D., Jiao, P., Ye, X., Xia, L.: An image-based model of the whole human heart with detailed anatomical structure and fiber orientation. Comput. Math. Methods Med. **2012**, 1–16 (2012)

177. Farah, J., Broggio, D., Franck, D.: Creation and use of adjustable 3d phantoms: application for the lung monitoring of female workers. Health Phys. **99**(5), 649–661 (2010)
178. Potse, M., Dube, B., Richer, J., Vinet, A., Gulrajani, R.: A comparison of monodomain and bidomain reaction-diffusion models for action potential propagation in the human heart. IEEE Trans. Biomed. Eng. **53**(12), 2425–2435 (2006)
179. Utkualp, N., Ercan, I.: Anthropometric measurements usage in medical sciences. Biomed. Res. Int. **2015**, 1–7 (2015)
180. Tanaka, G., Kawamura, H., Griffith, R., Cristy, M., Eckerman, K.: Reference man models for males and females of six age groups of Asian populations. Radiat. Prot. Dosim. **79**(1), 383–386 (1998)
181. Lee, C.S., Lee, J.K.: Computational anthropomorphic phantoms for radiation protection dosimetry: evolution and prospects. Nucl. Eng. Technol. **38**(3), 239–250 (2006)
182. Tusscher, K.H.W.J.T., Hren, R., Panfilov, A.V.: Organization of ventricular fibrillation in the human heart. Circ. Res. **100**, e87–e101 (2007)
183. Fleureau, J., Garreau, M., Donal, E., Leclercq, C., Hernández, A.: A hybrid tissue-level model of the left ventricle: application to the analysis of the regional cardiac function in heart failure. In: International Conference on Functional Imaging and Modeling of the Heart, pp. 258–267. Springer, Berlin, Heidelberg (2009)
184. Pravdin, S.: A mathematical spline-based model of cardiac left ventricle anatomy and morphology. Computation **4**(4), 42 (2016)
185. Krishnamurthy, A., Gonzales, M., Sturgeon, G., Segars, W., McCulloch, A.: Biomechanics simulations using cubic Hermite meshes with extraordinary nodes for isogeometric cardiac modeling. Comput. Aided Geom. Des. **43**, 27–38 (2016)
186. Loewe, A., Krueger, M.W., Holmqvist, F., Dössel, O., Seemann, G., Platonov, P.G.: Influence of the earliest right atrial activation site and its proximity to interatrial connections on P-wave morphology. EP Europace, **18**(suppl_4), iv35–iv43 (2016)
187. Chang, K., Trayanova, N.: Mechanisms of arrhythmogenesis related to calcium-driven alternans in a model of human atrial fibrillation. Sci. Rep. **6**(1) (2016)
188. Harrild, D.M., Henriquez, C.S.: A computer model of normal conduction in the human atria. Circ. Res. **87**, E25–E36 (2000)
189. Jacquemet, V., Virag, N., Ihara, Z., Dang, L., Blanc, O., et al.: Study of unipolar electrogram morphology in a computer model of atrial fibrillation. J. Cardiovasc. Electrophysiol. **14**, S172–S179 (2003)
190. Reumann, M., Bohnert, J., Seemann, G., Osswald, B., Dossel, O.: Preventive ablation strategies in a biophysical model of atrial fibrillation based on realistic anatomical data. IEEE Trans. Biomed. Eng. **55**(2), 399–405 (2008)
191. Plank, G., Prassl, A.J., Wang, J.I., Seemann, G., Scherr, D., et al.: Atrial fibrosis promotes the transition of pulmonary vein ectopy into reentrant arrhythmias. Heart Rhythm **5**, S162–S163 (2008)
192. Kim, H., Vignon-Clementel, I., Coogan, J., Figueroa, C., Jansen, K., Taylor, C.: Patient-specific modeling of blood flow and pressure in human coronary arteries. Ann. Biomed. Eng. **38**(10), 3195–3209 (2010)
193. Biglino, G., Cosentino, D., Steeden, J., De Nova, L., Castelli, M., Ntsinjana, H., et al.: Using 4D cardiovascular magnetic resonance imaging to validate computational fluid dynamics: a case study. Front. Pediatr. **3** (2015)
194. Chen, K., Liu, J.: MRI image segmentation based on watershed algorithm and WKFCM algorithm. J. Electron. Meas. Instrum. **25**(6), 516–521 (2011)
195. Cutroneo, G., Bruschetta, D., Trimarchi, F., Cacciola, A., Cinquegrani, M., Duca, A., et al.: In vivo CT direct volume rendering: a three-dimensional anatomical description of the heart. Pol. J. Radiol. **81**, 21–28 (2016)
196. Mueller, D.C.: Direct volume illustration for cardiac applications. Doctoral dissertation, Queensland University of Technology (2008)

Multimodal Medical Image Fusion as a Novel Approach for Aortic Annulus Sizing

N. A. Hamzah, Z. Omar, M. Hanafi and L. I. Izhar

Abstract Securing the safety and efficacy of transcatheter aortic valve implantation (TAVI) on the human heart largely depends upon an accurate size estimate of the aortic annulus. Correct aortic annulus sizing is needed in order for the clinician to select a suitable prosthetic valve, where incorrect selection can lead to aortic regurgitation. To date, echocardiography and computed tomography (CT) are two imaging modalities that have been used extensively for measuring annulus size. Our contention is that fusion of both modalities may merge complementary information and produce an altogether more wholesome image of higher quality, thus facilitating a more precise measurement of aortic annulus. This paper proposes the novel idea of Echocardiography-CT image fusion for the purpose of aortic annulus sizing, and further goes to review trends and advancements in the field of medical image fusion. Research shows that fusion of two imaging modalities may provide a more accurate and complete image for diagnosis purposes. In the case of aortic annulus sizing, image fusion can potentially decrease uncertainty and minimize redundancy in the output while maximising relevant information. In the paper, we have introduced the exploratory study of using medical image fusion as an alternative method of measuring aortic annulus, as well as its challenges and related works. We found that the use of pixel-level image fusion has been effective in medical image fusion for optimising diagnostics and intervention. A number of image fusion algorithms were reviewed; from those, we can conclude that curvelet-based methods score the highest using quality assessment metrics that best associate with the human visual system (HVS), and are therefore preferred when performing image fusion.

N. A. Hamzah (✉) · Z. Omar
IJN-UTM Cardiovascular Engineering Centre, Faculty of Biomedical Engineering, Universiti Teknologi Malaysia, Johor Bahru, Malaysia
e-mail: nuraqilahhamzah@gmail.com

Z. Omar
Faculty of Electrical Engineering, Universiti Teknologi Malaysia, Johor Bahru, Malaysia

M. Hanafi
Faculty of Engineering, Universiti Putra Malaysia, Serdang, Malaysia

L. I. Izhar
Department of Electrical and Electronic Engineering, Universiti Teknologi Petronas, Seri Iskandar, Malaysia

D. E. O. Dewi et al. (eds.), *Cardiovascular Engineering*, Series in BioEngineering,
https://doi.org/10.1007/978-981-10-8405-8_4

Keywords Image fusion · Multimodality · TAVI · Aortic stenosis

1 Introduction

A recent epidemiological study shows that one out of every eight people aged over 75 years suffers moderate or severe heart valve disease, including aortic stenosis (AS) [1]. This disease is most common in developed countries, especially in Europe and North America [2]. Stenosis is present when the cardiac valve area is smaller or equal to 1.5 cm^2, restricting the valve opening and not letting all blood to pass through. Prior to the year 2002, patients with severe AS are treated by surgical aortic valve replacement (AVR) which is an open-heart surgery that replaces the existing aortic valve with a prosthetic valve. Surgical AVR is a risky procedure for patients over 75 years old when taking into account the possibility of operative mortality [3]. Due to this, Cribier et al. [4] have introduced an alternative technique that invasively implants the prosthetic valve via transfemoral or transpical procedures, known as transcatheter aortic valve implantation (TAVI). TAVI was introduced to perform valve replacement on selective patients who are at high or prohibitive to surgical risk.

The selection of patients is based on the risk score of the Society of Thoracic Surgeons-Predicted of Mortality score (STS-PROM) and the European System for Cardiac Operative Risk Evaluation (EuroScore) [5–7]. These scores determine the risk factor of the patient for cardiac surgery. However, other conditions that are not included in the risk score that needs to be considered are disabling arthritis, liver cirrhosis, porcelain aorta, chest radiation, dementia, recurrent pulmonary emboli, right ventricular failure, cancer, cachexia, and frailty. More recently, Valve Academic Research Consortium (VARC-2) has standardized the definition of patient criteria for TAVI [8].

Edward SAPIEN and CoreValve are two examples of approved prosthetic valve for TAVI [9]. Both show haemodynamic results at mid-term follow-up and are good for clinical practice [10]. These types of prosthetic valve come in several sizes. The choice of the prosthetic valve depends on the aortic annulus size of the particular patient. It is very important to determine the accurate size of the aortic annulus. An incorrect sizing of the prosthetic valve may lead to aortic regurgitation (AR), valve embolization, coronary obstruction, left bundle branch block or aortic annular rupture [11, 12]. It has been reported that 11.7% of patients that undergo TAVI experience moderate-to-severe aortic regurgitation (AR), thereby showing the risk involved in the procedure. The most common AR that occurs is paravalvular AR, which means that AR occurs between the native annulus with the prosthetic valve [12]. To alleviate this concern, it is essential that an accurate measure of the size of the aortic annulus is obtained.

A number of modalities can be used to measure sans-operatively the aortic annulus size, namely echocardiography, computed tomography (CT) and magnetic resonance imaging (MRI). However due to its high procedural costs, MRI is less commonly utilised. Both echocardiography and CT are therefore the modalities of choice, specifically at the National Heart Institute (IJN) in Malaysia. Further descriptions of image acquisition on both modalities shall be explained in the following section. Although each modality offers a sufficient view of the aortic annulus, the gold standard in its measurement is yet to exist. This is because echocardiography and CT have their own advantages and limitations, and at times their measurements may contradict each other [13].

Echocardiography can provide the size of the aortic annulus, the geometry of the left ventricular outflow tract (LVOT) and the size of aortic root (Fig. 2) [14]. The images that are provided by echocardiography tend to have low spatial resolution. On the other hand, CT can deliver more detailed information on the shape and length of the aortic annulus (Fig. 3) [15]. Though CT images are prone to the calcification on the aortic valve. This obstructs clinicians from detecting the borderline of the aortic annulus, thereby resulting in incorrect measurement of aortic annulus size, either by overestimation or underestimation. Echocardiography and CT shows important characteristics of the aortic annulus, but both display inconsistent measurement of the aortic annulus. In some cases, when there is a conflicting measurement of aortic annulus sizing due to asymmetrical calcifications or eccentric leaflets, clinicians resort to using balloon aortic valvuloplasty (BAV) as to confirm any uncertainties [16]. However, the disadvantage of BAV is it being a maximally-invasive method of determining the aortic annulus size, and therefore not preferred.

Thus, we suggest the use of image fusion between echocardiography and CT to overcome inconsistent measurements of aortic annulus sizing. This solution will expectedly minimise the need and usage of BAV. The aim of image fusion is to provide a superior image containing comprehensive information by merging multiple input images. Image fusion has been used in a wide array of fields such as remote sensing, microscopic imaging, computer vision, robotics and medical diagnosis. The application of image fusion has been proven to be an excellent tool in combining information from different images into a singular output image [17–19], thereby increasing diagnostic accuracy.

The rest of the article are arranged as follows. The modalities used to measure aortic annulus sizing are discussed in the next section. We provide an in-depth understanding of the methods of measuring the aortic annulus sizing using echocardiography, CT and BAV. Furthermore, related works pertaining the use of fusion in the medical imaging field for localisation and detection are elaborated in the subsequent chapter. We discuss the merits and limitations of each technique, and will attempt to conclude the best method suited for Echocardiography-CT fusion on aortic annulus sizing tasks. Lastly, assessment techniques of the fused image are elaborated, before we summarise the paper in the final section.

2 Modalities Used in Aortic Annulus Sizing

The importance of measuring the accuracy of aortic annulus for TAVI has been long stressed in literature [11, 20–22]. When it comes to measuring the aortic annulus, a surgical AVR is straightforward, as opposed to TAVI. In surgical AVR the aortic annulus is measured once the valve has been declassified [23]. Meanwhile in clinical practices that utilise TAVI, echocardiography and CT are the main imaging modalities used to measure the aortic annulus size (Fig. 1).

The term 'sizing' indicates the choice of prosthesis that is in the range of available sizes that best fit into the native valve root. This measurement is considered difficult due the dynamic changes of aortic annulus in the patient during the systolic and diastolic phases [11]. To obtain the size of the aortic annulus, the most conventional method is to calculate the aortic root and the geometry of LVOT.

The available sizes for an Edwards SAPIEN prosthetic valve are 23, 26 and 29 mm. As for CoreValve, sizes that are available are 23, 26, 29 and 31 mm. The selection of this prosthetic valve is dependent on the measurement obtained by echocardiography and CT (Tables 1, 2 and 3), hence why they are so important.

2.1 *Echocardiography*

The role of echocardiography is to identify patients for intervention and provide intra-procedural monitoring [22]. Echocardiography gives real-time examination of

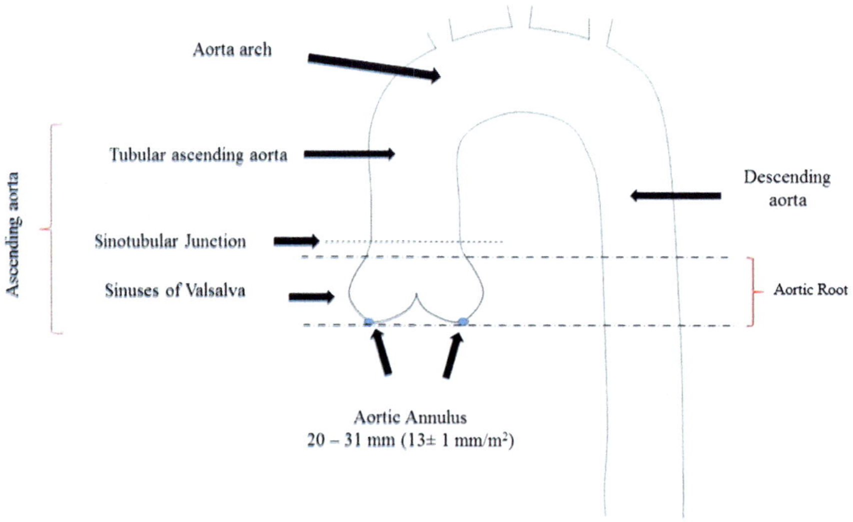

Fig. 1 View of the aorta, including ascending aorta, descending aorta and sinuses

Table 1 Prosthetic valve selection for Edward SAPIEN[a] [16]

Edward SAPIEN (mm)	23	26
Echocardiography (mm)	18–21	22–24 (25[a])
CT mean diameter (mm)	19–22	23–25

[a]Recommended by the manufacturer. For uncertain cases, use BAV for confirmation

Table 2 Prosthetic valve selection for Edward SAPIEN XT[a] [16]

Edward SAPIEN XT (mm)	23	26	29
Echocardiography (mm)	18–21	22–24	25–27
CT mean diameter (mm)	19–22	23–25	26–28

[a]Recommended by the manufacturer. For uncertain cases, use BAV for confirmation

Table 3 Prosthetic valve selection for CoreValve[a] [16]

CoreValve (mm)	23	26	29	31
Echocardiography (mm)	17–19	19–22	22–26	25–28
CT mean diameter (mm)	18–20	20–23	23–27	26–29

[a]Recommended by the manufacturer. For uncertain cases, use BAV for confirmation

the highly mobile aortic leaflets. The use of echocardiography in the detection of AS has shown to be prominent, as proven by studies in [24, 25].

In indicating the aortic annulus size using echocardiography, transesophageal echocardiogram (TEE) and transthoracic echocardiogram (TTE) are used. Cerillo et al. [14] have reported that the size obtained by echocardiography plays an essential role in choosing the correct prosthetic valve. This measuring technique is as illustrated in Fig. 2.

The size of the aortic annulus is measured by the geometry of the LVOT and the aortic root. Based on the guidelines for TTE [26], the aortic annulus is measured by the distance between the insertion of the adjacent leaflets on the parasternal long axis view (refer to Fig. 2). Using TEE, it is measured on the midesophageal long-axis view of the ascending aorta and aortic valve at the end-systolic phase. However, the measurement of aortic annulus using TTE or TEE are prone to underestimation of its size. Even though TEE and TTE are able to acquire the image, the probability of under- or over-estimating the aortic annulus size is high due to its low image quality. Another limitation of TEE and TTE is that the measurement of aortic diameter is based on a single annular plane, and assumes a circular annular orifice. A single plane may not capture the whole structure of the aortic annulus, therefore clinicians may face uncertainty in obtaining the correct size of the patient's aortic annulus.

When it comes to measuring aortic annulus using echocardiography, the measurement must be done in the systolic phase of the heart and zooming to the LVOT [10]. This measurement should be performed using an enlarged view of the midesophageal long axis, referred to as the "3 chamber view". From this image, clinicians should be able to obtain a clear view of an aligned image of the left ventricular chamber,

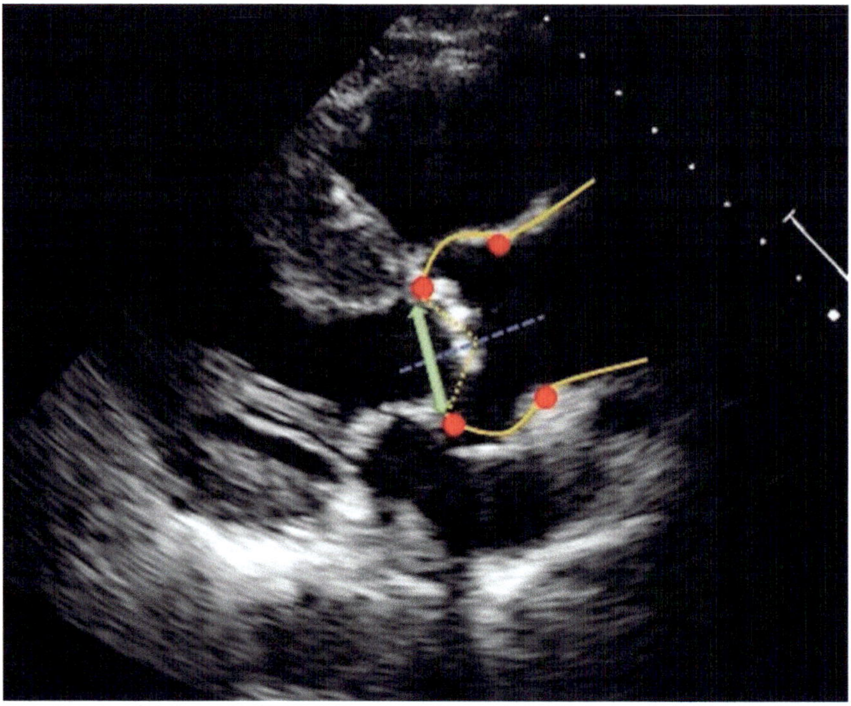

Fig. 2 Measurement of aortic annulus in echocardiography (green arrow). Red dots at the reference plane. Aortic annulus is measured between the hinge point of the cusps

outflow tract and ascending aorta. This is done to ensure that the sagittal plane bisects the maximal diameter of the annulus. The aortic annulus should be measured from the edge of the sinus to the hinge point of the right coronary cusp perpendicular to the long axis of the aorta, as shown in Fig. 2 [16].

2.2 Computed Tomography

For valve pathology, CT-based modalities may provide certain information, for instance the view of valve calcification. The cardiac morphology and function are also obtained by using CT. The technique provides imagery of all valves and can therefore assist in calculating the geometric orifice area. This is important, as the calcification of valves correlates with the AS [27]. In the CT image, the residual valve area is significantly similar to the valve area obtained by the echocardiography image. Further, evaluation of leaflet and cusp are done in cine-loops [28].

The use of CT can provide a detailed understanding of the anatomy in predicting the final shape of the prosthetic valve. Furthermore, CT gives an understanding of the

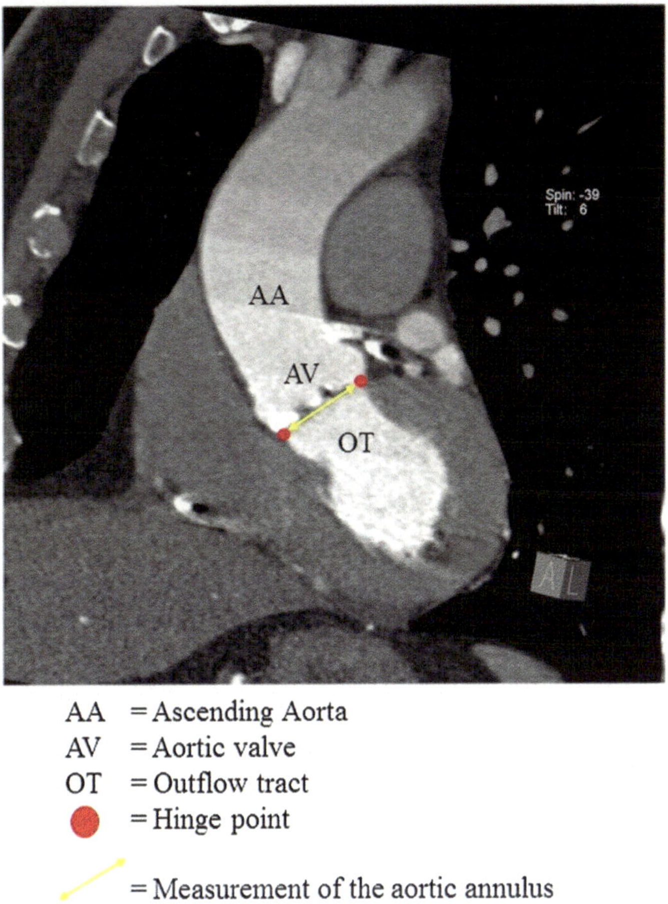

Fig. 3 The measuring of aortic annulus in CT with the presence of bulky calcification

complex three-dimensional aortic root, which includes the crown shaped anatomic aortic annulus [14]. However, the CT images will also include the presence of bulky calcification at the aortic root. This is undesirable as it will lead to uncertainty in estimating the aortic annulus size.

The annulus sizing is measured in CT geometry of the LVOT and the aortic root similar to that in echocardiography. The measurement of the aortic annulus by CT can be seen in Fig. 3. A notable difference is that the aortic annulus measurement in CT is measured both in diastolic and systolic phases. The data is obtained by tracing the LVOT contour at 20 and 70% points of the R-R interval [29].

In TAVI, it is imperative that the height of the coronary ostia is measured relative to the aortic annulus [10]. This however can be considerably hampered by the presence of bulky calcification (Fig. 3), which increases the risk of gaps between the external surface of prosthetic valve and the host native valve.

Subsequently, this may affect the determination of the borderline of the aortic annulus that is necessary for sizing the prosthetic valve.

2.3 Balloon Aortic Valvuloplasty

In certain cases, there could be uncertainties in selecting the appropriate prosthetic valve. Often, this forces the clinician to resort to using BAV to determine the aortic size. BAV is an invasive procedure where the surgeon inserts a catheter into the artery and through to the heart; once the tube reaches the aortic valve the balloon is inflated. The inflation of the balloon is carefully measured to indirectly confirm the size of the aortic annulus. The indication of the balloon size is gathered once the balloon has no movements at the aortic valve. Another indication is when the centre of the balloon is at the level of the annulus. Lastly, the absence of regurgitation in the aortic valve may also indicate the size [16]. Despite all these however, some literature have reported the size obtained by BAV can still be misleading, as noted in Pacchioni et al. [30].

3 Multimodal Medical Image Fusion

As previously stated, a single modality may not be sufficient in providing the correct measurement of the aortic annulus. The limitations of each modality often impedes the process to obtain the correct size of the aortic annulus. A multimodal approach is therefore suggested to provide a more reliable imagery of the subject [24, 31]. The approach denotes that echocardiography, coupled with CT, can be used to overcome this deficiency. It has been stated that when two images from different modalities are combined, they are expected to produce a much more reliable image that ideally features useful information of both original images [32].

For this reason, engineers have come up with a solution to improve medical diagnosis by fusing images between different modalities [32]. In a nutshell, image fusion may be divided into three levels that have been established by the scientific community, as seen in Fig. 4.

Our hypothesis is that the fusion of echocardiography and CT images shall result in an improved image for more accurate annulus sizing, as stated in the original aim. More specifically, the use of image fusion will increase the confidence of measuring the aortic annulus, thus eliminating the need for an invasive method, which can be painful and cumbersome.

The definition of image fusion in scientific terms is "the process of combining multiple input images into a single composite" [19]. On the whole, there are several techniques in implementing image fusion. The techniques belong either to the spatial or to the frequency domain methods. Spatial domain methods manipulate every pixel value in the input images in order to achieve the desired result. On

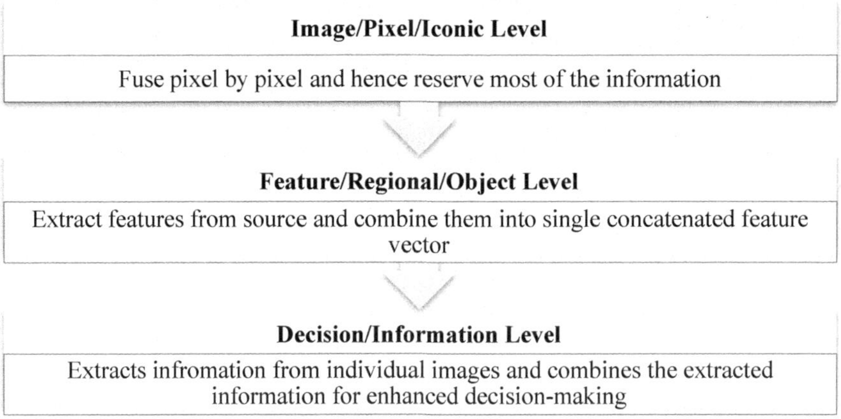

Fig. 4 Levels of fusion

Fig. 5 Classification of image fusion technique

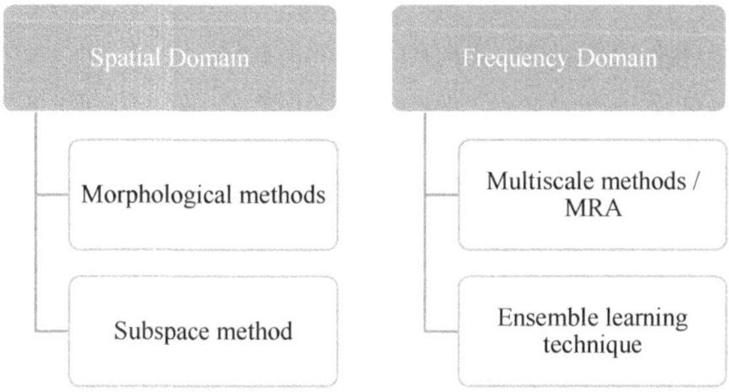

the other hand, frequency domain techniques first transform the image into another domain before implementing fusion [33]. The rationale is that salient image features are more visible in the transform domains, and can therefore be more easily manipulated. Based on literature, the various image fusion techniques are suitably classified into morphological methods, multiscale or multiresolution analysis (MRA) methods, subspace methods and ensemble learning techniques. A more detailed explanation on each of these are carried out in the section to come. Further, the categories can be denominated into either spatial or frequency domain as summarized in Fig. 5 [34].

Image fusion has been applied to many fields, most notably in satellite imaging and also in medical imaging. For medical imaging, the purpose of fusion is to improve the confidence in clinical diagnosis [35]. Medical image fusion is usually performed by fusing the structural and functional information of the same organ or anatomical region. The use of image fusion in the medical domain also helps to increase the reading efficiency and to compute the difference between clinical scans, where it acts as a guide for clinicians to plan for intervention.

In recent years multimodal medical image fusion (MMIF) has developed to become a viable solution to integrate information between multiple imaging devices that can produce a more comprehensive and precise description of the subject organ [13]. Scientists have shown that the use of image fusion in the medical context can lead to additional clinical information that is not visible in the separate images [36]. This allows the clinicians to evaluate the disease on a single image rather than on multiple images, thereby reducing complexity and error. MMIF also has the advantage of reducing the storage space, as the pertinent information from several input images are stored in a single image. Overall, we believe that MMIF is an advanced image processing technique that can provide high quality, reliable and comprehensive information to support the detection of the aortic annulus. The ability of evaluating the progression of the disease can be improved by MMIF [37].

3.1 Challenges of Multimodal Cardiac Image Fusion

Despite the many advances in medical image fusion, the field is still open for further research and development. More pertinently, significant challenges need to be reviewed in applying cardiac image fusion, including medical imaging technologies, applied methods and medical datasets.

One of the major challenges of the fusion process is borne by the modality used for fusing. Aortic annulus sizing is readily measured via echocardiography and CT. It is known that echocardiography is a procedure that is entirely human dependant, where the probe is handled, and its subsequent images are determined, by the user. Taken into perspective the angles of the image that are obtained by echocardiography, CT thereby requires appropriate slicing of its 3D model to fit the same viewpoint and render both images ‘registered’, so to speak. This is a major challenge that needs to be considered prior to the fusion process.

The computational complexity and the overall performance are some of the challenges that may prevent the implementation of various proposed works in the real world. While it is important to select the best fusion technique that preserves all the important information, due consideration must also be given to their processing speed and parameter requirements. Given that the main aim of fusing echocardiography and CT is to obtain a better view of the aortic annulus, the fusion method must be capable of performing adequately in real-time.

Finally, a crucial issue is the dynamic movements of the aortic annulus which changes at the systolic and diastolic phases. Practitioners and engineers who develop the fusion system must be clear which phase that best represents the aortic annulus. Following that, both the CT and echocardiography images must be correlated at the same phase. All these require processing and methods involving alignment, enhancement and transformation, beyond the conventional procedures associated with image fusion alone.

Table 4 Example studies of medical image fusion for localisation and measurement

Author	Organ	Fusion technique	Modalities	Purpose
Mazaheri et al. [39]	Cardiac	Hybrid fusion (1) Discrete wavelet transform (DWT) (2) Principal component analysis (PCA)	Echocardiography + echocardiography	Obtain measurement of the cardiac structure
Wong et al. [40]	Cardiac	Ensemble learning technique	MRI + body surface potential maps (BPSM)	Subject-specific cardiac physiology and pathology
Tavard et al. [41]	Cardiac	Ensemble learning technique	Echocardiography + electro-anatomical mapping (EAM) + CT	Left ventricle (LV) measurement for cardiac resynchronization therapy
Ambily et al. [42]	Brain	Multiscale method	Magnetic resonance imaging (MRI) + CT	Detection of tumour size and location
Hu et al. [43]	Sacroiliac joints (SIJ)	Software (virtual navigator 4.2)	Ultrasound + CT	Locate abnormal vascularization

3.2 *Related Works*

The application of medical image fusion has been widely used for diagnostic and analysis purposes [38]. A literature study by James and Dasarathy [38] has concisely summarised the application of image fusion for brain, breast, prostate, lungs and other organs, where fusion was found to substantially increase the clinical confidence in diagnostics and analysis. Further, we find that medical image fusion can be characterised by the techniques used, image modalities involved and organs under study. In specific cases, researchers rely on medical image fusion in order to determine the location and measurement of an organ, region or abnormalities present in the organ. This forms the basis of our approach, where we will be using fusion to gain a more accurate measure of the aortic annulus. We derive this motivation from a number of works in previous literature, as encapsulated in Table 4.

The table signifies various works which attest to the viability of using medical image fusion for identifying various structures of the human organ. With regards to the technique, the choice is diverse. A number of applications were noted where medical image fusion is used to obtain the structure of the heart [39–41]. Elsewhere, fusion has been successfully utilised to detect the precise location and size of a tumour [42], and also the location of abnormal vascularization [43].

Previous researches have proven that image fusion can be used to produce a clear view of an organ, as well as locating and determining a disease. This provides a strong foundation to our proposed idea, where image fusion is implemented to determine aortic annulus sizing. In fusing an image, the most important task is to produce an image that copes with human view. As stated earlier in this article, image fusion is used to provide an image that contains all the information from the input images. An important aspect of image fusion is to preserve the information associating with HVS.

3.3 Image Fusion Techniques

Image fusion techniques can be categorized into two groups, namely spatial domain and frequency domain. Spatial domain image fusion is based on direct manipulation of pixels of the image. The simplest form of a fusion technique is in the spatial domain and compares the corresponding pixel coordinates of both source images. In a weighted-average method, the idea is to calculate the average weight between the pixels of the two sources. Though this technique may reduce the signal-to-ratio (SNR) of the resulting image, the approach has shown itself to be unreliable. Fusion via weighted-average alone tends to introduce artefacts within the output [44]. This impedes the original aim of fusion, which is to improve the image quality so that further diagnosis, detection and decision-making may benefit from it [45].

Apart from weighted-average, other techniques have also been introduced for image fusion [44, 46]. Techniques are mostly chosen based on the characteristics of the input images and the resulting fused image [19, 37, 38, 44, 47, 48]. In our case, the main objective of fusion between echocardiography and CT images is to produce a clear view of the aortic annulus. For this reason, several factors need to be considered prior to selecting the fusion technique—the ability to preserve edge and textural details of the aortic annulus, and the ability to purge artefacts.

In recent years, image fusion has become a trend whereby its approaches are adept at increasing the practitioner's confidence in diagnosing and planning for clinical intervention. Among those, several technique have been proposed in order to fuse images in the medical context at pixel level. The techniques can be classified by:

(1) Morphological Methods

Morphological methods are the straightforward forms of pixel-level image fusion, where fusion is performed via mathematical operations [38, 47]. The operations include simple arithmetic operators, minimum, maximum, median and rank [19]. Markov Random Field (MRF) is considered a morphological method, in which it identifies a model by using probabilistic means. Given that this class is a straightforward method of fusion, it often faces certain limitations including contrast reduction of the image.

(2) Multiscale Methods/Multiresolution Analysis

The idea of wavelet based-methods for fusion is to extract the detailed information from one image and inject it into another [38]. The input images are transformed to the frequency domain so that they can be represented in a multiscale manner, which can highlight particular image features at different scales. Multiresolution analysis (MRA) decomposes the image into several components, in which information is captured at a given scale [19, 49]. The use of MRA as an image processing tool correlates well with the ability to capture information by the human visual system (HVS).

For pixel-level image fusion, the input images are decomposed using MRA. At every level of decomposition the fusion rule, such as max-abs, weighted average, entropy and others, is applied as to construct a composite MRA representation. The fused image is then obtained by applying the inverse transform on the composite MRA representation [50–53]. Various incarnations of MRA exist which have since been applied on image fusion tasks [19], these include wavelet transform [32, 36, 54–57], curvelet transform [58, 59], contourlet transform [60–63] and ripplet transform [64].

(3) Subspace Method

Subspace method is a collection of statistical techniques that remove the correlation between the input images [19, 47]. This method is done at the spatial domain, where the idea is to project a high dimensional input image into a lower dimensional space. Principal component analysis (PCA) [65], intensity-hue-saturation (IHS) [66], independent component analysis (ICA) [67], and canonical correlation analysis (CCA) [68] all fall into this category.

The main reason of using subspace method on medical images is the ability to visualise the intrinsic structure of the input data, once it has been reduced to a lower dimensionality. In addition, the method also boasts faster manipulation of the image and low memory requirements.

(4) Ensemble Learning Technique

The central aim of this method is to construct accurate predictors or classifiers through assembling weak predictors or classifiers [47]. Ensemble learning represents the fusion of K images I_k, $k \in \{1, 2, \ldots, k\}$, that are derived from the same base image. In obtaining I_k, different K transformation methods can be applied to the base image. Fusion using ensemble learning will be successful if I_k is independent, where it should highlight different characteristics of the base image.

Notwithstanding the above four classes, often there are cases where using just one image fusion technique is inadequate for medical images. In this case, a hybrid image fusion method is needed, for example to remove the ringing artifacts and the noise in the source images [35].

(5) Hybrid Image Fusion

Hybrid image fusion of medical images is an advanced approach that uses a combination of different fusion techniques to gain better results. By using hybrid image fusion minimised mean square error (MSE) value and maximised SNR [69] can be obtain. In Omar and Stathaki [44], Agarwal and Bedi [45] have demonstrated the approach's viability by combining the curvelet transform and WT in order to fuse the image between magnetic resonance imaging (MRI) and CT. The result displays minimal error and produces a higher image quality. Another solution is to combine the fusion rules among the different fusion techniques before integrating the results together to obtain a single image. Thus far, hybrid fusion seems the most attractive approach as it enables us to merge the advantages of various fusion algorithms. This will therefore be taken into consideration in developing the fusion framework for aortic annulus size detection.

3.4 Proposed Image Fusion Method

As mentioned above, the objective of fusion is to combine two images into one single image that can merge information from both inputs. In this, the fusion algorithm plays an essential part in extracting salient features of the input images.

In recent times, wavelet transform has grown to become the most popular image fusion technique in medical imaging and indeed other imaging tasks [13, 32, 36, 70, 71]. However it is not without its limitations. Despite the ability of wavelet transform in representing point singularities, it tends to ignore geometric properties of the structures and therefore loses the ability to exploit regularity of edges [51]. To overcome this limitation, researchers have performed modifications on the wavelet transform.

Candes and Donoho [72] introduced the first generation curvelet transform to overcome the directionality problem faced by wavelets. In comparison, the curvelet transform uses ridgelets to obtain the sub images. The limitation to this first generation curvelet transform however is its heavy time consumption. In 2006, Candes et al. [73] subsequently developed the fast discrete curvelet transform to eliminate the use of ridgelet transform and deliver faster performance.

Another major parameter that affects the fusion performance is the fusion rule. The majority of research thus far rely on simple fusion rules, for instance maximum absolute value, weighted average and maximum selection. Though, these rules inherently do not consider the relations between neighbouring pixels in the input image.

In this research we propose using a hybrid approach comprising two established fusion techniques: PCA and curvelet transform. The echocardiography and CT input images are each initially decomposed using the curvelet transform. Thereafter PCA

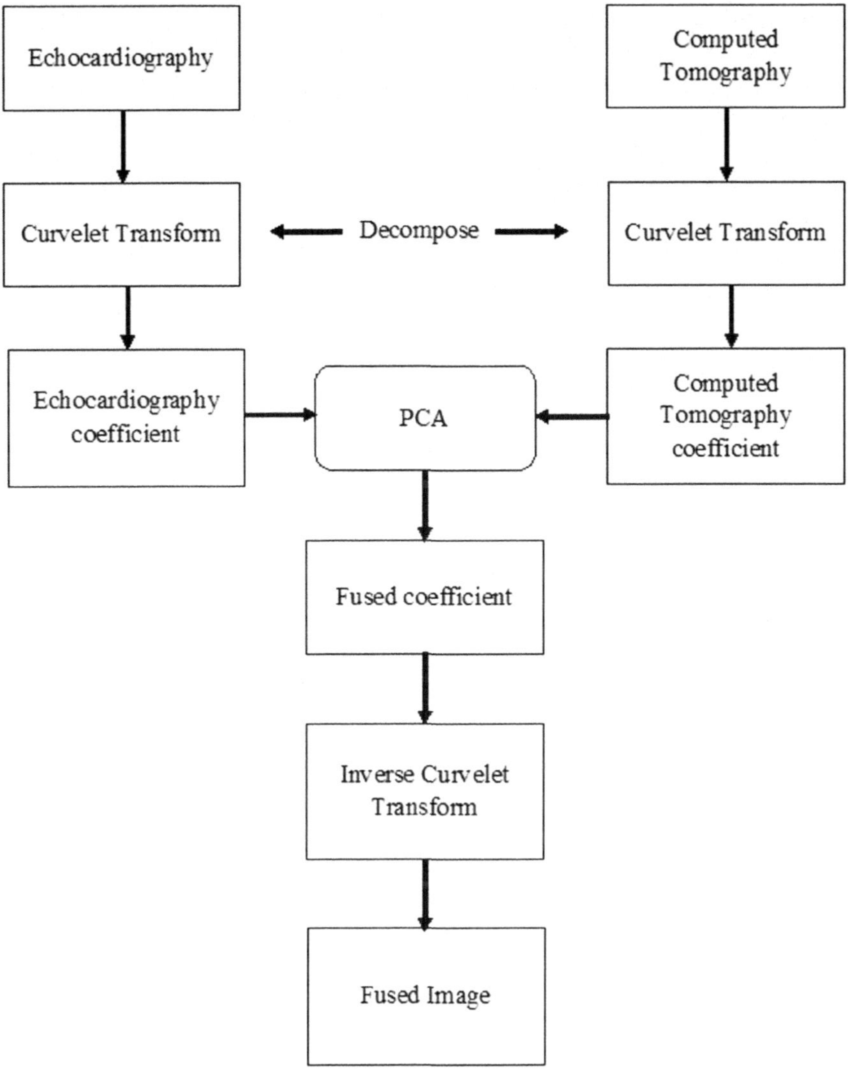

Fig. 6 Proposed image fusion workflow

is used as the fusion rule between the Echo and CT coefficients. The high correlative property of PCA is projected to yield a better fusion result, as compared to regular rules. Finally the inverse curvelet transform is applied to the fused coefficients to generate the fused output image. The proposed framework is illustrated in Fig. 6.

3.5 Image Evaluation

An important criterion of the overall image fusion system is the performance measure, which evaluates the fusion output both qualitatively and quantitatively. Based on the conventions used by the image fusion community, the evaluation metric should ideally meet the following requirements [74, 75]:

(a) To accurately perceive the most important information from either one, or both, input images.
(b) To measure the capacity of the fusion process in extracting information from the input image to the output image as accurately as possible.
(c) To accord with the HVS.

Given the highly specific nature of this research endeavour, where we aim to fuse Echo and CT cardiac images to improve detection of aortic annulus size, an exclusive dataset is desired. Specifically, we require dataset from patients that have gone through TAVI where their prosthetic valve size acts as the ground truth. For this purpose, we will be collaborating with clinicians at the Malaysian National Heart Institute (IJN) to provide the patient data as needed. In turn, to measure the performance, the size obtained by the fused image shall be compared with the prosthetic valve size that has been implanted in the patient, and the accuracy is recorded.

Multiple techniques have been established to identify the quantitative score of the fused image. The evaluation of the images is done to assess how good a fusion algorithm is at retaining salient features, whilst purging artefacts and withstanding image degradation.

The most commonly used image quality assessment are entropy, structural similarity image measure (SSIM), mutual information (MI) and universal image quality index (UIQI) [18, 32, 36, 39, 40, 42, 43, 53, 57–59, 61, 63, 64, 76, 77]. Though these quality assessments clearly possess the ability to measure the fusion algorithm, they do not necessarily associate with HVS. The significance of the human visual system is readily known. The human eyesight is more adept at discerning edges, outlines and contrasting features within a scenery. As such, metrics that emphasise on this particular feature is regarded superior and are thus preferred.

To overcome this problem, visible difference (VD) has been introduced. This method has shown to be useful in measuring the probability of signal difference between input images and fused image that can be identified by the human visual system (HVS) [78]. They are used to indicate how much information the fused image conveys.

The concept of VD is widely used to assess image fidelity that determines the probability of a spectator in detecting the difference between two forms of two signals [79]. VD is used to compare between the input images and the fused image by detecting the visual information transferred between them. Furthermore, VD is useful in detecting the presence of any fusion artefacts in the fused image [78]. Evaluating the fusion performance is done by using a simple measurement of average probability in identifying the difference between the input images and the fused image. A low probability value indicates that the fusion process has been successful. VD has shown to be the most reliable technique for fusion quality assessment that correlates well with HVS. Indeed, the use of VD in evaluating fused images in different fields, such as remote sensing, has shown to be prominent.

4 Conclusion

The aim of this paper has been to explore and investigate the potential use of medical image fusion in obtaining the correct aortic annulus sizing. As has been mentioned earlier, correct measurement of aortic annulus size is important as it is used for selecting the prosthetic valve for TAVI. This may alleviate sizing error, which can potentially complicate the patient's health.

Image fusion has shown to be successful in many fields, where it can improve upon uncertainties that may be present when the images are analysed individually. With regards to aortic annulus size estimation, evaluating the echocardiography and CT images separately may at times produce an inadequate result. The merging of images from different modalities can maximise the information that users require. The combination of echocardiography and CT modalities retained under a single image will help clinicians to clearly observe the aortic annulus and measure its size, thus facilitating the choice for a suitable prosthetic valve. Additionally, fusion will be able to alleviate the dependency on BAV for aortic annulus size confirmation. Overall, medical image fusion has the potential to improve areas pertaining diagnosis, planning, intervention and monitoring in clinical use.

We thus contend hybrid image fusion to be the preferred solution to aortic annulus sizing tasks. More specifically, the combination of curvelet transform in the decomposition step and PCA as the fusion rule within an enhanced hybrid fusion framework is expected to generate higher quality fusion results. This ensures the preservation of image edges and absolves the presence of artefacts and blurring, therefore better facilitating the correct measurement of the aortic annulus.

This fusion algorithm should have the tendency to preserve important information from echocardiography and CT, so as to produce a clearer view of the aortic annulus that correlates highly with HVS. Both qualitative and quantitative evaluation of the fused image shall be done by means of perceptive comparison with ground truth, as well as analysis using the VD metric.

Acknowledgements The research was made possible by the fundings of the Ministry of Higher Education (MOHE) Malaysia and Universiti Teknologi Malaysia (UTM) under the Research University Tier 1 Grant (vote 12H72). We would also like to acknowledge the help of Dr. Al Fazir Omar (IJN) and Professor Christine Pohl (UTM) for their immense help during the early phase of this study.

References

1. Nkomo, V.T., Gardin, J.M., Skelton, T.N., Gottdiener, J.S., Scott, C.G., Enriquez-Sarano, M.: Burden of valvular heart diseases: a population-based study. Lancet **368**, 1005–1011 (2006)
2. Vahanian, A., Iung, B., Himbert, D.: Transcatheter aortic valve implantation: a treatment we are going to need!. J. Am. Coll. Cardiol. **62**, 1013–1014 (2013)
3. Takano, H., Miyata, H., Motomura, N., Yamauchi, T., Shirakawa, Y., Takamoto, S.: Analysis of risk factors for mortality and morbidity of surgical aortic valve replacement for aortic stenosis: risk models from a Japanese database. Circulation **132**, A16999 (2015)
4. Cribier, A., Eltchaninoff, H., Bash, A., Borenstein, N., Tron, C., Bauer, F., Derumeaux, G., Anselme, F., Laborde, F., Leon, M.B.: Percutaneous transcatheter implantation of an aortic valve prosthesis for calcific aortic stenosis first human case description. Circulation **106**, 3006–3008 (2002)
5. Arai, T., Lefèvre, T.: Who is the right patient for TAVI? J. Cardiol. **63**, 178–181 (2014)
6. Bax, J.J., Delgado, V., Bapat, V., Baumgartner, H., Collet, J.P., Erbel, R., Hamm, C., Kappetein, A.P., Leipsic, J., Leon, M.B.: Open issues in transcatheter aortic valve implantation. Part 1: patient selection and treatment strategy for transcatheter aortic valve implantation. Eur. Heart J. **35**, 2627–2638 (2014)
7. Piazza, N., Lange, R., Martucci, G., Serruys, P.W.: Patient selection for transcatheter aortic valve implantation: patient risk profile and anatomical selection criteria. Arch. Cardiovasc. Dis. **105**, 165–173 (2012)
8. Kappetein, A.P., Head, S.J., Généreux, P., Piazza, N., van Mieghem, N.M., Blackstone, E.H., Brott, T.G., Cohen, D.J., Cutlip, D.E., van Es, G.-A., Hahn, R.T., Kirtane, A.J., Krucoff, M.W., Kodali, S., Mack, M.J., Mehran, R., Rodés-Cabau, J., Vranckx, P., Webb, J.G., Windecker, S., Serruys, P.W., Leon, M.B.: Updated standardized endpoint definitions for transcatheter aortic valve implantation: the valve academic research consortium-2 consensus document. J. Am. Coll. Cardiol. **60**, 1438–1454 (2012)
9. Bourantas, C.V., Farooq, V., Onuma, Y., Piazza, N., Van Mieghem, N.M., Serruys, P.W.: Transcatheter aortic valve implantation: new developments and upcoming clinical trials. EuroInterv. J. EuroPCR Collab. Work. Group Interv. Cardiol. Eur. Soc. Cardiol. **8**, 617–627 (2012)
10. Zamorano, J.L., Gonçalves, A., Lang, R.: Imaging to select and guide transcatheter aortic valve implantation. Eur. Heart J. **35**(24), 1578–1587(2014)
11. Jurencak, T., Turek, J., Kietselaer, B.J.H., Mihl, C., Kok, M., van Ommen, V.V.A., van Garsse, L.F.M., Nijssen, E., Wildberger, J., Das, M.: MDCT evaluation of aortic root and aortic valve prior to TAVI. What is the optimal imaging time point in the cardiac cycle? Eur. Radiol. 1–9 (2015)
12. Bax, J.J., Delgado, V., Bapat, V., Baumgartner, H., Collet, J.P., Erbel, R., Hamm, C., Kappetein, A.P., Leipsic, J., Leon, M.B.: Open issues in transcatheter aortic valve implantation. Part 2: procedural issues and outcomes after transcatheter aortic valve implantation. Eur. Heart J. **35**, 2639–2654 (2014)
13. Bhatnagar, G., Wu, Q.J., Liu, Z.: Human visual system inspired multi-modal medical image fusion framework. Expert Syst. Appl. **40**, 1708–1720 (2013)
14. Cerillo, A., Mariani, M., Berti, S., Glauber, M.: Sizing the aortic annulus. Ann. Cardiothorac. Surg. **1**, 245–256 (2012)
15. Messika-Zeitoun, D., Serfaty, J.-M., Brochet, E., Ducrocq, G., Lepage, L., Detaint, D., Hyafil, F., Himbert, D., Pasi, N., Laissy, J.-P., Iung, B., Vahanian, A.: Multimodal assessment of the aortic annulus diameter: implications for transcatheter aortic valve implantation. J. Am. Coll. Cardiol. **55**, 186–194 (2010)
16. Kasel, A.M., Cassese, S., Bleiziffer, S., Amaki, M., Hahn, R.T., Kastrati, A., Sengupta, P.P.: Standardized imaging for aortic annular sizing: implications for transcatheter valve selection. JACC Cardiovasc. Imaging **6**, 249–262 (2013)
17. Cao, W., Li, B., Zhang, Y.: A remote sensing image fusion method based on PCA transform and wavelet packet transform. In: International Conference on Neural Networks and Signal Processing, 2003. Proceedings of the 2003, pp. 976–981. IEEE (2003)

18. Gaemperli, O., Schepis, T., Valenta, I., Husmann, L., Scheffel, H., Duerst, V., Eberli, F.R., Luscher, T.F., Alkadhi, H., Kaufmann, P.A.: Cardiac image fusion from stand-alone SPECT and CT: clinical experience. J. Nucl. Med. **48**, 696–703 (2007)
19. Mitchell, H.B.: Image Fusion: Theories, Techniques and Applications. Springer (2010)
20. Khalique, O.K., Kodali, S.K., Paradis, J.M., Nazif, T.M., Williams, M.R., Einstein, A.J., Pearson, G.D., Harjai, K., Grubb, K., George, I., Leon, M.B., Hahn, R.T.: Aortic annular sizing using a novel 3-dimensional echocardiographic method use and comparison with cardiac computed tomography. Circ. Cardiovasc. Imaging **7**, 155–163 (2014)
21. Leipsic, J., Gurvitch, R., LaBounty, T.M., Min, J.K., Wood, D., Johnson, M., Ajlan, A.M., Wijesinghe, N., Webb, J.G.: Multidetector computed tomography in transcatheter aortic valve implantation. JACC Cardiovasc. Imaging **4**, 416–429 (2011)
22. Zamorano, J.L., Badano, L.P., Bruce, C., Chan, K.-L., Gonçalves, A., Hahn, R.T., Keane, M.G., La Canna, G., Monaghan, M.J., Nihoyannopoulos, P., Silvestry, F.E., Vanoverschelde, J.-L., Gillam, L.D.: EAE/ASE recommendations for the use of echocardiography in new transcatheter interventions for valvular heart disease. J. Am. Soc. Echocardiogr. **24**, 937–965 (2011)
23. Kempfert, J., Van Linden, A., Lehmkuhl, L., Rastan, A.J., Holzhey, D., Blumenstein, J., Mohr, F.W., Walther, T.: Aortic annulus sizing: echocardiographic versus computed tomography derived measurements in comparison with direct surgical sizing. Eur. J. Cardiothorac. Surg. **42**, 627–633 (2012)
24. Leong, D.P., Joseph, M.X., Selvanayagam, J.B.: The evolving role of multimodality imaging in valvular heart disease. Heart **100**, 336–346 (2014)
25. Douglas, P.S., Khandheria, B., Stainback, R.F., Weissman, N.J., Brindis, R.G., Patel, M.R., Alpert, J.S., Fitzgerald, D., Heidenreich, P., Martin, E.T.: ACCF/ASE/ACEP/ASNC/SCAI/SCCT/SCMR 2007 appropriateness criteria for transthoracic and transesophageal echocardiography: a report of the American College of Cardiology Foundation Quality Strategic Directions Committee Appropriateness Criteria Working Group, American Society of Echocardiography, American College of Emergency Physicians, American Society of Nuclear Cardiology, Society for Cardiovascular Angiography and Interventions, Society of Cardiovascular Computed Tomography, and the Society for Cardiovascular Magnetic Resonance endorsed by the American College of Chest Physicians and the Society of Critical Care Medicine. J. Am. Coll. Cardiol. **50**, 187–204 (2007)
26. Shanewise, J.S., Cheung, A.T., Aronson, S., Stewart, W.J., Weiss, R.L., Mark, J.B., Savage, R.M., Sears-Rogan, P., Mathew, J.P., Quiñones, M.A.: ASE/SCA guidelines for performing a comprehensive intraoperative multiplane transesophageal echocardiography examination: recommendations of the American Society of Echocardiography Council for Intraoperative Echocardiography and the Society of Cardiovascular Anesthesiologists Task Force for Certification in Perioperative Transesophageal Echocardiography. J. Am. Soc. Echocardiogr. **12**, 884–900 (1999)
27. Qanadli, S.D.: Cardiac CT: practical approach to integrate appropriate indications in daily practice. Presse Med. **40**, e413–e423 (2011)
28. Schuijf, J.D., Kroft, L.J., de Roos, A., Bax, J.J.: Classic images: cardiac computed tomography. Curr. Probl. Cardiol. **34**, 277–295 (2009)
29. Clavel, M.-A., Malouf, J., Messika-Zeitoun, D., Araoz, P.A., Michelena, H.I., Enriquez-Sarano, M.: Aortic valve area calculation in aortic stenosis by CT and Doppler echocardiography. JACC Cardiovasc. Imaging **8**, 248–257 (2015)
30. Pacchioni, A., Penzo, C., Reimers, B., Saccà, S.: Size (or sizing) matters! In: 34th National Congress Italian Society of Invasive Cardiology (2013)
31. Rajani, R., Khattar, R., Chiribiri, A., Victor, K., Chambers, J.: Multimodality imaging of heart valve disease. Arq. Bras. Cardiol. **103**, 251–263 (2014)
32. Alfano, B., Ciampi, M., Pietro, G.: A wavelet-based algorithm for multimodal medical image fusion. In: Falcidieno, B., Spagnuolo, M., Avrithis, Y., Kompatsiaris, I., Buitelaar, P. (eds.) Semantic Multimedia, pp. 117–120. Springer, Berlin, Heidelberg (2007)
33. Sahu, D.K., Parsai, M.: Different image fusion techniques—a critical review. Int. J. Mod. Eng. Res. (IJMER) **2**, 4298–4301 (2012)

34. Singh, S., Rajput, R.: A comparative study of classification of image fusion techniques. Int. J. Eng. Comput. Sci. **3**, 7350–7353 (2014)
35. Kavitha, C., Chellamuthu, C.: Medical image fusion based on hybrid intelligence. Appl. Soft Comput. **20**, 83–94 (2014)
36. Yong, Y., Dongsun, P., Shuying, H., Nini, R.: Medical image fusion via an effective wavelet-based approach. EURASIP J. Adv. Sign. Process. **2010**, 579341 (2010)
37. Constantinos, S.P., Pattichis, M.S., Micheli-Tzanakou, E.: Medical imaging fusion applications: an overview. In: Conference Record of the Thirty-Fifth Asilomar Conference on Signals, Systems and Computers, vol. 1262, pp. 1263–1267 (2001)
38. James, A.P., Dasarathy, B.V.: Medical image fusion: a survey of the state of the art. Inf. Fusion (2014)
39. Mazaheri, S., Sulaiman, P.S., Wirza, R., Dimon, M.Z., Khalid, F., Moosavi Tayebi, R.: Hybrid pixel-based method for cardiac ultrasound fusion based on integration of PCA and DWT. Comput. Math. Methods Med. (2014)
40. Wong, K.C., Wang, L., Zhang, H., Liu, H., Shi, P.: Physiological fusion of functional and structural images for cardiac deformation recovery. IEEE Trans. Med. Imaging **30**, 990–1000 (2011)
41. Tavard, F., Simon, A., Leclercq, C., Donal, E., Hernández, A.I., Garreau, M.: Multimodal registration and data fusion for cardiac resynchronization therapy optimization. IEEE Trans. Med. Imaging **33**, 1363–1372 (2014)
42. Ambily, P., James, S.P., Mohan, R.R.: Brain tumor detection using image fusion and neural network (2015)
43. Hu, Z., Zhu, J., Liu, F., Wang, N., Xue, Q.: Feasibility of US-CT image fusion to identify the sources of abnormal vascularization in posterior sacroiliac joints of ankylosing spondylitis patients. Sci. Rep. **5** (2015)
44. Omar, Z., Stathaki, T.: Image fusion: an overview. In: 2014 5th International Conference on Intelligent Systems, Modelling and Simulation (ISMS), pp. 306–310. IEEE (2014)
45. Agarwal, J., Bedi, S.S.: Implementation of hybrid image fusion technique for feature enhancement in medical diagnosis. Hum.-centric Comput. Inf. Sci. **5**, 1–17 (2015)
46. Mengyu, Z., Yuliang, Y.: A new image fusion algorithm based on fuzzy logic. In: 2008 International Conference on Intelligent Computation Technology and Automation (ICICTA), pp. 83–86. IEEE (2008)
47. El-Gamal, F.E.-Z.A., Elmogy, M., Atwan, A.: Current trends in medical image registration and fusion. Egypt. Inform. J. (2015)
48. Galande, A., Patil, R.: The art of medical image fusion: a survey. In: 2013 International Conference on Advances in Computing, Communications and Informatics (ICACCI), pp. 400–405. IEEE (2013)
49. Piella, G.: A general framework for multiresolution image fusion: from pixels to regions. Inf. Fusion **4**, 259–280 (2003)
50. Pajares, G., Manuel de la Cruz, J.: A wavelet-based image fusion tutorial. Pattern Recogn. **37**, 1855–1872 (2004)
51. Ma, J., Plonka, G.: The curvelet transform. Sig. Process. Mag. IEEE **27**, 118–133 (2010)
52. Do, M.N., Vetterli, M.: The contourlet transform: an efficient directional multiresolution image representation. IEEE Trans. Image Process. **14**, 2091–2106 (2005)
53. Yang, J., Ma, Y., Yao, W., Lu, W.: A spatial domain and frequency domain integrated approach to fusion multifocus images. In: 2008 Proceedings of Commission VII ISPRS Congress Beijing (2008)
54. Wang, A., Sun, H., Guan, Y.: The application of wavelet transform to multi-modality medical image fusion. In: Proceedings of the 2006 IEEE International Conference on Networking, Sensing and Control, 2006 (ICNSC '06), pp. 270–274. IEEE (2006)
55. Hong, Z., Lei, L., Nan, L.: A novel wavelet medical image fusion method. In: International Conference on Multimedia and Ubiquitous Engineering, 2007 (MUE '07), pp. 548–553 (2007)
56. Singh, R., Srivastava, R., Prakash, O., Khare, A.: Mixed scheme based multimodal medical image fusion using Daubechies complex wavelet transform. In: 2012 International Conference on Informatics, Electronics & Vision (ICIEV), pp. 304–309. IEEE (2012)

57. Xiao, Z.-S., Zheng, C.-X.: Medical image fusion based on an improved wavelet coefficient contrast. In: 3rd International Conference on Bioinformatics and Biomedical Engineering, 2009 (ICBBE 2009), pp. 1–4. IEEE (2009)
58. Himanshi, V.B., Krishn, A., Sahu, A.: Medical image fusion in curvelet domain employing PCA and maximum selection rule. In: Proceedings of the Second International Conference on Computer and Communication Technologies (IC3T 2015), p. 1. Springer (2015)
59. Ali, F., El-Dokany, I., Saad, A., Abd El-Samie, F.: A curvelet transform approach for the fusion of MR and CT images. J. Mod. Opt. **57**, 273–286 (2010)
60. Bhatnagar, G., Wu, Q.J., Liu, Z.: A new contrast based multimodal medical image fusion framework. Neurocomputing (2015)
61. Yang, L., Guo, B.L., Ni, W.: Multimodality medical image fusion based on multiscale geometric analysis of contourlet transform. Neurocomputing **72**, 203–211 (2008)
62. Yang, Y., Tong, S., Huang, S., Lin, P.: Log-Gabor energy based multimodal medical image fusion in NSCT domain. Comput. Math. Methods Med. **2014** (2014)
63. Azzawi, N.A.A., Sakim, H.A.M., Abdullah, A.K.W.: An efficient medical image fusion method using contourlet transform based on PCM. In: IEEE Symposium on Industrial Electronics & Applications (ISIEA), pp. 11–14. IEEE (2009)
64. Das, S., Chowdhury, M., Kundu, M.K.: Medical image fusion based on ripplet transform type-I. Prog. Electromagnet. Res. B **30**, 355–370 (2011)
65. He, C., Liu, Q., Li, H., Wang, H.: Multimodal medical image fusion based on IHS and PCA. Procedia Eng. **7**, 280–285 (2010)
66. Daneshvar, S., Ghassemian, H.: MRI and PET image fusion by combining IHS and retina-inspired models. Inf. Fusion **11**, 114–123 (2010)
67. Mitianoudis, N., Stathaki, T.: Pixel-based and region-based image fusion schemes using ICA bases. Inf. Fusion **8**, 131–142 (2007)
68. Correa, N.M., Li, Y.-O., Adali, T., Calhoun, V.D.: Canonical correlation analysis for feature-based fusion of biomedical imaging modalities and its application to detection of associative networks in schizophrenia. IEEE J. Sel. Top. Sign. Process. **2**, 998–1007 (2008)
69. Zaveri, T., Makwana, I., Zaveri, M.: Hybrid multispectral image fusion method. In: World Congress on Nature & Biologically Inspired Computing (NaBIC), pp. 930–935. IEEE (2009)
70. Kushwaha, A., Khare, A., Prakash, O., Song, J.-I., Jeon, M.: 3D medical image fusion using dual tree complex wavelet transform. In: 2015 International Conference on Control, Automation and Information Sciences (ICCAIS), pp. 251–256. IEEE (2015)
71. Shuaiqi, L., Jie, Z., Peng, G., Xiuling, L., Yuchao, S.: Medical image fusion based on non-subsampled direction complex wavelet transform. Int. J. Appl. Math. Mach. Learn. **1**, 21–34 (2014)
72. Candes, E.J., Donoho, D.L.: Curvelets: a surprisingly effective nonadaptive representation for objects with edges. DTIC Document (2000)
73. Candes, E., Demanet, L., Donoho, D., Ying, L.: Fast discrete curvelet transforms. Multiscale Model. Simul. **5**, 861–899 (2006)
74. Li, C., Yang, X., Chu, B., Lu, W., Pang, L.: A new image fusion quality assessment method based on contourlet and SSIM. In: 2010 3rd IEEE International Conference on Computer Science and Information Technology (ICCSIT), pp. 246–249. IEEE (2010)
75. Omar, Z., Stathaki, T.: GLCM-based metric for image fusion assessment. In: 2012 15th International Conference on Information Fusion (FUSION), pp. 376–381. IEEE (2012)
76. Julow, J., Major, T., Emri, M., Valalik, I., Sagi, S., Mangel, L., Németh, G., Tron, L., Várallyay, G., Solymosi, D.: The application of image fusion in stereotactic brachytherapy of brain tumours. Acta Neurochir. **142**, 1253–1258 (2000)
77. Montagner, J., Barra, V., Boire, J.-Y.: Synthesis of a functional information with anatomical landmarks by multiresolution fusion of brain images. In: 27th Annual International Conference of the Engineering in Medicine and Biology Society (IEEE-EMBS 2005), pp. 6547–6550. IEEE (2006)

78. Petrović, V., Xydeas, C.: Evaluation of image fusion performance with visible differences. In: Pajdla, T., Matas, J. (eds.) Computer Vision—ECCV 2004: 8th European Conference on Computer Vision, Prague, Czech Republic, 11–14 May 2004. Proceedings, Part III, pp. 380–391. Springer, Berlin, Heidelberg (2004)
79. Daly, S.J.: Visible differences predictor: an algorithm for the assessment of image fidelity. In: SPIE/IS&T 1992 Symposium on Electronic Imaging: Science and Technology, pp. 2–15. International Society for Optics and Photonics (1992)

Part II
Cardiovascular Informatics

Introduction of Telecardiology and Challenges in Developing Countries

Kui Lin Kam, Hoe Tung Yew and Yuan Wen Hau

Abstract Telecardiology is the delivery of cardiac care services by using information and communication technology (ICT) at distance. Research has proven telecardiology is useful for improving cardiac diseases management and reducing unnecessary hospitalizations and travelling. Although telecardiology is not a new cutting edge technology, yet the infrastructure is still not well established in most of the developing countries. This chapter reviews the development of telecardiology system in some of the developing countries such as Malaysia and Indonesia, the related research works, and the challenges of telecardiology deployment in terms of human resource, financial, technology, user acceptance and policy.

Keywords Developing countries · Issues and challenges · Telecardiology deployment

1 Introduction

Cardiovascular diseases are the leading cause of non-communicable disease death in the world. In 2015, cardiovascular diseases killed 17.7 million people globally and caused over three quarters of death in developing countries [1]. As one of the developing countries, cardiovascular diseases have become the main cause of death in Malaysia for last two decades [2]. Among the reported death due to cardiovascular diseases in 2015, it was estimated that 7.4 million were due to ischaemic heart disease. Ischaemic heart disease, or also known as coronary heart disease, is the heart disease that caused by narrowed arteries which affect the blood supply to the heart and lead to heart attack [3]. For the patient who is having a silent ischemia, they may not realise about it until the heart attack occurred. Therefore, in order to reduce the

K. L. Kam · Y. W. Hau (✉)
UTM-IJN Cardiovascular Engineering Centre, School of Biomedical Engineering and Health Sciences, Faculty of Engineering, Universiti Teknologi Malaysia, Johor Bahru, Johor, Malaysia
e-mail: hauyuanwen@biomedical.utm.my

H. T. Yew
Faculty of Engineering, Universiti Malaysia Sabah, Kota Kinabalu, Sabah, Malaysia

D. E. O. Dewi et al. (eds.), *Cardiovascular Engineering*, Series in BioEngineering,
https://doi.org/10.1007/978-981-10-8405-8_5

risk of fatality, the continuous monitoring of cardiac activity and rapid initiation of treatment are required for patients with suspected or actual ischaemic heart disease.

The scarcity of doctors and specialists is not a new issue to the developing countries. In Indonesia, there were around 1,000 cardiologists available in year 2017 or ratio of one cardiologist to 250,000 populations [4], whereas the recommended density of cardiologists by British Cardiac Society is between the range of 52.7–84.2 per million populations [5]. Similarly in Malaysia, referring to Health Indicator from Ministry of Health (MoH) Malaysia [6], the ratio of doctor to population in year 2016 was 1:656, which is lower than the recommended ratio of 1:600 from World Health Organization (WHO). Apart from that, there were only 247 cardiologists throughout the country in year 2013, which was equivalent to overall density of 0.83 per 100,000 populations [7].

In addition, concentration of cardiologists in urban city has caused rural population seriously lacking of cardiologists. For example, in Malaysia, there are 25.7% of the population lived in the rural area and majority of them are from East Malaysia (Sabah and Sarawak) [6]. However, based on the National Healthcare Establishment and Workforce Statistics [7], most of the cardiologists, that is 73 or 29.6%, are located at Kuala Lumpur. In contrast, Sabah and Sarawak have only 6 (2.4%) and 17 (6.9%) cardiologists, respectively, in the states. This shows a very poor cardiologist to population ratio, especially in rural area.

Furthermore, based on the survey outcome of National Cardiovascular Disease Database—Percutaneous Coronary Intervention (NCVD—PCI) Registry 2007–2009, 27% of patients who underwent PCI does not follow-up the cardiac treatment and rehabilitation after discharged from hospitals [8, 9]. These may due to the reason of unwillingness to long distance travelling, long queuing time and monetary constraints [10].

In order to overcome the barrier of healthcare services such as distance, time, cost, and effectiveness of health care services, deployment of telecardiology is necessary. Thus, area with less developed healthcare facilities, for example, rural area in East Malaysia, can capitalize on the expertise available in the West Coast Peninsular by centrally managing patients across the nation through telecardiology. This would help to improve the overall healthcare quality across the whole country.

Research has proven that telecardiology is useful in the management of acute coronary syndrome, atrial fibrillation, and syncope, and even beneficial in implementing strategies of cardiovascular primary care [11]. It also reduces the door-to-balloon time of ST-elevation myocardial infarction (STEMI) [11–16]. In addition, PausJenssen [17] had proven that telecardiology is feasible for cardiovascular patients living in remote region and improve the follow up rate. The other benefits of telecardiology are improvement in patient compliance with medication [18], reduce mortality and morbidity [12–16], better healthcare resource utilization such as ambulance and tertiary beds, less travel of both consultants and patients [19], improve hospitals' clinical administration workflow, and environment friendly (paperless, less travel).

2 Review of Telecardiology System

Telecardiology is a mechanism that delivers cardiovascular services by using ICT at distance. The most common part of telecardiology is dealing with transmission of electrocardiography (ECG) signal and echocardiography image over different types of telecommunication technologies from home or clinics to cardiology specialty centres. These ECG signal and echocardiography image can be transmitted in two ways, which are store-and-forward and real-time transmission.

The store-and-forward is the user sends the pre-recorded health data (e.g. ECG data, vital signs) to hospital information system, so that healthcare professionals could check the patients' electronic health record some time later. In this application, sender and receiver does not need to present simultaneously. On the other hand, the real-time telecardiology requires both parties (patients and healthcare professionals) have to present at the same time. One of the examples of real-time telecardiology application is the transmission of ECG data from ambulance to the hospital emergency department. The store-and-forward has been widely used in telecardiology to follow-up the chronic cardiac disease due to its expediency and high data integrity. However, the real-time telecardiology service is crucial for the patient in critical or emergency cases. It could help patients to get treatment promptly and also can be a strong educational component for the remote practitioners [20].

Telecardiology applications can be categorised into three, which are pre-hospital, in-hospital and post-hospital. ECG obtained during pre-hospital period is useful for early detection of myocardial infarction and this information is transferred to the receiving emergency physician before the arrival of the patient [21]. In-hospital telecardiology is mainly used for the communication between small hospitals in rural areas and main hospitals in urban city for information exchange [22]. The applications of post-hospital include tele-consultation between physicians and cardiologists [23], home monitoring for chronic cardiac diseases [24], and the diagnosis of arrhythmias [25].

Since the advent of telemedicine in the early 1900s [26, 27], there are a lot of research have been done on the topic of telecardiology, either for the invention of portable bio-signal acquisition device that enables the transmission of medical data via different types of network for remote monitoring, or development of telecardiology application to further enhance the functionality of existing commercial medical devices for medical data sharing purpose.

Figure 1 shows the architecture of wireless telecardiology system. The telecardiology device collects the patient health data from the electronic medical sensors via wireless personal area network (WPAN). The collected health data is then forwarded to hospital server via internet. The healthcare professionals can remote monitor their patient's health data history either off-line or in real-time by accessing the hospital server.

A portable telecardiology monitoring system using wireless local area network (WLAN) is presented in [28]. This system transmits user or patient ECG (electrocardiogram) signal to telecardiology centre in the hospital via WLAN. The main

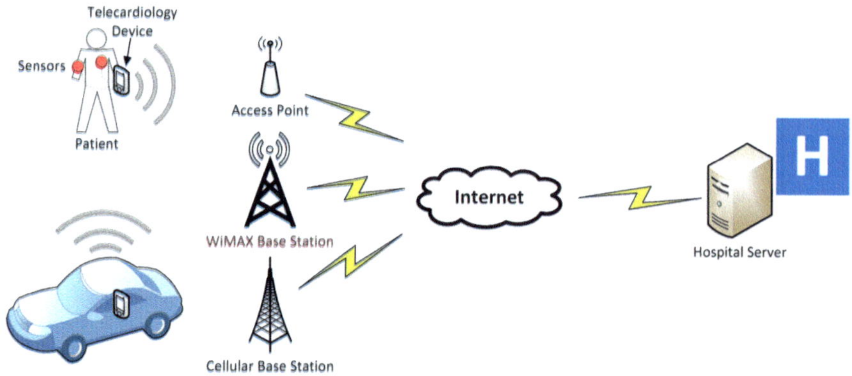

Fig. 1 Architecture of wireless telecardiology system

drawback of this system is that patient movement is restricted by the small coverage of WLAN cell. However, the use of cellular network, mobile health (m-health), can overcome the coverage issue. Research on m-health has rapidly increased in recent years with the help of advances in mobile smart phone and cellular network technologies. For example, Mobile-care unit [29] is a portable and wearable home monitoring medical device that enables continuous collection and analysis of medical signals from patients. This device applies both real-time and store-and-forward modes. The real-time mode is automatically activated when the device detects abnormal vital signs or ECG signal. It will send an emergency alert to the remote server via cellular network [Global System for Mobile Communications (GSM)/General Packet Radio Service (GPRS)] in real-time. For non-emergency case, the health data is sent through internet by store-and-forward mode. A similar work is also presented in [30, 31] which uses GSM/GPRS cellular network to monitor patient's ECG signal remotely. The GSM/GPRS based telecardiology systems are useful in rural areas that have limited coverage of high bandwidth networks such as Worldwide Interoperability for Microwave Access (WiMAX), 3G and 4G cellular networks. These GSM and GPRS based telecardiology systems are suitable for telecardiology services such as vital signs, ECG and voice which require less bandwidth. However, they may have insufficient bandwidth to support certain telecardiology service such as the high quality real-time video transmission.

The high bandwidth 3G and 4G cellular network could partially overcome the issue faced by GSM and GPRS networks. WE-CARE [32] is a real time 7-lead ECG monitoring system using 3G cellular network and WLAN mainly for indoor application. The ECG device continuously collects the user's ECG data, and then forwards the collected data to the telecardiology service provider over 3G network. Healthcare professionals could read the patients' real-time or historical ECG data via the web interface. Furthermore, Chin-Teng et al. [33] presented an intelligent ECG device that can be connected to a mobile and ubiquitous real-time display platform. The acquired ECG signals are transmitted to mobile devices through Bluetooth and then processed by the system. An alert signal is sent to the remote database server automatically via WLAN or cellular networks once an abnormal ECG is detected.

To guarantee the quality of telecardiology services, authors in [34, 35] proposed WiMAX technology for telecardiology application. WiMAX has large coverage, high bandwidth and support high mobility. WiMAX network could compensate the imperfection of WLAN (limited cell coverage) and cellular networks (insufficient bandwidth). However, the WiMAX network coverage is limited because most network service providers are ceasing development of WiMAX [36]. As a result, the application of the WiMAX based telecardiology system is limited in future.

For areas which are out of the coverage of ground networks, satellite communication becomes the only possible solution. Satellite communication is suitable for offshore oil platform [37] and military application [38]. AIDMAN [39] was a telecardiology pilot project that used a 512 Kbps satellite network to transmit real-time video and ECG data. The link is only capable of carrying a single video transmission at a time, and to ensure high quality, data should not be carried over the link at the same time. The satellite communication is not widely used in telecardiology because of high cost, high delay and poor quality especially during poor weather.

Most of the existing telecardiology systems are relying on a single wireless technology. These systems are unable to guarantee users remain continuously connected to the telecardiology service provider due to the imperfection of network coverage and poor network quality. A disconnection of a link will disrupt the telecardiology services and lead to misdiagnosis from healthcare professional. The study of Bergrath et al. [40] presented that only 73% of the still images had good signal quality that was helpful for teleconsultation. The reasons of failed transmission or transmission problems were attributed to connection breakdown and poor network condition [41]. The next generation of telecardiology system should be capable of selecting the best network from among the available candidate networks to guarantee the quality of telecardiology services. With heterogeneous wireless access, the telecardiology user can connect to the telecardiology service providers at anywhere.

Apart from that, there are also some research have been done on the topic of telecardiology which focus on the ECG interoperability. Since many ECG device manufacturers employ a protective scheme by having their own proprietary ECG data format, the telecardiology industry has been bombarded with a mixture of custom and preferred standard storage formats. Although the specifications of the standard formats are openly available, none has been universally adopted to be used with all ECG machines. Thus remains a lack of global interoperability of ECG information [42]. Indeed, the problem is not the lack but the existence of multiple standard ECG formats.

A web-based telecardiology framework has been developed by Mahesh et al. [22] for the diagnosis of cardiac patients in rural areas. The proposed system only requires a computer with a web browser and internet network connection to enable the uploading of Philips Extensible Markup Language based ECG (XML-ECG) and ECG waveforms display for easier diagnosis. Besides, the system also allows patients post diagnostic information to be stored in the server for future reference. Hsieh and Hsu [43] presented a cloud computing based ECG tele-consultation and tele-diagnosis system that supports three ECG data formats such as Philips XML-ECG, HP compatible Standard Communications Protocol for Computer-assisted Electrocardiography (SCP-ECG) and Mortara Digital Imaging and Communications in Medicine based ECG (DICOM-ECG). The data is first transcoded from vendor-specific format to an XML format before transmitting. This service greatly improve medical service quality by partially overcome the problem of vendor-dependent of most of the telecardiology application. Hsieh and Lo [44] proposed a DICOM-based 12-lead ECG information system that transforms SCP-ECG and XML-ECG to DICOM-ECG. The system provides an easier management of ECG where the DICOM-ECG can be integrated in the Picture Archiving and Communication System (PACS) with medical images. Moreover, a Hypertext Markup Language 5 (HTML5) powered web ECG management system is proposed in [45] for telecardiology application. The system enables the general practitioners or technicians to share XML-ECG with cardiologists remotely through a HTML5 enabled web browser without any additional plugins.

3 Telecardiology in Developing Countries

According to a global survey on eHealth by WHO [46], there are 28 countries providing telecardiology service with different levels of development in 2009 as summarized in Table 1. The levels of development were classified into three: 'established' where the service is continuously supported by funds from government or other sources, 'pilot' where the services is still under testing and evaluation stage, and 'informal' where the service is not part of an organized programme. Among those 28 countries, there are 17 countries having an established service, 9 under pilot study, one providing informal service and one do not mention the stage of development.

In Malaysia, telecardiology system was commenced in year 2000 as pilot project. At that time, most of the hospitals were lack of digital electronic medical devices. In addition, physicians were facing difficulty to apply telecardiology over a low internet speed. Hence, majority hospitals were inactive in telecardiology network. Finally, this telecardiology pilot project was terminated after 30 months operations due to technological problem, insufficient ICT infrastructure, short of funding and demand inconsistency [47].

The new telecardiology system was launched in February 2010, which provided healthcare services between specialist hospitals and non-specialist hospitals. According to the analysis from the MoH [48], there are only total of 409 cases have been sent through telecardiology from 2010 to 2012, in which the usage is still considered underutilized compared to other disciplines such as neurosurgery and radiology. Moreover, referring to Table 2, the number of hospitals involved in new telecardiology system is very little and there are no rural clinics involve in this new system due to poor internet connection, lack of ICT infrastructure, and unstable electricity supply in rural area [49].

In Indonesia, telecardiology (phone-based) was started in year 2001. In 2012, a pilot project on telecardiology was implemented to provide patient management

Table 1 Countries provide telecardiology service [46]

Country	Stage
Austria	Pilot
Belarus	Established
Belgium	Established
Bhutan	Established
Burundi	Pilot
Cape Verde	Established
Colombia	Established
Croatia	Established
Czech Republic	Pilot
Estonia	Established
Greece	Pilot
Iceland	Pilot
Indonesia	Established
Latvia	Informal
Malaysia	Pilot
Mali	No stage provided
Mexico	Established
Mongolia	Established
Mozambique	Pilot
Nepal	Established
Norway	Established
New Zealand	Established
Pakistan	Established
Paraguay	Pilot
Peru	Established
Senegal	Established
Singapore	Pilot
Turkey	Established

Table 2 Telecardiology stations in Malaysia [48]

Receiving station (Specialist hospital with cardiologist)	Sending station (Medical officer)
Hospital Pulau Pinang	• Hospital Tuanku Fauziah, Perlis • Hospital Sultanah Bahiyah, Kedah • Hospital Taiping, Perak
Hospital Queen Elizabeth, Sabah	• Hospital Duchess of Kent, Sandakan • Hospital Tawau

services. The system is only used for emergency cases where the 12 leads ECG data is transmitted to Cardiac Centre Harapan Kita Hospital in Jakarta using 3G connection for real-time transmission. In the pilot project, there are 3 sites in Banten Province, 4 sites in Jakarta Province, 3 sites in Bandung Province with all sites involves teleradiology. However, there are some technical problems encountered in the pilot project, for example, the high operational cost for ECG machine and unstable 3G connection or even 3G blank spot in telecardiology sites. Apart from that, the policy and regulation are not yet established and there is lack of awareness to ICT technology in human resources. Furthermore, the system is product-dependent in which the system is designed based on the functionality provided by the product in-used instead of user needs [50].

4 Challenges of Telecardiology Implementation at Developing Countries

The challenges of telecardiology deployment at developing countries could be categorized into human resource, financial, technology, user acceptance and policy. Table 3 summarizes the challenges and possible solutions of telecardiology implementation at developing countries.

Table 3 Challenges and possible solutions of telecardiology implementation at developing countries

	Challenge	Solution
Human resource	• Scarcity of cardiologist and doctors	• Experience nurses and medical assistants can be trained to support telecardiology application
Financial	• Be governed by Ministry of Health (government)	• Propose a sustainable business model
Technology	• Lack of ICT infrastructure in rural area • Lack of universal ECG viewer	• Propose a telecardiology system with a handover mechanism in heterogeneous wireless networks • Encourage the implementation of open and standard ECG data format for computerized ECG
User acceptance	• Concerns about reliability of telecardiology • Lack of ICT skill and knowledge	• Ease-to-use and reliable technology • One-to-one training program • User awareness program
Policy	• Lack of enforcement	• Encourage healthcare professional to participate in telecardiology

4.1 Human Resource

An inadequate number of physicians may result an extremely high workload fall on this limited number of physicians. Thus, it will become a barrier for physicians to involve in telecardiology. To overcome this resistance, experience healthcare professional such as nurses and medical assistants can be trained to give support on telecardiology system 24 hours per day, 7 days per week by shift. They could monitor the patients' electronic health record and give preliminary consultation to the patients who are needed. Therefore, training programs for healthcare professionals are necessary to get them familiar with telecardiology equipment, indicate their roles, and their job scopes in telecardiology system.

4.2 Finance

Financial issue is another barrier to implement or sustain the telecardiology in developing countries. Many telecardiology projects were failed due to running out of project funding [51, 52]. In developing countries, the sustainability of telecardiology is highly depending on the government fund. In Malaysia, MoH is the only financial supporter to telecardiology system.

Telecardiology system has been proven able to reduce the traveling cost. However, it requires cost to purchase the medical equipment, maintenance, internet tariff charge, training programs, and incentives for the healthcare professional who participated in telecardiology. To be financially sustainable, a telecardiology business model is proposed in Fig. 2.

In this business model, a Telecardiology Centre is a centre in a tertiary hospital which provides cardiology specialty healthcare services to the outpatients via ICT. This centre is partnership with medical device manufacturers, internet service

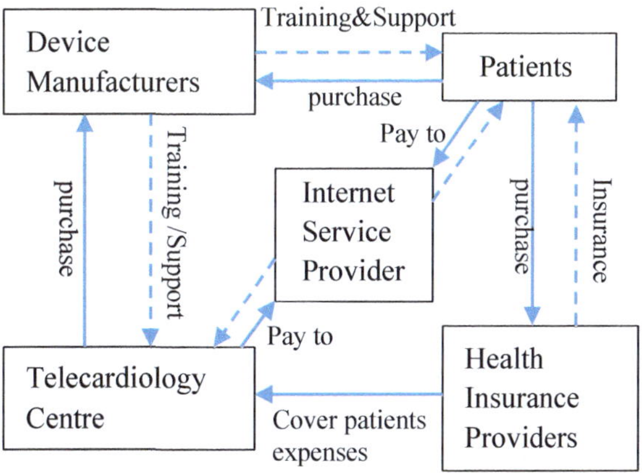

Fig. 2 Telecardiology business model

providers, and insurance companies. The medical device manufacturers are mainly supplying telecardiology related monitoring devices to telecardiology centre and patients. These manufacturers will provide free application training courses, warranty and technical support on their products.

In this model, both telecardiology centres and patients have to bare their own cost on purchasing a monitoring devices, telecommunication equipment and internet service. To reduce the financial burden, the internet service providers could offer a special tariff for telecardiology users. Furthermore, government should suggest private insurance companies to cover the telecardiology expenses in their health insurance policy. This suggestion has been proposed and start implement in United State [53].

The proposed telecardiology business model could avoid the scenario of high dependency on government fund and also reduce the patients' financial burden especially for low-income population in rural area. Furthermore, it shows the finance contribution to local economic.

4.3 Lack of ICT Infrastructure

The growth of telecardiology is very much in line with the development of ICT infrastructure. In Malaysia, cellular network [Universal Mobile Telecommunications System (UMTS)] coverage has reached to about 95% of populated areas [54]. However, the high bandwidth wireless networks such as WiMAX and Long Term Evolution (4G-LTE) are still under deployment and have coverage holes even in urban.

The high ICT infrastructure installation cost and low subscription rate have restricted internet service providers to expand the broadband internet service to sub-urban and rural areas [55]. Thus, the application of telecardiology in sub-urban and rural areas in developing countries is still a major problem due to the inadequate ICT infrastructures. The bandwidth limitation of cellular network has restricted the amount of data that can be transmitted in real-time [52], could lead to poor medical image quality and prohibit the use of real-time videoconferencing [38, 56]. In addition, the imperfection of high bandwidth network (WiMAX and LTE) coverage has put some limitation on user mobility.

To resolve the limitations of existing telecardiology system, next generation telecardiology system (as shown in Fig. 3) is expected integrates with heterogeneous wireless technologies to allow patients always connected to wireless network at anytime and anywhere with guarantee of services quality [57]. This system will be integrated with vertical handover mechanism which is able to search for the available networks and connect to the most appropriate network based on user requirements to guarantee telecardiology service quality and always connected to health care service providers.

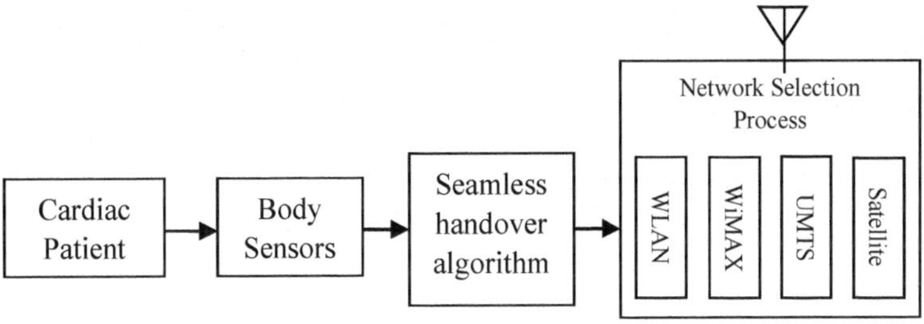

Fig. 3 Telecardiology system in heterogeneous networks

4.4 Lack of Universal ECG Viewer

There are heterogeneous, either standard or proprietary, ECG data format available nowadays. According to Bond et al. [42], there are at least nine types of ECG data format, including SCP-ECG [58], DICOM-ECG [59], Health Level Seven Annotated Electrocardiogram (HL7 aECG) [60], Electrocardiogram Markup Language (ecgML) [61], Medical Waveform Format Encoding Rules (MFER) [62], Philips XML-ECG [63], XML-ECG [64], Mobile Electrocardiograph Markup Language (mECGml) [65] and ECG markup language for ambulatory telemonitoring (ecgAware) [66]. Due to the fact that no current existing formats satisfy all the needs from potential users, researchers or medical device vendors continue to create their own proprietary ECG data formats. This heterogeneous ECG data format decreases the interoperability of various ECG records from multiple device vendors as well as increases the difficulty of integration between ECG devices and telecardiology system. Hence it is a challenge in developing a universal ECG viewer which compatible for all type of ECG data format. Interoperability of ECG is one of the important elements in developing a telecardiology system in terms of system integration and management.

Nowadays, there are many computer-based or web-based commercial products or ECG management systems that provide telecardiology functionality, such as Welch Allyn CardioPerfect Workstation [67], Mortara Pyramis ECG Management [68] and Schiller SEMA ECG Management System [69]. However, most these ECG management systems have limitation of system interoperability where they can only support specific ECG data format, which greatly increase the difficulty of the system integration with ECG devices from various manufacturers.

Interoperability of digital and computerized ECG is important for data exchange among different devices, applications, systems and hospitals. The best way to achieve interoperability is through the use of standards. In 2002, a European project called OpenECG was started to promote the SCP-ECG standard by providing information and technical support [70]. After the OpenECG project, there is a shift from closed and proprietary solution towards the development of open source [71], where tutorials for implementation of SCP-ECG standard as well as specifications and data sets of different ECG data formats are available for the users and developers [72].

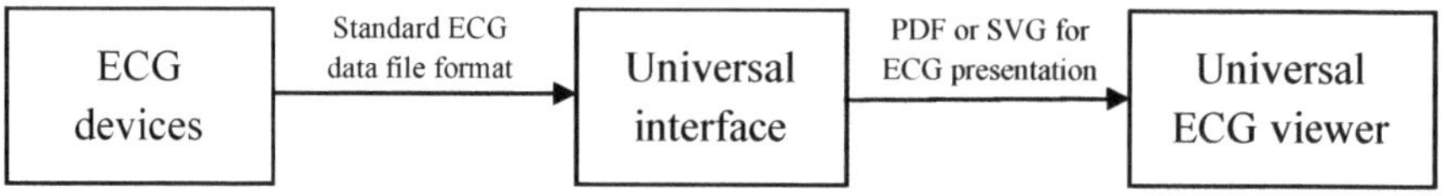

Fig. 4 Telecardiology universal ECG viewer

As an effort to improve the ECG interoperability, Philips Medical Systems, one of the ECG manufacturers, supports for OpenECG project by opening their ECG file format XML-ECG in 2003 [63]. Besides, Philips also provides a set of software tools for XML-ECG waveform manipulation and display, as well as set of converters to convert the XML-ECG into other formats such as HL7 aECG, XML with plain text decompressed waveform values, and Scalable Vector Graphics (SVG).

Integrating the Healthcare Enterprise (IHE) is an initiative designed to assist the integration of the information systems by providing a technical framework for the implementation of existing standards. IHE Cardiology was formed in 2003 to address issues specific to clinical workflow, information sharing and improved patient care in the clinical domain of cardiology. There are five integration profiles specified in the IHE Cardiology Technical Framework: Cardiac Catheterization Workflow (CATH), Echocardiography Workflow (ECHO), Retrieve ECG for Display (ECG), Displayable Reports (DRPT), and Evidence Documents (ED). For the integration profile of ECG, it provides framework for displaying the information such as ECG waveforms and report without requiring specialized cardiology software or workstations by using general purpose computer applications [73].

As a solution for the problem of heterogeneous ECG data format, the government can encourage the use of open and standard ECG data format by establishing a guideline of developing and purchasing an ECG device. Before purchasing an ECG device, the users have to check with the manufacturer for the supported ECG data format to make sure the interoperability of the device. By reducing the usage of proprietary ECG data format, a universal ECG viewer can be developed for a telecardiology system. The universal ECG viewer should able to read all type of standard ECG data file formats and display the data and waveform either in Portable Document Format (PDF) or SVG format as shown in Fig. 4.

4.5 User Acceptance

The successful and sustainability of telecardiology application is very much depending on user acceptance. The telecardiology users can be divided into two parties, patient and healthcare professional. In Malaysia, 65.8% of population are using the internet. The broadband penetration rate is 13.5 per 100 inhabitants [54]. The low internet user and broadband penetration rate may cause the application of telecardiology in sub-urban and rural area difficult to be implemented [53]. The patients in the rural area are mostly lack of ICT skills and knowledge [74]. Consequently, they are lack of confidence in technology and more comfortable with traditional face-to-face meeting with doctors.

For healthcare professional, they felt telecardiology does not reduce their workload and viewed telecardiology as an extra duty [75]. In addition, telecardiology will interrupt their daily work routine and shift the focus away from direct contact with patients [76]. The failure of initial telecardiology pilot project also gave a negative impact to healthcare professional on the reliability and sustainability of telecardiology [77]. Healthcare professional also thought that applying telecardiology system is too much dealing with technical equipment. They felt problematic for them to install, operate and maintain telecardiology equipment [77].

The suggestions to improve the patient acceptance on telecardiology system included user-friendliness of telecardiology technology [19] and one-to-one training program. The telecardiology equipment must be as simple as possible in application. For example, user can do installation based on the colour matching method such as red cable connects to red colour port on the equipment. Besides that, one-to-one training program should be provided to the users who are lack of ICT skill and knowledge. By one-to-one training, trainer will know well about the strengths and weaknesses of an individual user and improve the weaknesses of user to ensure every user is able to install and operate the telecardiology equipment competently.

On the other hand, training and continual professional education should be provided to healthcare professionals to improve their confidence in telecardiology system, get familiar with telecardiology technology, better collaborative with patients and know how to response to the patient in critical. Aside from that, the aims, visions and usefulness of telecardiology must be very clear to healthcare professional. This is to make healthcare professional feel that they are being taken seriously [78]. Chau and Hu [79] presented that perceived usefulness is the most significant factor to improve the acceptance of healthcare professional in telecardiology technology.

Furthermore, Ministry of Health should promote the contributions and benefits of telecardiology in health care by given successful examples of telecardiology system in developed countries. This is to increase the awareness of all levels and understanding what the telecardiology is.

4.6 Policy

Telecardiology policy is established and adopted in most of the developing countries. This policy included laws, regulations and reimbursement. According to Law of Malaysia (Act 564, Telemedicine Act 1997), only person who has fully registered or licensed medical practitioner holding a valid practising certificate may practise telecardiology. The complicated and time consuming licensing process has become a resistance for healthcare professional to participate in telecardiology application. The suggest solution is to make telemedicine course as a compulsory course for all healthcare professionals to enforce them to participate in telecardiology application. Moreover, an attractive incentive should be given to the qualified telecardiology medical practitioner to motivate more experienced healthcare professionals participate in telecardiology service.

5 Conclusions

A sustainable telecardiology system in developing countries is essential to overcome the scarcity of cardiology healthcare services especially for the population in sub-urban and rural areas. However, to implement a sustainable telecardiology system is not an easy task in developing countries due to the problems such as shortage of healthcare professionals, inadequate funding, lack of ICT infrastructure, lack of universal ECG viewer, low user acceptance and policy issue. The potential solutions of these challenges are discussed.

First is expansion of healthcare professional by including experienced nurses and medical assistants. They can give a preliminary consultation to the patients who are not in critical case. This could reduce the burden of medical doctors and cardiac specialists. Next, a sustainable business model is proposed to overcome the monetary issue. The proposed model could bring an economic contribution and reduce the reliance on government fund.

A next generation telecardiology integrates with multiple wireless technologies and come with vertical handover mechanism is suggested to overcome the shortage of ICT infrastructure in sub-urban and rural areas. The next generation telecardiology system will guarantee the service quality and also ensure the users are continuously connected to healthcare service providers. In order to generating a universal ECG viewer, the implementation of open and standard ECG data format for computerized ECG has to be encouraged. In term of user acceptance on telecardiology, training and supporting programs are necessary. Lastly, a policy to enforce the healthcare professional to participate in telecardiology application is needed.

Acknowledgements This research is supported and funded by Ministry of Higher Education (MoHE) Trans-Disciplinary Research Grant Scheme (TRGS) with Grant no. TRGS/1/2015/UTM/02/3/3 (UTM vote no. R.J130000.7845.4L842) and Universiti Teknologi Malaysia Research University Grant with vote no. Q.J130000.2545.16H83.

References

1. World Health Organization (WHO): Fact sheet: Cardiovascular Diseases (CVDs). http://www.who.int/mediacentre/factsheets/fs317/en/ (2017). Accessed on 4 Jan 2018
2. Ministry of Health (MoH) Malaysia: Health Facts Malaysia 2001–2017. Health Informatics Centre, Planning and Development Division, Ministry of Health (MoH) Malaysia, Malaysia (2017)
3. American Heart Association: Silent Ischemia and Ischemic Heart Disease. http://www.heart.org/HEARTORG/Conditions/HeartAttack/TreatmentofaHeartAttack/Silent-Ischemia-and-Ischemic-Heart-Disease_UCM_434092_Article.jsp#.WXRe2oiGPIV (2016). Accessed 28 Jan 2018

4. Harian Kompas: Jumlah Ahli Kardiovaskular Belum Memadai. https://kompas.id/baca/ilmu-pengetahuan-teknologi/2017/11/18/jumlah-ahli-kardiovaskular-belum-memadai/ (2017). Accessed on 4 Jan 2018
5. Hackett, D.: Cardiac Workforce Requirements in the UK. British Cardiac Society, London (2005)
6. Ministry of Health (MoH) Malaysia: Health Indicators 2016: Indicator for Monitoring and Evaluation of Strategy Health for All. Health Informatics Centre Planning Division, Ministry of Health (MoH) Malaysia, Malaysia (2016)
7. Sivasampu, S., Foo, C.Y., Aimi, N.J., et al.: National Healthcare Establishment and Workforce Statistics (Hospital) 2012–2013. The National Healthcare Statistics Initiative (NHSI), National Clinical Research Centre, Ministry of Health Malaysia, Malaysia (2015)
8. Ahmad, W.A., Ali, R.M., Khanom, M., et al.: The journey of Malaysia NCVD-PCI (national cardiovascular disease database-percutaneous coronary intervention) registry: a summary of three years report. Int. J. Cardiol. **165**(1), 161–164 (2013)
9. Ahmad, W.A., Sim, K.H.: Annual Report of the NCVD-PCI Registry 2007–2009. National Cardiovascular Disease Database, National Heart of Association Malaysia and Clinic Research Centre, Ministry of Health (MoH) Malaysia, Malaysia (2011)
10. Raju, P.K., Prasad, S.G.: Telemedicine and cardiology—decade of our experience. J. Indian Coll. Cardiol. **2**(1), 4–16 (2012)
11. Backman, W., Bendel, D., Rakhit, R.: The telecardiology revolution: improving the management of cardiac disease in primary care. J. R. Soc. Med. **103**(11), 442–446 (2010)
12. Brown, J.P., Mahmud, E., Dunford, J.V., et al.: Effect of prehospital 12-lead electrocardiogram on activation of the cardiac catheterization laboratory and door-to-balloon time in ST-segment elevation acute myocardial infarction. Am. J. Cardiol. **101**(2), 158–161 (2008)
13. Brunetti, N.D., De Gennaro, L., Amodio, G., et al.: Telecardiology improves quality of diagnosis and reduces delay to treatment in elderly patients with acute myocardial infarction and atypical presentation. Eur. J. Cardiovas. Prev. Rehabil. **17**(6), 615–620 (2010)
14. Brunetti, N.D., De Gennaro, L., Dellegrottaglie, G., et al.: Fast and furious: telecardiology in acute myocardial infarction triage in the emergency room setting. Eur. Res. Telemed. **2**(2), 75–78 (2013)
15. Chaudhry, S.I., Barton, B., Mattera, J., et al.: Clinical trials: methods and design, randomized trial of telemonitoring to improve heart failure outcomes (Tele-HF): study design. J. Cardiac Fail. **13**(9), 709–714 (2007)
16. Sorensen, J.T., Clemmensen, P., Sejersten, M.: Telecardiology: past, present and future. Rev. Esp. Cardiol. **66**(3), 212–218 (2013)
17. PausJenssen, A.M., Spooner, B.A., Wilson, M.P., et al.: Cardiovascular risk reduction via telehealth: a feasibility study. Can. J. Cardiol. **24**(1), 57–60 (2008)
18. Adeogun, O., Tiwari, A., Alock, J.R.: Models of information exchange for UK telehealth systems. Int. J. Med. Inform. **80**(5), 359–370 (2011)
19. Maarop, N., Win, K.T.: The interplay of environmental factors in the acceptance of teleconsultation technology: a mixed methods study. Open Int. J. Inform. (OIJI) **1**(1), 46–48 (2012)
20. Wootton, R.: Realtime telemedicine. J. Telemed. Telecare **12**(7), 328–336 (2006)
21. Campbell, P.T., Patterson, J., Cromer, D., et al.: Prehospital triage of acute myocardial infarction: wireless transmission of electrocardiograms to the on-call cardiologist via a handheld computer. J. Electrocardiol. **38**(4), 300–309 (2005)
22. Mahesh, V., Kandaswamy, A., Venkatesan, R.: Telecardiology for rural health care. Int. J. Recent Trends Eng. (IJRTE) **2**(1), 6–9 (2009)
23. Ozen, N., Karlik, B.: A telecardiology system design with real-time diagnosis and teleconsultation. In: The First International Conference on the Applications of Digital Information and Web Technologies (ICADIWT), pp. 500–506 (2008)
24. Kong, K.Y., Ng, C.Y., Ong, K.: Web-based monitoring of real-time ECG data. Comput. Cardiol. **2000**, 189–192 (2000)
25. Kohila, S., Gowri, K.: A novel real-time intelligent tele cardiology system using wireless technology to detect cardiac abnormalities. Glob. J. Comput. Sci. Technol. **11**(12), 51–57 (2011)

26. Wright, D., Androuchko, L.: Telemedicine and developing countries. J. Telemed. Telecare **2**(2), 63–70 (1996)
27. Armstrong, I.J., Haston, W.S.: Medical decision support for remote general practitioners using telemedicine. J. Telemed. Telecare **3**(1), 27–34 (1997)
28. Cai, K., Liang, X.: Development of WI-FI based telecardiology monitoring system. In: 2010 2nd International Workshop on Intelligent Systems and Applications (ISA), pp. 1–4 (2010)
29. Abo-Zahhad, M., Ahmed, S.M., Elnahas, O.: A wireless emergency telemedicine system for patients monitoring and diagnosis. Int. J. Telemed. Appl. **2014**, 4 (2014)
30. Mitra, S., Mitra, M., Chaudhuri, B.B.: Rural cardiac healthcare system—a scheme for developing countries. In: TENCON 2008—2008 IEEE Region 10 Conference, pp. 1–5 (2008)
31. Elena, M., Quero, J.M., Tarrida, C.L., et al.: Design of a mobile telecardiology system using GPRS/GSM technology. In: Engineering in Medicine and Biology, 2002. 24th Annual Conference and the Annual Fall Meeting of the Biomedical Engineering Society EMBS/BMES Conference, 2002. Proceedings of the Second Joint, Houston, Texas, USA, pp. 1859–1860 (2002)
32. Anpeng, H., Chao, C., Kaigui, B., et al.: WE-CARE: an intelligent mobile telecardiology system to enable mHealth applications. IEEE J. Biomed. Health Inform. **18**(2), 693–702 (2014)
33. Chin-Teng, L., Kuan-Cheng, C., Chun-Ling, L., et al.: An intelligent telecardiology system using a wearable and wireless ECG to detect atrial fibrillation. IEEE Trans. Inf. Technol. Biomed. **14**(3), 726–733 (2010)
34. Niyato, D., Hossain, E., Diamond, J.: IEEE 802.16/WiMAX-based broadband wireless access and its application for telemedicine/e-health services. IEEE Wirel. Commun. **14**(1), 72–83 (2007)
35. Chorbev, I., Mihajlov, M.: Wireless telemedicine services as part of an integrated system for e-medicine. In: MELECON 2008. The 14th IEEE Mediterranean Electrotechnical Conference, pp. 264–269 (2008)
36. Pareit, D., Lannoo, B., Moerman, I., et al.: The history of WiMAX: a complete survey of the evolution in certification and standardization for IEEE 802.16 and WiMAX. IEEE Commun. Surv. Tutor. **14**(4), 1183–1211 (2012)
37. Mair, F., Fraser, S., Ferguson, J.: Telemedicine via satellite to support offshore oil platforms. J. Telemed. Telecare **14**(3), 129–131 (2008)
38. de Barbara, A.H.A.: The Spanish ministry of defence (MOD) telemedicine system. In: Advances in Telemedicine: Technologies, Enabling Factors and Scenarios. InTech (2011)
39. Clarke, M., Jones, R.W., Lioupis, D.: AIDMAN-telecardiology over a high-speed satellite network. Comput. Cardiol. **2000**, 657–660 (2000)
40. Bergrath, S., Rortgen, D., Rossaint, R., et al.: Technical and organisational feasibility of a multifunctional telemedicine system in an emergency medical service—an observational study. J. Telemed. Telecare **17**(7), 371–377 (2011)
41. Sejersten, M., Sillesen, M., Hansen, P.R., et al.: Effect on treatment delay of prehospital teletransmission of 12-lead electrocardiogram to a cardiologist for immediate triage and direct referral of patients with ST-segment elevation acute myocardial infarction to primary percutaneous coronary intervention. Am. J. Cardiol. **101**(7), 941–946 (2008)
42. Bond, R.R., Finlay, D.D., Nugent, C.D., et al.: A review of ECG storage formats. Int. J. Med. Inform. **80**(10), 681–697 (2011)
43. Hsieh, J.C., Hsu, M.W.: A cloud computing based 12-lead ECG telemedicine service. BMC Med. Inform. Decis. Mak. **12**(1), 77 (2012)
44. Hsieh, J.C., Lo, H.C.: The clinical application of a PACS-dependent 12-lead ECG and image information system in E-medicine and telemedicine. J. Digit. Imaging **23**(4), 501–513 (2010)
45. Kumar, M.A., Srinivasan, A., Bussa, N.: HTML5 powered web application for telecardiology: a case study using ECGs. In: 2013 IEEE Point-of-Care Healthcare Technologies (PHT), pp. 156–159 (2013)
46. World Health Organization (WHO): Telemedicine: Opportunities and Developments in Member States: Report on the Second Global Survey on eHealth 2009. In: Global Observatory for eHealth Series, vol. 2. WHO, Geneva (2010)

47. Maarop, N., Win, K.T., Masrom, M., et al.: Exploring factors that affect teleconsultation adoption: in the case of Malaysia. In: 15th Pacific Asia Conference on Information Systems (PACIS 2011): Quality Research in Pacific, pp. 1–12 (2011)
48. Ministry of Health (MoH) Malaysia: Teleconsultation Portal. http://tc.moh.gov.my (2015). Accessed on 7 May 2015
49. Supriyanto, E., Yew, H.T., Satria, H.: Telecardiology for rural area in developing countries: challenges. In: WSEAS Proceeding of the 18th International Conference on Circuits, Systems, Communications and Computers (CSCC '14), Greece, pp. 124–128 (2014)
50. Thomas, A.I., Mars, A.C.: Telemedicine in Indonesia: country experiences. Meeting presentation at WHO SEA Regional Consultation on Telemedicine, Pyongyang, DPRK. http://www.searo.who.int/entity/health_situation_trends/events/12_Indonesia_Telemedicine.pdf?ua=1 (2013). Accessed on 4 Jan 2018
51. Wootton, R.: Telemedicine support for the developing world. J. Telemed. Telecare **14**(3), 109–114 (2008)
52. Yellowlees, P.M.: Successfully developing a telemedicine system. J. Telemed. Telecare **11**(7), 331–335 (2005)
53. LeRouge, C., Garfield, M.J.: Crossing the telemedicine chasm: have the U.S. barriers to widespread adoption of telemedicine been significantly reduced? Int. J. Environ. Res. Public Health **10**(12), 6472–6484 (2013)
54. Yew, H.T., Satria, M.H., Hau, Y.W., et al.: A telecardiology framework for rural area. In: WSEAS Proceeding of the 13th International Conference on Applied Computer and Applied Computational Science (ACACOS '14), Kuala Lumpur, pp. 140–148 (2014)
55. Pejovic, V.: Adaptive and resource-efficient rural area wireless networks. Ph.D. Thesis, University of California (2012)
56. Moffatt, J.J., Eley, D.S.: Barriers to the up-take of telemedicine in Australia—a view from providers. Rural Remote Health **11**(1581), 1–6 (2011)
57. Yew, H.T., Aditya, Y., Satria, H., et al.: Telecardiology system for fourth generation heterogeneous wireless networks. ARPN J. Eng. Appl. Sci. **10**(2), 600–607 (2015)
58. International Standards Office: Health Informatics—Standard Communication Protocol—Part 91064: Computer-Assisted Electrocardiography (ISO 11073-91064:2009). ISO, Geneva (2009)
59. DICOM Standards Committee, Working Group 1—Cardiac and Vascular Information: Digital Imaging and Communications in Medicine (DICOM) Supplement 30: Waveform Interchange. ftp://medical.nema.org/medical/dicom/final/sup30_f2.pdf (2000). Accessed 23 July 2017
60. Brown, B.D., Badilini, F.: HL7 aECG Implementation Guide. http://www.amps-llc.com/website/documents/UsefulDocs/aECG_Implementation_Guide.pdf (2005). Accessed 23 July 2017
61. Wang, H., Azuaje, F., Jung, B., et al.: A markup language for electrocardiogram data acquisition and analysis (ecgML). BMC Med. Inform. Decis. Mak. **3**(1), 4 (2003)
62. Kimura, E., Norihiko, T., Ishihara, K.: Development MFER (medical waveform format encoding rules) parser. In: AMIA 2006 Annual Symposium Proceedings, p. 985 (2006)
63. Helfenbein, E.D., Gregg, R., Zhou, S.: Philips medical systems support for open ECG and standardization efforts. Comput. Cardiol. **2004**, 393–396 (2004)
64. Lu, X., Duan, H., Zheng, H.: XML-ECG: An XML-based ECG presentation for data exchanging. In: 2007 1st International Conference on Bioinformatics and Biomedical Engineering (ICBBE), pp. 1141–1144 (2007)
65. Fang, Q., Sufi, F., Cosic, I.: A mobile device based ECG analysis system. In: Data Mining in Medical and Biological Research, pp. 209–226. InTech (2008)
66. Gonçalves, B., Pereira Filho, J.G., Andreão, R.V.: ECGWARE: an ECG markup language for ambulatory telemonitoring and decision making support. In: International Conference on Health Informatics (HEALTHINF), pp. 37–43 (2008)
67. Welch Allyn: Welch Allyn CardioPerfect Workstation. https://www.welchallyn.com/content/welchallyn/americas/en/products/categories/cardiopulmonary/cardiopulmonary-data-management/cardioperfect-workstation.html (2015). Accessed on 4 Jan 2018

68. Mortara: Mortara Pyramis ECG Management. http://www.mortara.com/products/healthcare/hospitals/data-management-ehr-connectivity/pyramis-ecg-management/ (2015). Accessed on 4 Jan 2018
69. Schiller: SEMA-200 Data Management Solution Product Brochure. http://www.schillerservice.com/v/vspfiles/files/repertoire/SEMA_LQ.pdf (2015). Accessed on 4 Jan 2018
70. Chronaki, C.E., Chiarugi, F., Lees, P.J., et al.: Open ECG: A European project to promote the SCP-ECG standard, a further step towards interoperability in electrocardiography. In: Computers in Cardiology, pp. 285–288 (2002)
71. Lees, P.J., Chronaki, C.E., Chiarugi, F.: Standards and interoperability in digital electrocardiography. The OpenECG project. Hell. J. Cardiol. **45**, 364–369 (2004)
72. Chronaki, C.E., Chiarugi, F., Macerata, A., et al.: Interoperability in digital electrocardiography after the openECG project. In: Computers in Cardiology, pp. 49–52 (2004)
73. Integrating the Healthcare Enterprise (IHE): IHE Cardiology (CARD) Technical Framework, Volume 1 (CARD TF-1): Integration Profiles. http://www.ihe.net/uploadedFiles/Documents/Cardiology/IHE_CARD_TF_Vol1.pdf (2013). Accessed on 4 Jan 2018
74. Imani, B., Hajalizadeh, A., Jahangiri, A., et al.: The challenges of ICT development in rural area case study: Village Aleni, Meshkin Shahr in Ardebil Province. Aust. J. Basic Appl. Sci. **6**(9), 674–682 (2012)
75. Sharma, U., Barnett, J., Clarke, M.: Clinical users' perspective on telemonitoring of patients with long-term conditions: understood through concepts of Giddens's structuration theory & consequence of modernity. Stud. Health Technol. Inform. **160**(1), 545–549 (2010)
76. Brewster, L., Mountain, G., Wessels, B., et al.: Factors affecting frontline staff acceptance of telehealth technologies: a mixed-method systematic review. J. Adv. Nurs. **70**(1), 21–33 (2014)
77. Mair, F., Hiscock, J., Beaton, S.C.: Understanding factors that inhibit or promote the utilization of telecare in chronic lung disease. Chronic Illn. **4**(2), 110–117 (2008)
78. Buck, S.: Nine human factors contributing to the user acceptance of telemedicine application: a cognitive-emotional approach. J. Telemed. Telecare **15**(2), 55–58 (2009)
79. Chau, P.Y.K., Hu, P.J.H.: Investigating healthcare professionals' decisions to accept telemedicine technology: an empirical test of competing theories. Inf. Manag. **39**(4), 297–311 (2002)

Clinical Pathway in Cardiovascular Disease Management

Rania Hussien Al-Ashwal and Eko Supriyanto

Abstract This chapter review the clinical pathway evolution in cardiovascular disease management and its implementation in Malaysia. We conducted a literature review on the clinical pathway use in cardiovascular disease management. In the nest sections we will give an insight into the situation in Malaysia mainly. In summary, we found that the available evidence on the clinical pathway is scarce and does not help in building strong evidence for evaluating the clinical pathway in Malaysia. The clinical pathway is not a new field in Malaysia. However, interest in reporting the contemporary scientific evidence is scanty. It is necessary to engage the health care researchers on clinical pathway design, implementation and even analysis.

Keywords Clinical pathway · Cardiovascular · Malaysia

1 Introduction

The way we treat and diagnose the diseases is changing rapidly. There is a long list of unsatisfactory treatments with the presence of a variety of medications, surgeries techniques, and even facility. Many cases either not respond or face complications to the process of care, which lead to increase in medical errors and side effects.

In the new era, as highlighted by Lodewijckx et al. [1], the clinical pathway has become a central issue for quality improvement that is used to provide better quality with the latest evidence.

Clinical pathways come as a solution to standardise the care, decrease the medical error and improve the quality of care. At the time clinical path started to be used it has gained much attractiveness with poor understanding to what and how to use it [2]. That has come after clinical pathway has proven in some trials as a useful tool

R. H. Al-Ashwal (✉)
School of Biomedical Engineering and Health Science, University Technology Malaysia, Skudai, Malaysia
e-mail: rania@utm.my

E. Supriyanto
Universiti Teknologi Malaysia, Skudai, Malaysia
e-mail: eko@utm.my

D. E. O. Dewi et al. (eds.), *Cardiovascular Engineering*, Series in BioEngineering,
https://doi.org/10.1007/978-981-10-8405-8_6

to improve efficiency and effectiveness of healthcare. However it evolves with better understanding within the course of time, and numerous studies started to evaluate its impact and criteria for success and failure.

Clinical pathway research evolves as a consequence of health care research popularity in modern health system development; the results of health care researchers work assisted the practitioners and policy makers in decision-making for better care.

In practice, the high level of evidence from the research (randomised clinical trial) makes the main arm in developing clinical path as one trustable source of definite proof. Besides, more organised and standardised care has been promised by the clinical pathway. So, within the course of time, the clinical pathway has been used to standardise the attention of various diseases (acute and chronic care) especially for the most costly, high volume and complex conditions such as cardiovascular medicine. As the demand for more standard care with high effectiveness and efficiency increased and due to the increasing trend of morbidity and mortality rate of the cardiovascular diseases globally and locally, the use the clinical pathway in the hospitals became important for better care.

The Clinical pathway evolution in cardiovascular disease management and its implementation in Malaysia will be the focus of the overview in the following sections. Thus, in this book chapter, we will highlight on the clinical pathway use in cardiovascular disease management, and we will give an insight into the situation in Malaysia mainly.

2 Significant of Cardiovascular Disease

Cardiovascular diseases(CVDs) is one of the most common causes of death worldwide in which 50% of them attributed to Asia [3]. According to World Health Organization (WHO) fact sheet [4], up to 17.5 million people died from CVDs in 2012. At least three-quarter of CVD deaths in developing countries compared to people in developed countries due to the difficulties in these countries to access the integrated primary health care programs. Consequently, morbidity among productive people increased, with high-cost burden due to high expenditure on treatments.

Similarly, in Malaysia CVDs among the first causes of death due to the ever-growing ageing population and the lifestyle changes that certainly have most of the cardiovascular diseases risk factors [5].

According to the study conducted on Malaysian population by Ghazali et al. [6], the risk factors for cardiovascular diseases are trending up. Therefore, the prevalence of cardiovascular diseases will be more in the future.

The influence of the CVDs significance leads the 'Global Non- Communicable Disease Action Plan 2010–2030' to assign two of their targets toward cardiovascular disease prevention and control 'the sixth and eight targets' (WHO, 2013). As the prevalence of cardiovascular diseases increased and the efforts in the healthcare industry trending up to provide the best quality care, managers have incorporated critical pathways in practice.

Due to the increasing competition with others to provide the service that clinical pathway has promised such as reducing the variation in care in term of cost, resources, outcome and quality [1]. However, the promise that fulfilled by other industries when using the Standard Operation Procedures(SOP) could not achieve the same success in healthcare industries.

That might be due to the human factor that plays a role in variation occurrence [7]. The health care system is one of organised complexity in which all the components are in dynamic interaction.

3 Clinical Pathway

The care pathway concept first evolved at the New England Medical Center in Boston in 1985 [8]. Clinical pathways inspired by the Standard Operating Procedures (SOPs) for high-quality work to be done on time. Clinical pathways have many names "care pathways, critical pathways, integrated care pathways or care maps" [9].

The critical pathway is a tool provides us with activities and processes of care but not warrant its success [7]. Clinical protocol is the umbrella term for standard care. Protocols also use the guideline as a base for designing the treatment recommendations. Comparable to the clinical pathway, the goal of the clinical protocol is to decrease treatment variations with focus on guideline compliance where not necessary to consider the continuous monitoring and data-evaluation component.

Clinical pathways are being used worldwide, more in developed countries, while sporadic evidence in Asia about the clinical pathway deployment and studies on its impact on outcome and quality of care is found and even evidence about their usefulness has been unclear [10, 11]. Evidence has shown that there is variation in the use of health services for conditions and circumstances that are quite similar. At present, pathways are used with a minority of patients, mainly in acute hospital trusts. Cardiovascular is one area where the clinical pathway has been used and proof success.

4 The Need for Clinical Pathway

In the era of increasing demand for more coordinated, organised care and growing competition in medical care, using the standards care comes as a solution method for providing a high quality of healthcare service. Therefore, clinical guidelines have appeared as one of the most popular new initiatives. The clinical guideline would provide information to the practice of the disease management based on current evidence from relevant clinical trial data and published literature. According to 'American College of Cardiology Foundation and American Heart Association [12]', guidelines: "reflect a consensus resulted from a review of the available, current scientific evidence and are intended to improve patient care".

The management of cardiovascular diseases patients in guidelines such as the 'American College of Cardiology Foundation/American Heart Association and the European Society of Cardiology and Malaysia MOH 2001, 2007, and 2014' are accessible for everyone. Therefore, the healthcare system has adopted guidelines to reduce clinical practice error, improve the quality of cardiovascular care, improve cost-effectiveness, and find areas of further research needs. However, that will happen only if been followed appropriately. Thus by having the proper guideline, we can enhance the outcome of patient care [13].

A gap between the guidelines and clinical practice exist which may be due to documentation and statement nature of the guideline. The gap between knowledge and practice noticed although the guideline has shown a positive impact on the practice in many settings such as the stroke prevention in atrial fibrillation disease which shown improvement in the disease outcomes [14]. At the same time, we still find the acute coronary disease, for instance, as high as previously with less change in the incident rate or mortality trend. Therefore, another approach to ease the use of guideline and improve adherence to its standard has been used "Clinical pathway".

The use of clinical guidelines in the design of clinical pathways comes as a guarantee that the content is evidence-based and following the latest definite evidence. However, variation in care still happened. It provides an excellent reference for clinicians regardless the lengthy, abstract and static (not dynamic) nature that requires searching and reading, therefore become underused [15].

5 Clinical Pathway in Cardiovascular Diseases

What makes clinical pathway attractive for cardiovascular disease is the capability to cope with the complexity of the multitasks care process in cardiovascular diseases. It facilitates the multidisciplinary patient care by moving the processes of care from stage to another sequentially, within specified time frames, to achieve the desired outcome. The medical application of this concept adopted by nursing professionals in the 1980s and now it became well established in the literature and widely used worldwide [16, 17].

In recent years, an increasing weight placed on the delivery of standard management and quality health. There are initiatives shift from organisation structure to care process and clinical and non-clinical outcomes in health care [18]. However, that does not guarantee the success as the previous studies have shown that a clinical pathway is a tool which highlights the details process of care and inefficiencies of care regardless the presence of warrants to change in those process [7].

The observed evolve in clinical pathway use, was due to its impact on the time, cost, quality [19], clinical outcomes [9, 20–22].

In most of health care program's endeavours, the quality is a significant milestone. In fact, the clinical pathway furthers the consolidation and strengthening effort toward high-quality care. Many researchers worldwide proved that such as: Aniza et al. [10, 11, 19, 23–32].

The increasing trend of chronic diseases became the main problem that has to be tackled the majority of health-care systems care [1]. Despite the presence of various evidence-based prevention and management guidelines and the presence of recommendations for cardiovascular disease, significant gaps between the evidence and its implementation still exist ('evidence-practice gap').

It is important to understand the treatment strategies of cardiovascular diseases, and quality of life and risk evaluation and especially control of risks has essential economic and operational implications. Heart disease nowadays is one of the most common health cost burdening disease.

The delivery of high quality for this patients not only depending on the treatment service but it is a complete process that may cost high if malpractice. The link between the quality of healthcare delivery and the preventable factors of cardiovascular disease has been emphasised in many initiatives for health improvement [33, 34]. Coronary heart disease is, however, just one part of the picture and the most significant cause of heart-related ill health and death Evidence-based medicine needs to be complemented with evidence-based implementation.

Thus, the prevention and optimal treatment of time-sensitive diseases for example 'obstructive Coronary artery disease (CAD)' represents a significant public health imperative. Considering the importance of the time to the heart muscle safety, proper management does not mean only using the best choice of treatment but also giving the timely reperfusion to prevent necrosis from happening. Therefore it is obligatory to decrease the gap from waiting times to treatment for instance at the Emergency Department (ED) until the patient gets the first treatment correctly [35].

That could occur when we use one of the management tools(guidelines, protocols, clinical pathway) that will minimise the crowding and improve the patient flow throughout the organisation [36, 37]. The issue of time has been tackled in many efforts, such as "the percent of heart attack patients receiving treatment within the recommended time window has increased" as shown in Southon et al. [38]. Therefore, clinical pathways need to be used at Hospital to deploy and enhance clinical standards use. Patients on clinical pathway have a lower rate of complications, readmission and mortality rate compared patients, not on clinical pathway. The Impact of the clinical pathway has been shown in term of length of stay (average duration of the treatment process for a particular disease). It also helps in minimising the medical errors(inappropriate or incorrect treatment) during the treatment process through the recommendation given through consensus in guidelines show high compliance [39], clinical pathway could help in managing the care in a timely, sequenced and organised manner. There are numerous studies that show positive impact of clinical pathway on different CVD such as heart failure, Myocardial infarctions, cardiac surgeries, aortic surgeries, and PCI, and event nutritional management and prevention measures [40–45].

Also, recent data from the National Health Survey of Indonesia showed that cerebrocardiovascular disease is the leading cause of death in Indonesia too (MOH Indonesia [46].

Few studies on CP in Asia and fewer carried out in Malaysia tackling the myocardial infarction mainly in the field of cardiovascular diseases such as Aniza Ismail et al. [47, 48] study. Despite the significance of the CVDs in Malaysia, in which they remain as an "important cause of death accounting for 20–25% of all deaths in government hospitals from 2000–2005" [49].

6 Clinical Pathway in Malaysia

Malaysia has an area of 330,000 km^2 and a population of 27 million, comprising three significant ethnic populations: Malay, Chinese and Indian [37]. In 2010, the proportional mortality was 32% out of total population 28,401,017 just for CVD. According to Ministry of Health Malaysia, one of 10 leading causes of mortality in 2009 was due to heart diseases and diseases of pulmonary circulation where the number of deaths was 7738 out of 48,091. Primary provider for health care service is Ministry of Health (MOH) for most of the sectors such as Ministry of Education, Ministry of Defend, Ministry of Housing and Local Government Ministry of Home Affairs and Ministry of human resource. Healthcare professionals and workers have used Clinical pathway as a clinical management tool to define the best treatment, procedures and duration of treatment for patients with specific diagnoses according to evidence-based medicine (EBM). Which aims to minimise time interval between the clinically proven treatments and their use in everyday medical practice by using dynamic tools such as the clinical pathways [50, 51] (Table 1).

The 9th Malaysia plan insisted on development of the Health Information Management. Ministry of health has addressed in its reports the safety and quality of the patient in health care by various efforts to improve the quality of healthcare by the use of clinical pathway as a tool for management of care. However, the academic focus on the clinical pathway and the research that cover the rigour of the implemented programmes are limited.

Various studies that tackled the different type of diseases in Malaysia exist such as the 'dental pathway' that has been developed by the Oral Health Division in collaboration with the University of Malaya Community Dentistry Department and 'neonatal jaundice pathway [34]. Some papers found to discuss the quality in term of cost mainly for UKM researchers [47, 48].

The author of this chapter has conducted thorough literature to scan the situation of clinical pathway research in Malaysia. Among the only fourteen peer-reviewed papers currently have been found; a few studies found that cover the clinical pathway in Malaysia and mostly focus on development, while few studies found to discuss the impact and effectiveness. Most of the study design was on pre- post-implementation comparative studies.

The central theme in most of 14 articles was about the development then implementation of the clinical pathway for improving the outcome. The majority of the publication show success in arthritis, knee replacement, myocardial infarction and dental clinical pathway outcome [47, 48, 52, 59]. However, mostly using the pre-post-implementation study design which applies to the first publication found about the nutritional clinical pathway [53]. After that, active research conducted and pub-

Table 1 Demographic indicators, Malaysia, 2002–2003

Indicator	2002		2003	
	Number (thousands)	% of Total population	Number (thousands)	% of Total population
Population of males	12,487.1	50.9	12,751.9	50.9
Population of females	12,039.4	49.1	12,296.4	49.1
Population of youth (below 20 years old)	10,648.5	43.4	10,784.1	43.1
Population of elderly (above 60 years old)	1,554.1	6.3	1599.7	6.4
Economically-productive population (aged 15–64 years)	15,318.3	62.5	15,702.3	62.7
Economically-dependent population (aged below 15 years and above 64 years)	9208.0	37.5	9346.0	37.3
Urban population	15,310.7	62.4	15,687.0	62.6
Rural population	9215.8	37.6	9361.3	37.4

Source Malaysia Health Report [34]

lished in 2006 in the same trend with others in the other countries. However, the ups and downs in reporting studies are evident as shown in Fig. 1. Every two years there was a study published in Malaysia from 2006 until 2014. That might be due

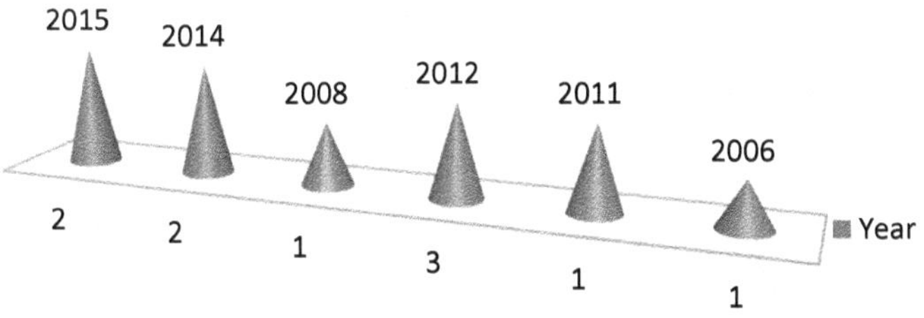

Fig. 1 Clinical pathway related publication in Malaysia after [53] publication. Adapted from AL-Ashwal and Supriyanto [58]

to the prospective nature and consensus studies that take a long time to be executed. However, more studies have been published in 2014 and 2015 consecutively. Most probably due to the exploratory nature of the last studies [47, 48, 52, 54–57, 59].

In summary, the clinical pathway is not a new field in Malaysia. However, interest in reporting the contemporary scientific evidence is scanty.

7 Conclusion

Health care research is new especially for a clinical pathway in Malaysia, and like any new programs or innovation, it takes time to be accepted and researched. The global evidence supports clinical pathway advantage, while scarce known from Malaysia literature on the clinical pathway. Adapting plan from international context to local might not be successful [7]. Therefore more research and studies need to be conducted toward its effectiveness in the local context. The available evidence on the clinical pathway is scarce and does not help in building strong evidence for evaluating the clinical pathway in Malaysia. There is a need for more engagement of health care researchers on clinical pathway design, implementation and even analysis.

Acknowledgements This book chapter has been supported by the Ministry of Higher Education (MOHE) with university technology Malaysia(UTM) by TRGS grant reference number PY/2016/05857.

References

1. Lodewijckx, C., et al.: Eight-step method to build the clinical content of an evidence-based care pathway: the case for COPD exacerbation. Trials **13**, 229 (2012)
2. Pearson, S.D., Goulart-Fisher, D., Lee, T.H.: Critical pathways as a strategy for improving care: problems and potential. Ann. Intern. Med. **123**(12), 941–948 (1995)
3. Ohira, T., Iso, H.: Cardiovascular disease epidemiology in Asia. Circ. J. **77**(7), 1646–1652 (2013)
4. WHO fact sheet.: Cardiovascular Diseases (CVDs). (2015)
5. Zambahari, R., et al.: 14 Annals of Internal Medicine Clinical Practice Guidelines Management of Acute ST-Segment Elevation Myocardial Infarction 2014, 3rd edn. Clinical Practice Guidelines, Kuala-Lumpur, Malaysia (2014)
6. Ghazali, S., et al.: Sociodemographic factors associated with multiple cardiovascular risk factors among Malaysian adults. BMC Publ. Health **15**(1), 68 (2015)
7. Every, N.R., et al.: Critical pathways a review. Circulation **101**, 461–465 (2000)
8. Grol, R., Grimshaw, J.: From best evidence to best practice: effective implementation of change in patients' care. Lancet **362**(9391), 1225–1230 (2003)
9. Ban, A., et al.: Impact of clinical pathway on clinical outcomes in the management of COPD exacerbation. BMC Pulm. Med. **12**, 27 (2012)
10. Cheah, J.: Clinical pathways–an evaluation of its impact on the quality of care in an acute care general hospital in Singapore. Singap. Med. J. **41**(7), 335–346 (2000)

11. Cheah, J.: Development and implementation of a clinical pathway programme in an acute care general hospital in Singapore. Int. J. Qual. Healthca. J. Int. Soc. Qual. Healthc. ISQua **12**(5), 403–412 (2000)
12. American College of Cardiology Foundation, American Heart Association.: Methodology manual and policies from the ACCF/AHA task force on practice guidelines. Med. Sci. Sports Exerc. 1–88 (2010)
13. Kontos, M.C., et al.: Lower hospital volume is associated with higher in-hospital mortality in patients undergoing primary percutaneous coronary intervention for ST-segment-elevation myocardial infarction: a report from the NCDR. Circ. Cardiovasc. Qual. Outcomes **6**(6), 659–667 (2013)
14. Lip G.Y.H. et al.: Improved outcomes with European society of cardiology guideline-adherent antithrombotic treatment in high-risk patients with atrial fibrillation: a report from the EORP-AF general pilot registry. Europace (January 2016): euv269 (2015)
15. Glasziou, P., Les, I., David, M.: Monitoring in chronic disease: a rational approach. BMJ Br. Med. J. **330**(7492), 644–648 (2005)
16. Bjurling-Sjöberg, P., et al.: Prevalence and quality of clinical pathways in Swedish intensive care units: a national survey. J. Eval. Clin. Pract. **20**(1), 48–57 (2014)
17. Hunter, B., Segrott, J.: Re-mapping client journeys and professional identities: a review of the literature on clinical pathways. Int. J. Nurs. Stud. **45**(4), 608–625 (2008)
18. Croucher, M.: An evaluation of the quality of integrated care pathway development in the UK national health service. J. Integr. Pathways **6**(9), 8 (2005)
19. Rotter, T., et al.: A systematic review and meta-analysis of the effects of clinical pathways on length of stay, hospital costs and patient outcomes. BMC Health Serv. Res. **8**, 265 (2008)
20. Gooch, K., et al.: Comparative effectiveness of alternative clinical pathways for primary hip and knee joint replacement patients: a pragmatic randomized, controlled trial. Osteoarthritis Cartilage **20**(10), 1086–1094 (2012)
21. Okamura, K., et al.: Changes in perioperative management of radical prostatectomy using clinical pathways according to a standardized care plan: a multi-institutional study. Int. J. Urol. **20**(3), 337–343 (2013)
22. Panella, M et al.: A cluster randomized trial to assess the effect of clinical pathways for patients with stroke: results of the clinical pathways for effective and appropriate care study. BMC Med. 10(1), 71 (2012)
23. Aniza, I., Saperi, S., Aljunid, S.M.: Clinical pathway as a strategy for improving healthcare quality and cost containment. Med. Health **3**(2), 239–246 (2008)
24. Cardozo, L., Aherns, S.: Assessing the efficacy of a clinical pathway in the management of older patients hospitalized with congestive heart failure. J. Healthc. Qual. **21**(3), 12–17 (1999)
25. Leigh, K.: Clinical pathway compliance and quality improvement. Nurs. Stan. **18**(18), 33–35 (2004). (Royal College of Nursing (Great Britain) : 1987)
26. Kitchiner, D.J., Bundred, P.E.: Clinical pathways. Med. J. Aust. **170**(2), 54–55 (1999)
27. Kucenic, M.J., Meyers, D.G.: Impact of a clinical pathway on the care and costs of myocardial infarction. Angiology **51**(5), 393–404 (2000)
28. Luc, De, Kathryn, E., Whittle, Claire: An integrated care pathway appraisal tool: a 'badge of quality'. J. Integr. Pathways **6**(1), 13–17 (2002)
29. Neuman, M.D., et al.: The relationship between short-term mortality and quality of care for hip fracture: a meta-analysis of clinical pathways for hip fracture. J. Am. Geriatr. Soc. **57**(11), 2046–2054 (2009)
30. Panella, M., Marchisio, S., Di Stanislao, F.: Reducing clinical variations in clinical pathways: do pathways work? Int. J. Qual. Healthc. J. Int. Soc. Qual. Healthc. ISQua **15**(6), 509–521 (2003)
31. Rotter, T., et al.: The quality of the evidence base for clinical pathway effectiveness: room for improvement in the design of evaluation trials. BMC Med. Res. Methodol. **12**(1), 80 (2012)
32. Schwarzbach, M., et al.: Effects of a clinical pathway on quality of care in kidney transplantation: a non-randomized clinical trial. Langenbeck's Arch. Surgery **395**(1), 11–17 (2010)

33. Krumholz, H.M., et al.: ACC/AHA clinical performance measures for adults with ST-elevation and non-ST-elevation myocardial infarction: a report of the American College of Cardiology/American Heart Association task force on performance measures (Writing Committee to Develop Perfo). J. Am. Coll. Cardiol. **47**(1), 236–265 (2006)
34. Ministry of Health Malaysia. Annual Report 2003, 294 (2003)
35. Amin, A.: Improving the management of patients after myocardial infarction, from admission to discharge. Clin. Ther. **28**(10), 1509–1539 (2006)
36. Rezai-Rad, M., Vaezi, R., Nattagh, F.: E-Health readiness assessment framework in Iran. Iran. J. Publ. Health **41**(10), 43–51 (2012)
37. Venkatesh, V., Davis, F.D.: A model of the antecedents of perceived ease of use: development and test. Decis. Sci. **27**(3), 451–481 (1996)
38. Southon, F.C.G., Sauer, C., Dampney, C.N.G.: Information technology in complex health services. J. Am. Med. Inform. Assoc. **4**(2), 112–124 (1997). (Kit)
39. Kazui, H., et al.: Effectiveness of a clinical pathway for the diagnosis and treatment of dementia and for the education of families. Int. J. Geriatr. Psychiatry **19**(9), 892–897 (2004)
40. Gurvitz, M., Marelli, A., Mangione-Smith, R., Jenkins, K.: Building quality indicators to improve care for adults with congenital heart disease. J. Am. Coll. Cardiol. **62**(23), 2244–2253 (2013)
41. Kul, S., et al.: Effects of care pathways on the in-hospital treatment of heart failure: a systematic review. BMC Cardiovasc. Disord. **12**(1), 81 (2012)
42. Sternberg, S. Influence of Stroke Clinical Pathway on Documentation. (December) (2007)
43. Sulch, D., Melbourn, A., Perez, I., Kalra, L.: Integrated care pathways and quality of life on a stroke rehabilitation unit. Stroke **33**(6), 1600–1604 (2002)
44. Taylor, R.S., et al.: Exercise-based rehabilitation for patients with coronary heart disease: systematic review and meta-analysis of randomized controlled trials. American J. Med. **116**(10), 682–692 (2004)
45. Vinen, J.: Time to initiation of thrombolysis after myocardial infarction: quality indicators. Int. Med. J. **32**(4), 125–126 (2002)
46. Jakarta, Ministry of Health R.I.: Zhurnal Eksperimental'noi i Teoreticheskoi Fiziki Indonesia Health Profile 2008. eds. Mkes Dra. Rahmaniar Brahim, Apt, MKes Sugito, SKM, Mkes Hary Purwanto, MKes, MMSi Hasnawati, SKM, and M.Epid Members Drg. Vensya Sitohang. Ministry of Health, Republic of Indonesia (2010)
47. Ismail, A., Sulung, S., AlJunid, S.M., Yahaya, N.H.M., Harunarashid, H., Maskon, O., Ban, A., Harun, R., Saiboon, I.M., Nor, I.M., et al.: Clinical pathways: development and implementation at a tertiary hospital in Malaysia. Int. J. Publ. Health Res. **2**(2), 153–160 (2012)
48. Ismail, A., et al.: Clinical pathways: development and implementation at a tertiary hospital in Malaysia. Int. J. Publ. Health Res. **2**(2), 153–160 (2012)
49. Malaysia Ministry of health. Clinical Practice Guideline. 07(2), 1–64 (2007)
50. Davis, F.D., Bagozzi, R.P., Warshaw, P.R.: User acceptance of computer technology: a comparison of two theoretical models. Manage. Sci. **35**(8), 982–1003 (1989)
51. Selvaraju, D.S.: Health information management : Malaysian experience. 1–7 (2005)
52. Aljunied, S.M., et al.: Development of clinical pathway for mild cognitive impairment and dementia to quantify cost of age-related cognitive. Malays. J. Publ. Health Med. **14**(3), 88–96 (2014)
53. Ramanujam, T.M.: The role of nutritional support in acute pancreatitis: a review and proposal of a clinical pathway for management. J. Univ. Malaya Med. … 2, 81–87 (1999)
54. Aljunid, Sm, Ismail, A., Sulong, S.: Can clinical pathways enhance the implementation of a casemix system? a case study in a teaching hospital in Malaysia. BMC Health Serv. Res. **11**(Suppl 1), A6 (2011)
55. Cheah, T.S.: Clinical pathways–the new paradigm in healthcare? Med. J. Malaysia **53**(1), 87–96 (1998)
56. Sulong, S., Osman, H.C.K., Ismail, A.: The evaluation of knowledge and practice on clinical pathways among health care workers at Universiti Kebangsaan. mjphm.org.my. 15(1), 69–76 (2015)

57. Tay, H.L., Raja Latifah, R.J., Razak, I.A.: Clinical pathways in primary dental care in Malaysia: clinicians' knowledge, perceptions and barriers faced. Asia-Pacific J. Publ. Health/Asia-Pacific Acad. Consortium Publ. Health **18**(2), 33–41 (2006)
58. AL-Ashwal, R., Supriyanto, E.: Evidence for the contemporary clinical pathway quality measures : literature review. In ICRIIS2015, 1–7 (2016)
59. Aljunid, S.M., Maimaiti, N., Ahmed, Z., Nur, A.M.: Development of clinical pathway for mild cognitive impairment and dementia to quantify cost of age-related cognitive. researchgate.net **14**(3), 88–96 (2014)

Square Root Design for Natural Frequency Module of Dynamic ECG Features—a Preliminary Study

Nurul Ashikin Abdul-Kadir, Norlaili Mat Safri, Mohd Afzan Othman and Fauzan Khairi Che Harun

Abstract In field-programmable gated array (FPGA) design platform, the device is invented into smaller scale instrument, such as a system-on-a-chip that is used for a specific application design. Not only that, the feasibility study of the natural frequency (ω) obtained from a second-order dynamic system (SODS) applied to an ECG signal was discovered recently. This new amazing discovery enabled the conversion of algorithms into hardware design (HD) which is an important earlier step to study the reliability of the algorithm for hardware application. The module developed in HD was a preliminary step to analyze the SODS algorithm and also the performance in register-transfer level (RTL) design. **Methods**: The selected algorithm was mapped into a specific module which was designed to reduce the scale or resources. A high-level (HL) synthesis approach was chosen for mapping the algorithm by using SystemVerilog hardware descriptive language. **Results**: The algorithm was successfully designed into hardware which utilized of 2432 logic elements and 16 digital signal processing blocks of chosen FPGA platform. Hence, the design was proven success as the output is equivalent to the truth table. **Conclusion**: This study found that algorithm ω was able to be implemented into hardware design platform.

Keywords ECG · Heart · Natural frequency · Second-order dynamic system

1 Introduction

The development of information and communication technology has improved health tele-care by providing more sophisticated analysis software to support a real-time processing of ECG signal [1]. A surge in the manufacture of small or integrated circuit system, or, scaled system module in medical diagnostic equipment for digital signal

N. A. Abdul-Kadir (✉) · N. Mat Safri · M. A. Othman · F. K. C. Harun
School of Electrical Engineering, Faculty of Engineering, Universiti Teknologi Malaysia, Johor Bahru, Malaysia
e-mail: ashikin.kadir@utm.my

N. A. Abdul-Kadir · F. K. C. Harun
IJN-UTM Cardiovascular Engineering Centre, Institute of Human Centered Engineering, Universiti Teknologi Malaysia, Johor Bahru, Malaysia

D. E. O. Dewi et al. (eds.), *Cardiovascular Engineering*, Series in BioEngineering,
https://doi.org/10.1007/978-981-10-8405-8_7

processing (DSP) applications can provide low cost ECGs with real-time processing, which is suitable for a monitoring and alert system [2, 3].

Software and hardware codesign of a reliable ECG detection and classification system can serve as a diagnostic tool that benefit patients' lives. Intense research has been devoted to the development of an affordable and reliable ECG for daily monitoring and outpatient usage [4–6]. There were studies implemented ECG systems and prototyped their designs by using Field-Programmable Gated Array (FPGA) platform [4, 5, 7] also in other medical application [6]. The flexibility of FPGA technologies make it possible to be used for prototyping the digital bio-signal processing system of ECGs for the detection of specific ECG features [8–10], ECG classification [10–12], and ECG data compression [4, 5].

In FPGA design platform, the device is invented into smaller scale instrument, for a specific application design such as a system-on-a-chip (SoC). SoC is an embedded system, that allows for parallel processing [13], the transfer rate was high as the bandwidth and the memory module were large, and the system was suitable for digital bio-signal processing [14]. Moreover, FPGA also is the most suitable platform to apply a specific application design, such as an ECG processor, in contrast to a microcontroller, which is suitable as a general purpose processor (multitasking) but has a lower processing speed [15]. The processing speed can be very fast, and multiple control loops can be run efficiently at different rates on each SoC [3]. Such inventions could highly benefit patients, i.e., increase in exertion does not have negative effect on the patients, not to mention they are able to maintain their regular activities. At the same time, their data can be wirelessly transmitted by mobile platform (i.e. android) to the central database, thus allowing an immediate analysis, lower cost, and efficient monitoring [7].

For real-time signal processing, an FPGA was used as a platform to detect the P- and T-waves of an ECG with a single lead [8]. The internal memory of the FPGA development board was used to store the ECG data. The study provided 30-ms latency for the detection of P- and T-waves. Instead of P- and T-waves detection, other studies detected the QRS complex in their FPGA-based systems [9, 10].

The classification and detection of ECG signals can be performed using various signal processing techniques based on time-domain analysis, frequency-domain analysis or statistical analysis. Time-domain features based on RR-interval [16–18], P-wave [19, 20], or QRS morphology [21] were explored. In the frequency domain, a wavelet transform [22] was examined. Another study used time and frequency domain features [23], or the time-frequency cross spectrum, namely the time-varying coherence function.

As mentioned by [9], an accuracy of 98% was achieved for diagnosing patients using the FPGA system. The study proposed the use of entropic measurements of fuzziness to map the set of all fuzzy subsets into a nonnegative real set [9]. The architecture of the neural network was 2-8-1, with input-hidden-output layers and with an inverse delayed function model for the neurons. The neural network was trained and tested on 1000 datasets. Tachycardia and bradycardia were indicated by the occurrence of QRS complexes at rates per minute that were greater than 100 and less than 60, respectively. If the QRS width was detected to be greater or less than

120 ms, then the tachyarrhythmia was classified as supraventricular or ventricular, respectively. A 12-bit binary counter was used to spot the rising edge of the QRS complex, and a 4-bit register was used when the falling edge of QRS complex was spotted. An 8-bit counter was used to detect bradycardia or tachycardia. The counter increased by one if a QRS was detected, and it was counted every minute. If the count reached more than 100 or was less than 60, then the signal was classified as tachycardia or bradycardia, respectively. A total of 4000 samples were collected, with a recording time of 8 s and 500 samples per second for each patient ($N = 200$) with 12 leads.

Another study used an FPGA to implement a detection algorithm that was 17-fold faster (sensitivity of 99%) than the system without FPGA [10]. The FPGA design platform did not need a high-performance microprocessor to prototype the targeted SoC. Instead of an ECG detection system, [10] classified an ECG as normal or abnormal based on the R-R interval with some alteration in the code for the ECG classification. The result showed a specificity of 95.54% for beat classification. The beat width and amplitude were compared to previously prior classified beat types. A further description of the R-R intervals patterns can be found in [10]. Other studies have proposed an ECG classification system using a neural network (NN) classifier [11, 12], and have utilised an FPGA in the prototype. The architecture of the back-propagation and feed-forward neural networks were designed using a single hidden-layer neuron, i.e., a reusable neuron architecture (RNA) in a three-layer (51-30-12) artificial neural network (ANN) [11]. The prototype system was realised using a Xilinx Virtex-5 FPGA board. The performance of the RNA developed by [11] led to a 98.7% area size savings, 99.1% dynamic power savings, 30-fold increase in time delay, 5000-fold increase in speed and 4000-fold reduction in energy consumption. In an RNA architecture, the multiplier(s) and adder(s) are configurable. Furthermore, the RNA design is more efficient than feed-forward or back-propagation neural network designs in which a multiplier is reused iteratively to multiply the input(s) by the corresponding bias(s) or reused to compute the updated weight(s).

The FPGA implementation of an RNA provides real-time ECG classification and allows ANNs to update the weight(s) and bias(s) unaided and without periodic retraining, which overcomes the limitation of smart phone-based ECGs for real-time analysis [24]. In contrast, [12] proposed the PNN classifier for distinguishing multiple cardiac arrhythmias and heartbeat recognition. The architecture of PNN consists of four layers (input, hidden, summation and output) with a network topology of 50-22-8-7. The ECG signal is classified using seven categories: normal beat, premature ventricular contraction, atrial premature beat, right bundle branch block beat, left bundle branch block beat, paced beat, and fusion of paced and normal beat; the classification was tested using the MIT-BIH Arrhythmias Database. An overall accuracy of 94% was achieved for all of the tested records. The proposed classifier showed outstanding computational efficiency, high precision, and flexibility for pattern recognition once implemented on the FPGA platform [12]. By using fuzzy neural signal processing system, [9] reported an FPGA-based embedded system for detecting the QRS complex of an ECG signal. The system performance reached an accuracy of 98% when tested with 200 patients who suffered from tachycardia and tachyarrhythmia.

Instead of prototyping ECG classification and detection using FPGA platform, prototypes for ECG data compression also can be implemented [4, 5]. An SoC that consists of ECG monitoring with real-time heart rate variability (HRV) analysis has been developed [4]. In the compression technique, the ECG data were packetised and multiplexed into a single data stream with a compression ratio of 2.50 and a maximum operation frequency of 32 MHz. In contrast, [5] achieved a maximum compression ratio (i.e., the ratio of the number of wavelet coefficients of the input relative to the number of the encoding wavelet coefficients) of 7.9 at a threshold of 10 and a decomposition level of 4.

Moreover, an SoC is potentially dependable acquiring ECG signals, perform automated analysis and wirelessly transmit the data [4]. This system is composed of an analog-to-digital converter (ADC) controller, HRV processor and compression engine. With the use of Lomb periodograms and a sliding window with 50% overlap for the time-frequency analysis of a 2 min window, the high- and low-frequency components of the HRV were successfully identified. The HRV processor normalised the spectrum (spectral analysis of HRV) to 1 Hz and only analyzed the frequency range from 0 to 0.5 Hz. SRAM (memory) was used to store the coefficients of the RR intervals from the HRV analysis. The architecture consists of an adjust unit, a context determiner, a differential pulse code modulation predictor, a context-based parameter estimator, a prediction memory array for 3-channel ECG, a set of context variable upkeep modules, a remap to unsigned unit, a Golomb-Rice entropy coder, and a 40-bit multistream multiplexer. Instead of time-domain analysis of HRV analysis, discrete wavelet transforms (DWTs) and a lossless encoding method were utilised by Ballesteros and coworkers for ECG compression in an FPGA platform [5]. The study showed excellent performance with a low error of quantisation and high signal quality, and the percent-root-mean-square-difference (PRD) satisfied the requirements for clinical applications.

However, none of the previously mentioned studies had used second-order dynamic system (SODS) approach towards ECG signals. The SODS method is a parametric method. SODS is still in the preliminary stage in ECG processing, therefore not much study have been done. Previous study had shown the feasibility of (SODS) to extract the features of ECG signals [25, 26]. Beforehand, [27, 28] introduced SODS by using a different name, which is semantic mining. Study [25] shows the SODS equation, namely second-order differential equation (SODE) for modeling the ECG signals as dynamic features. Additionally, [29–31] also described ECG signals as dynamical model which is also a nonlinear model [32]. SODE shows three features, they are natural frequency (ω), damping coefficient (ξ) and forcing input (u), which are used to describe the signals. Feature ω determines the ECG system oscillations speed during any transient response, feature ξ determines the ECG system oscillates as the response decays toward steady state, and feature u is the external input given to the ECG system. Besides that, the recognition of ECG signals by using feature ω has shown a comparable detection rate for AF [25], ventricular fibrillation (VF) and ventricular tachycardia (VT) conditions [27, 28] between the range of 95.0–97.9%, respectively.

Therefore, the algorithm of feature ω is chosen to be developed in hardware design to emphasize a specific module for ECG feature extraction according to SODS technique. This is a preliminary step to develop an application specific of ECG arrhythmias detection module based on SODS technique. Additionally, previous study [33] has showed the initial result of designing algorithm ω^2 and this current study focus on optimizing the function of ω^2 and designing a square root ($\sqrt{}$) function into hardware design. The performance and validation of the modules are further explored in this study.

2 Methods

The implementation of the DSP algorithm into hardware of the selected feature was performed by using the HL synthesis approach and the SystemVerilog (SV) hardware descriptive language (HDL). HL synthesis approach is the mapping of a behavioral description of a digital system into RTL design, or transforming the algorithm into an RTL design [34] HL synthesis approach is an initial step to develop hardware of own digital system in SoC design.

There are several techniques to map the algorithmic description in the form of data flow graph (DFG) for RTL design, which is suitable for hardware design. Figure 1 shows the steps for designing hardware using HL approach. Three phases are included; parsing, transformation and RTL synthesis, to implement the hardware design [35, 36].

Parsing phase is where less resources are needed by factoring the algorithm or removing redundant operations in the algorithm. For example, by replacing an unknown parameter into a fixed value or coefficient. The transformation phase is where DFG corresponding to the algorithm are allocated and scheduled to minimise resources and whole computation time. Finally, the RTL synthesis phase is where RTLs are coded and designed for a specific algorithm.

In this study, the specific algorithm is shown by Eq. (1) where $x(t)$ is the signals with respect to the variable t, d/dt is the slope, d/dt^2 is the velocity, d/dt^3 is the jerk, and d/dt^4 is the jounce of the signals $x(t)$ [37]. The selected algorithm was intended to be used in algorithm module (AM) which is shown in the flow chart of Fig. 2.

$$\omega = \sqrt{\frac{\frac{d^2}{dt^2}x(t)\cdot\frac{d^4}{dt^4}x(t) - \left(\frac{d^3}{dt^3}x(t)\right)^2}{\frac{d}{dt}x(t)\cdot\frac{d^3}{dt^3}x(t) - \left(\frac{d^2}{dt^2}x(t)\right)^2}} \qquad (1)$$

In addition, Fig. 2 shows several steps needed in AM design. AM is where the selected algorithm was decided either as a single module or several modules. As refer to Eq. (1), the feature ω is the product of a square root. Therefore, firstly, the value of

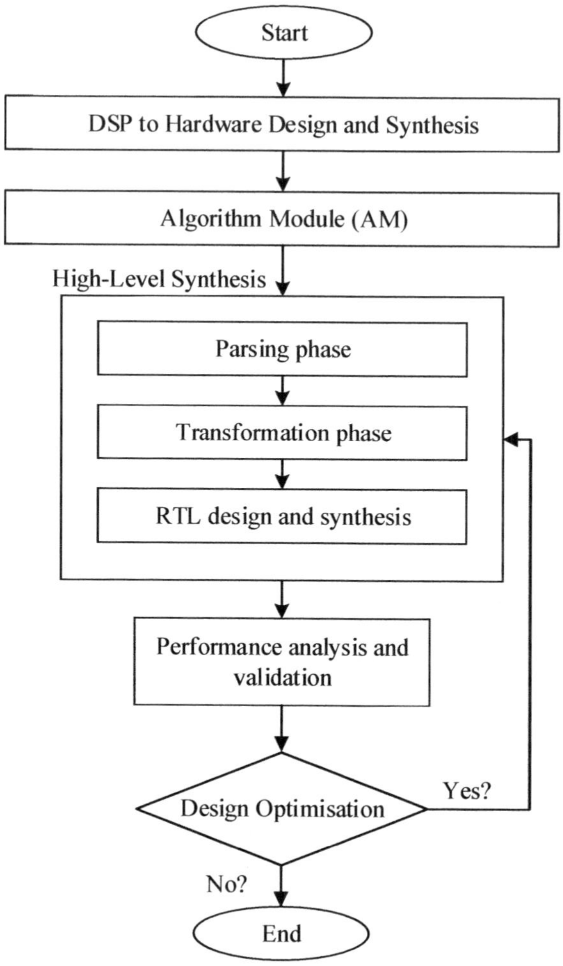

Fig. 1 The steps for designing hardware using HL approach

ω^2 (square algorithm) is determined, later the square root ($\surd(\omega^2)$) is defined. These led for designing two hardware modules, namely Module 1 (*NatFrequency2_opt*) and Module 2 (*SquareRoot*). Module 1 designs an optimization of square algorithm and Module 2 designs square root algorithm. The HL design synthesis is chosen for mapping the algorithm into hardware design. A design optimisation as well as the performance and validation are prepared for both designs. Further discussions are as follow.

According to Eq. (1), the right-hand side of the algorithm ω is rearranged as a new algorithm without square root function. Thus, producing left-hand side as ω^2, which is shown in Eq. (2). Feature ω is the natural frequency of a signal x. $x(t)$, dx/dt, d^2x/dt^2, d^3x/dt^3, and d^4x/dt^4 are time (t) derivatives of x for the first to forth derivation respectively.

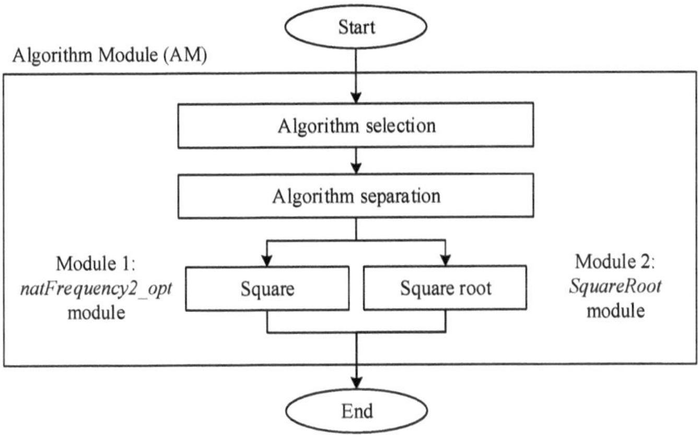

Fig. 2 Flow chart for algorithm module (AM) design

$$\omega^2 = \sqrt{\frac{\frac{d^2}{dt^2}x(t)\cdot\frac{d^4}{dt^4}x(t) - \left(\frac{d^3}{dt^3}x(t)\right)^2}{\frac{d}{dt}x(t)\cdot\frac{d^3}{dt^3}x(t) - \left(\frac{d^2}{dt^2}x(t)\right)^2}} \tag{2}$$

The value of ω^2 can be obtained by having all parameters of $dx/dt, d^2x/dt^2, d^3x/dt^3$, and d^4x/dt^4, which are considered as the values of signal x for five consecutive times. Additionally, Eqs. (3)–(6) show the algorithms respectively.

$$\frac{d}{dt}x(t) = \frac{x(t-1) - x(t)}{t} \tag{3}$$

$$\frac{d^2}{dt^2}x(t) = \frac{x(t-2) - 2x(t-1) + x(t)}{t^2} \tag{4}$$

$$\frac{d^3}{dt^3}x(t) = \frac{x(t-3) - 3x(t-2) + 3x(t-1) - x(t)}{t^3} \tag{5}$$

$$\frac{d^4}{dt^4}x(t) = \frac{x(t-4) - 4x(t-3) + 6x(t-2) - 4x(t-1) + x(t)}{t^4} \tag{6}$$

Parameter x in time delay of $x(t)$, $x(t-1)$, $x(t-2)$, $x(t-3)$, and $x(t-4)$, where t is 4. Therefore, the denominators for Eqs. (3)–(6) are changed to integer 4, 16, 64 and 256, respectively. Therefore, this study has separated the design of module ω into two modules as follow.

i. Module 1: *NatFrequency2_opt*—to compute the value of ω^2
ii. Module 2: *SquareRoot*—to compute the value of $\sqrt{\omega^2}$

3 Module 1—NatFrequency2_opt

There are five algorithms to be mapped into RTL design for finding the value of ω^2 which are shown as Eqs. (2)–(6). Equations (2)–(6) were included in a submodule and the square root value for Eq. (2) was set as a submodule of top-level design, namely, *NatFrequency2_opt* module.

Algorithms of Eq. (3) to (6) were transformed into DFGs as shown in Fig. 3. Firstly, the value of t is replaced with 4 to simplify the algorithms. Additionally, $x(t)$, $x(t-1), x(t-2), x(t-3)$, and $x(t-4)$, were changed to a, b, c, d, and e. These nodes are referring to five x values, which are needed to calculate dx/dt, d^2x/dt^2, d^3x/dt^3 and d^4x/dt^4. While the output for the time derivatives of x, were changed to $m, n, p,$

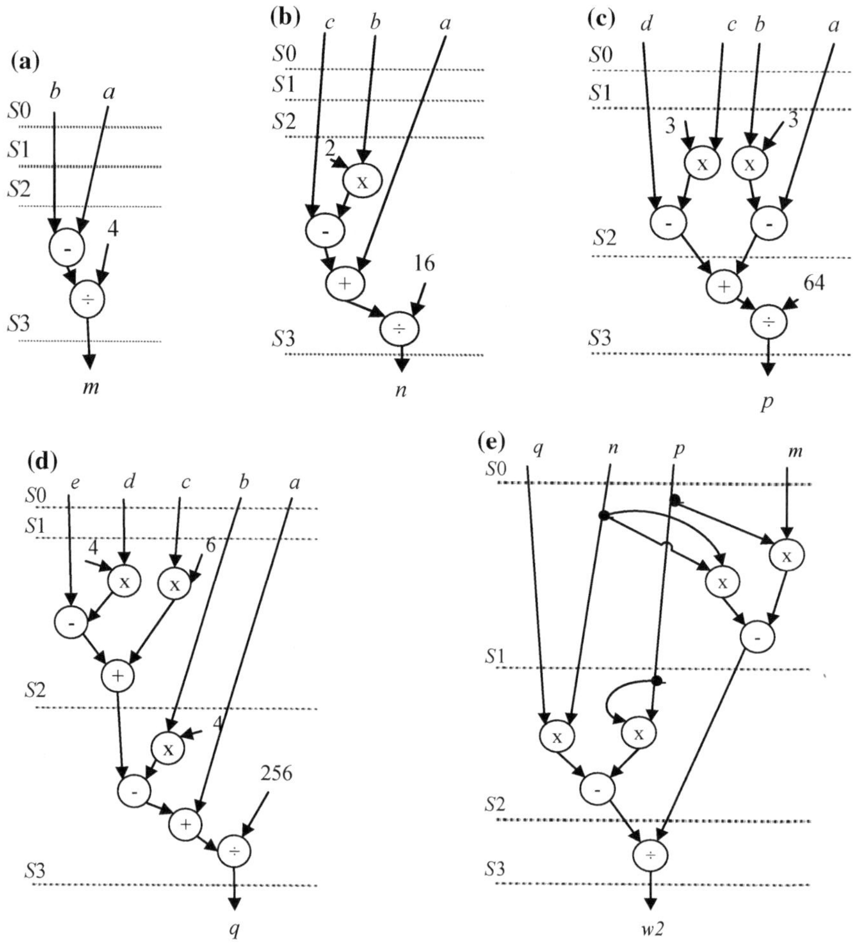

Fig. 3 The DFGs for module; **a** *xOne*, **b** *xTwo*, **c** *xThree*, **d** *xFour*, **e** *nfSquare*

and q. This step is to ensure the algorithms (Eqs. (3)–(6)) interpretation are correctly applied to HDL for hardware design. Moreover, this study wanted to determine the right-hand side of each algorithm is written in SV HDL coding properly. Equation (2) is produced using the output of Eqs. (3)–(6).

Figure 3 describes the DFGs of the equations by using only two multipliers/dividers and two adders/subtractors for each cycle. The algorithm for Eqs. (2)–(6) are labelled as *nfSquare-*, *xOne-*, *xTwo-*, *xThree-*, and *xFour*-module. Their outputs (m, n, p, q and $w2$) are produced after the third state. The outputs are the inputs to *nfSquare*-module. The output of *nfSquare*-module is $w2$.

3.1 Verification

The RTL code for *NatFrequency2_opt* design is shown in Table 1. The DFGs architecture of all modules were designed with four states, specifically, $S0$, $S1$, $S2$ and $S3$. State $S0$ for storing inputs. The RTL codes for *xOne-*, *xTwo-*, *xThree-*, *xFour-* and *nfSquare*-module were designed with up to $S3$ state.

According to Table 1, the design is a multi-cycle design, which needs several cycles to complete the algorithms' calculation. Based on the RTL code in Table 1, the SV codes were designed; of which the combinational logics for *xOne-*, *xTwo-*, *xThree-*, *xFour-*, and *nfSquare*-module by using SV HDL. The codes were compiled and simulated using QuartusII software [38].

Table 1 RTL code for *NatFrequency2_opt*-module

Module	RTL code
xOne	S0: R1 ← a; R2 ← b; state ← S1; S1: state ← S2; S2: state ← S3; S3: R12 ← (R2 − R1)/4; state ← S0; m ← R12;
xTwo	S0: R1 ← a; R2 ← b; R3 ← c; state ← S1; S1: state ← S2; S2: state ← S3; S3: R11 ← (R3 − 2*R2 + R1)/16; state ← S0; n ← R11;
xThree	S0: R1 ← a; R2 ← b; R3 ← c; R4 ← d; state ← S1; S1: state ← S2; S2: state ← S3; R6 ← 3*R2 − R1; R7 ← R4 − 3*R3; S3: R8 ← (R7 + R6)/64; state ← S0; p ← R8;
xFour	S0: R1 ← a; R2 ← b; R3 ← c; R4 ← d; R5 ← e; state ← S1; S1: state ← S2; S2: R9 ← R5 − 4*R4 + 6*R3; state ← S3; S3: R10 ← (R9 − 4*R2 + R1)/256; state ← S0; q ← R10;
nfSquare	S0: R1 ← m; R2 ← n; R3 ← p; R4 ← q; state ← S1; S1: R13 ← R1*R3 − R2*R2; state ← S2; S2: R14 ← R2*R4 − R3*R3; state ← S3; S3: R15 ← R14/R13; w2 ← R15;

**S—state*

3.2 Analysis

Table 2 shows the truth table for top-level design of *NatFrequency2* module. The output is calculated according to Eqs. (2)–(6).

After synthesising the RTL code of each module, the output waveforms were observed and monitored. Meanwhile, every input and output design was monitored at timing waveform, Fig. 4. All calculations based on algorithms of Eqs. (2)–(6), with inputs labeled as *a*, *b*, *c*, *d* and *e* while the outputs are *w2*, *m*, *n*, *p* and *q* [38].

The SV codes were simulated and the waveform window are analysed. Figure 4 shows the waveform for RTL codes from Table 1. The value of logic utilisation and the number of DSP blocks are obtained from the compilation results [38].

Submodules *m*, *n*, *p*, and *q* were designed using positive-edge clock-trigger while *nfSquare*-submodule used negative-edge clock-trigger. The *nfSquare*-module was set to use different clock cycles to show that differently triggered clock cycle could delay the output by half of a clock cycle.

The inputs are set for every four clock cycles of which 20 ps/cycle. From Fig. 4, the second and third inputs are changed after four clock cycles of the previous inputs, which are 80 ps. The inputs followed truth table in Table 2.

Aforementioned, Design 1 needs four clock-cycle to complete the concurrent modules, which are *m*-, *n*-, *p*-, and *q*-module. These modules, *m*–, *n*-, *p*-, *q*-module, must generate output and provide the input to *w*2 module. Then, another four clock cycles needs to complete the process for algorithm of Eq. (2). Every clock-cycle follows RTL code in Table 1. As shown in Fig. 4, the output of concurrent modules are produced after four clock cycles of the input. While the output of *w2* module is produced after four clock cycles of its input. Therefore, the output of *w*2 module is produced every eight clock cycles of the inputs to the *NatFrequency2_opt*-module. The control input (labeled as *done* signal) is introduced to the top-level module (*NatFrequency2_opt*-module). *w*2-module produces output when *done* signal is active or '1'.

Refer to Fig. 4, the first inputs ($a = 23$, $b = 42$, $c = 70$, $d = 11$, and $e = 23$) are fed into registers and produced submodules outputs, $m = 4.75$, $n = 0.5625$, $p = -1.5$, $q = 0.992188$. The first top-level (*NatFrequency2_opt*-module) output is 0.227362 after eight clock cycles of positive-edge triggered inputs. Here, the *done* control signal is active, or '1'. While the second inputs are set after 80 ps, such as $a = 54$, $b = 20$, $c = 35$, $d = 30$, and $e = 13$. The submodules outputs are $m = -8.5$, $n = 3.0625$, $p = -1.07813$, and $q = 0.300781$. The *NatFrequency2_opt*-module output, *w*2 is 1.122362 and the *done* signal is active, '1'. Furthermore, when given the third inputs ($a = 10$, $b = 34$, $c = 42$, $d = 50$, and $e = 19$) to the *NatFrequency2_opt* module, the submodules outputs are, $m = 6$, $n = -1$, $p = 0.25$, $q = -0.214844$. The top-level module produced output of $w2 = 0.304688$ when *done* signal is active or '1'.

It is shown that the waveform of Fig. 4 is the same when every number is rounded to two decimal points as truth table of Table 2. Therefore, RTL code for Design 1 was verified.

Table 2 The truth table for module *NatFrequency2_opt*

Input sequence	Input					Output				
	a	*b*	*c*	*d*	*e*	*m*	*n*	*p*	*q*	*w2*
1	23	42	70	11	23	4.75	0.56	−1.50	0.99	0.22
2	54	20	35	30	13	−8.50	3.06	−1.07	0.30	1.12
3	10	34	42	50	19	6.00	−1.00	0.25	−0.21	0.30

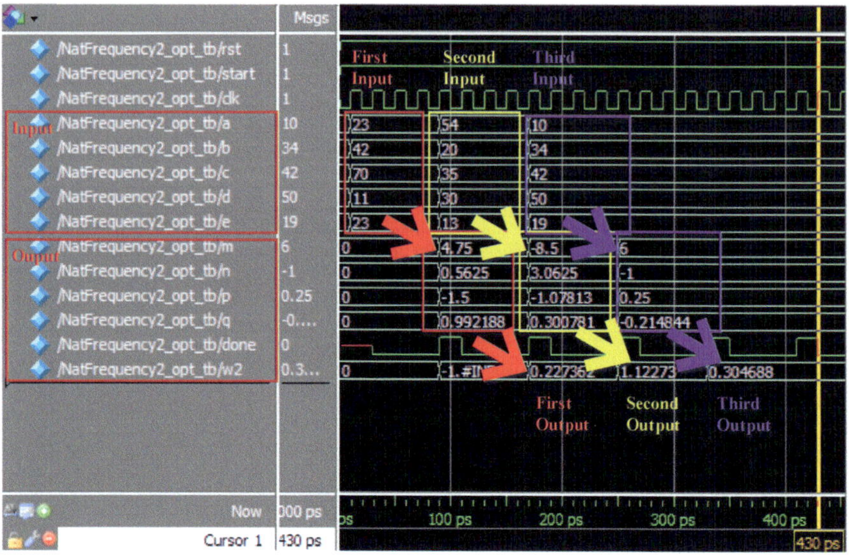

Fig. 4 Waveform of top-level module, *NatFrequency2_opt*

Table 3 System performance using Cyclone V 5CEFA731I7

Design	Cycle	Execution time for 1 complete operation	Logic utilisation	DSP block
1	Multi (8)	80 ps	1	0

The performance of each design is summarised in Table 3. The value of logic utilisation and DSP block were obtained from the QuartusII software after completed compiling the codes. A complete operation for Design 1 needs 80 ps. The result from QuartusII compilation shows total logic utilisation for Design 1 is only one and require none DSP blocks.

4 Module 2—SquareRoot

Square root of a number gives a value that could generate an output as either an integer or a decimal number. Therefore, two modules created as *A* and *B* for either integer or decimal output, respectively. Figure 5 shows the steps taken to calculate the value of square root which labeled as *NF*.

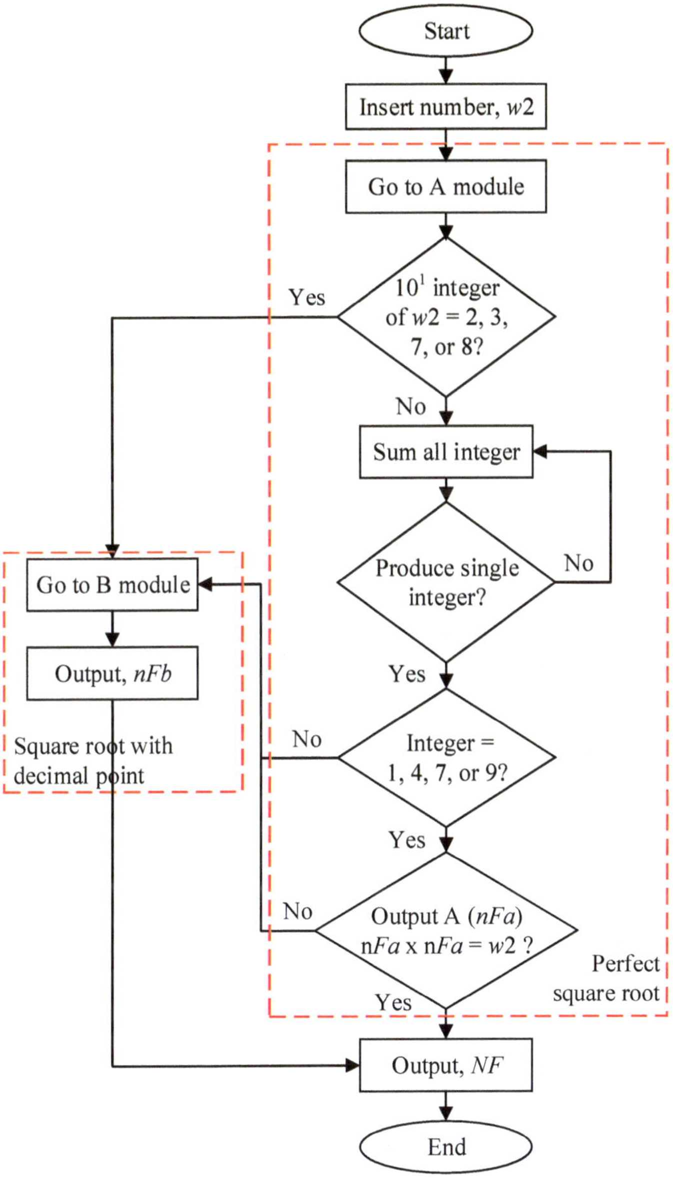

Fig. 5 Flow chart of top-level module, *SquareRoot*

4.1 Specification

The input is labeled as *w2*. There are two rules for determining the square root as an integer or decimal numbers (square root of *w*2 is an integer or a decimal number). First of all, a number (*w*2) end with 2, 3, 7, or 8 would never be a perfect square number. The second rule is, sum of all integers in number *w2* must be 1, 4, 7, or 9 [38]. A test condition was added to the steps as in Fig. 5. The test condition is to make sure the output of *A*, *nFa*, produced a squared value as same as *w*2. If the product of *nFa* squared is not equal to the input, *w*2, thus the output is a decimal number thus the module *B* can be proceeded. In depth of modules *A* and *B* are shown in flow chart as Fig. 6.

Both modules read input as 4-digit integer, $\{bcde\}$, i.e. $w2\{bcde\}$. Refer to Fig. 6a, the first and second digit of input $\{bc\}$ is read to determine the first output, $nFa[1]$. If $\{bc\}$ is equal to 1 and less than 4, output nFa [1] is 1. To determine further output $nFa[1]$ is through summarization in Fig. 4a. While the output of $nFa[0]$ is either *L* or *G* by comparing *w*2 with *m* ($nFa[1]$ multiply by 10 and add to 5). If *w*2 is greater, then $nFa[0]$ is *G*, vice versa $nFa[0]$ is *L*. *L* and *G* is a single integer. If $\{de\}$ is equal to 25, $nFa[0]$ is 5, and if $\{de\}$ is equal to 00, $nFa[0]$ is 0. The output of module *A* is $nFa\{1,0\}$.

While module *B* is to calculate the decimal point integer. First of all, the value of input, *w*2, *k*, *QA* and *R* are initialized. *QA* is the first digit of output integer, $nFb[3]$, while *k* and *R* is the coefficients to determine the remainder. *QA* is obtained through *m* from module *A* which has to minus 5 and divides by 10. The first divisor, *D*1, is equal to *QA*. The first remainder, *R*1, is equal to ($w2 - R$). Next is to determine *QB*. First of all is to set *QB* equal to 9 and calculate *D*2 and *R*1_*r*. Here, conditional loop is used. If *R*1_*r* is greater than *R*1, current *QB* value will be minus by 1. Then, D2 and R1_r recalculated until *R*1_*r* is less than or equal to *R*1. The same steps applied to find the value of *QC* and *QD*. Figure 7 shows the position of each parameter described previously.

The output of module *B* is *nFb*. This study support integer of single digit with up to three decimal places [38]. The integer is labeled as QA and the decimal points are labeled as *QB*, *QC* and *QD*. Therefore, the top-level design for Module *SquareRoot* as Fig. 8.

There are an input (*w*2) and an output (*NF*) for module *SquareRoot*. The submodules of module SquareRoot are labeled as *A_PerfectSquare* and *B_SquareRootDecimal* of which refer to module *A* and module *B*, respectively. The outputs of module *A_PerfectSquare* are *nFa*, *k* and *QA*. While the output for module *B_SquareRootDecimal* is *nFb*. The SV code for top-level design of *SquareRoot*-module is shown in Fig. 9.

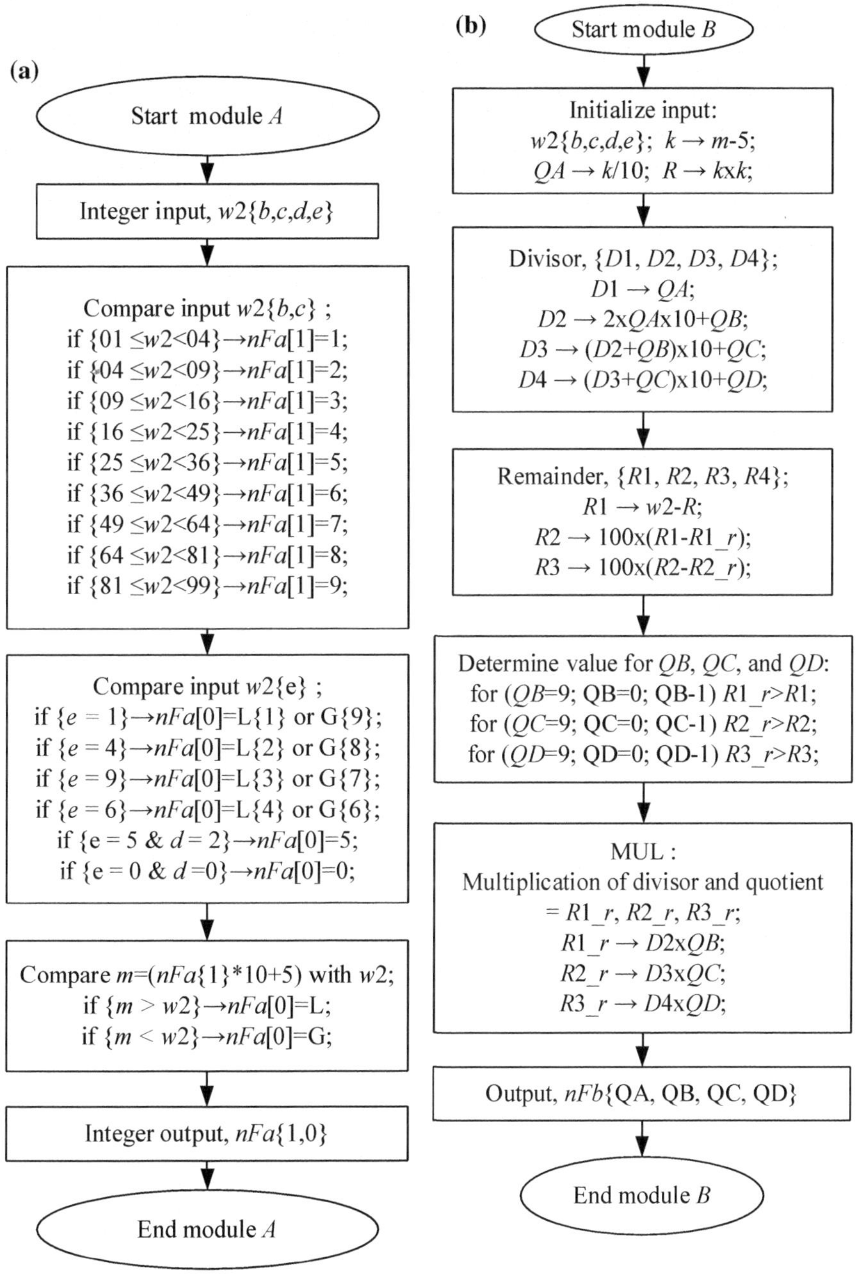

Fig. 6 Flow chart of submodules: **a** module A: perfect square root, **b** module B: square root with decimal point

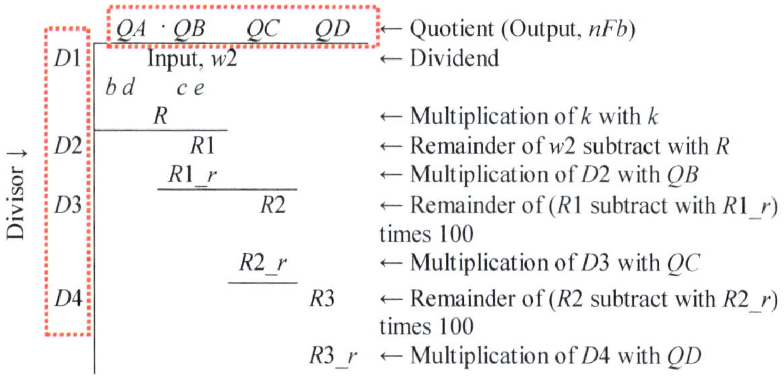

Fig. 7 The position of the input and output for *B*-module

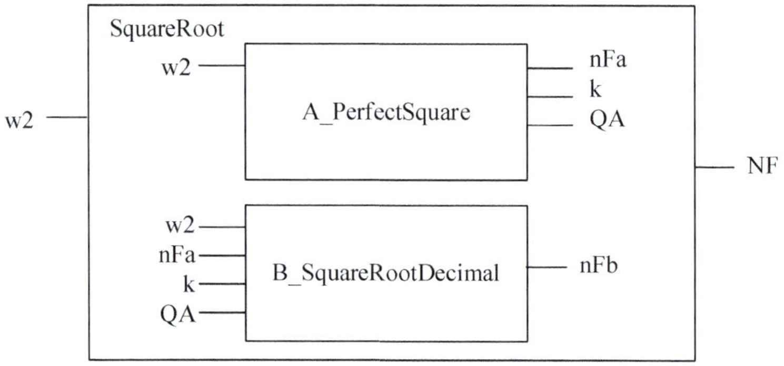

Fig. 8 Top-level design of *SquareRoot*-module

4.2 Verification

The design is a single cycle design. The truth table is shown in Table 4. The input for *SquareRoot*-module is *w2* and the output is *w*.

4.3 Analysis

The SV codes were compiled and simulated using QuartusII software. The waveform was analysed [38]. Figure 10 shows the waveform. Figure 10 shows the outputs (*NF*) provided by the inputs (*w2*) 5371, 7005, 1149, 7000, 9009, 6594 and 1055 are 73.28, 83.69, 33.89, 83.66, 94.91, 81.2, and 32.48, respectively. The first, fourth, fifth, and seventh outputs are valid after 11 ps, the second and third outputs are valid after 9 ps, and the sixth output is valid after 7 ps. However, the output is valid mostly at 11 ps, as 7 ps is only 4 ps quicker than 11 ps. The input and output follow the

```
module SquareRoot (
input rst, CLOCK_50, start,
input var real w2,
output real NF
);

real QA, nFb, nFa;
real k;

A_PerfectSquare  u1 (.*); // (.nFa(nFa), .k(k), .QA(QA), .w2(w2), .start(start),
.CLOCK_50(CLOCK_50), .rst(rst) );
B_SquareRootDecimal u2 (.*); //(.nFa(nFa), .nFb(NF), .k(k), .QA(QA), .w2(w2),
.start(start), .CLOCK_50(CLOCK_50), .rst(rst) );

always_ff@(posedge CLOCK_50)
begin
NF = nFb;
end

endmodule
```

Fig. 9 SV Code for top-level design of module-*SquareRoot*

Table 4 Truth table for *SquareRoot*-module

w2	w
5371	73.28
7005	83.69
1149	33.89
7000	83.66
9009	94.91
6594	81.20
1055	32.48

truth table of Table 4. Therefore, the RTL code of designing the module *SquareRoot* is verified. The performance of module *SquareRoot* is summarised in Table 5. A complete operation for module *SquareRoot* needs 11 ps. The result from QuartusII compilation shows total logic element utilisation for module *SquareRoot* which is 2432. Total DSP blocks are 16 blocks for this module *SquareRoot*.

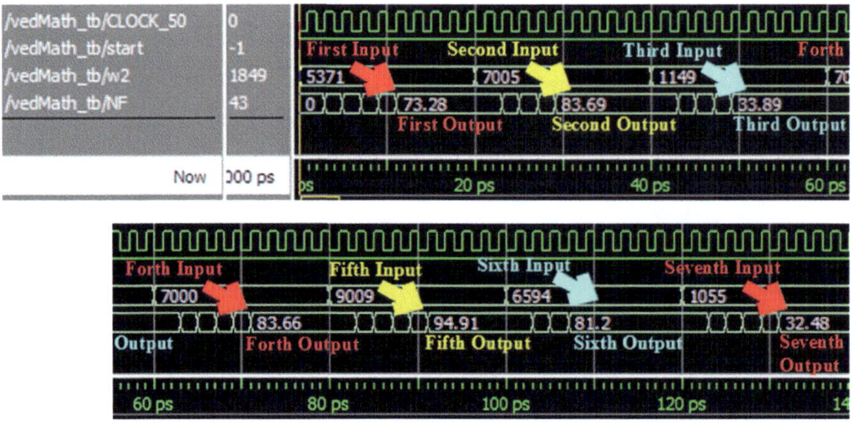

Fig. 10 Waveform of *SquareRoot*-module

Table 5 System performance using Cyclone V 5CEFA731I7 FPGA board

Cycle	Execution time for 1 complete operation	Logic element	DSP blocks
Single (1)	11 ps	2432	16

5 Conclusion

Based on the table, it is proven that the design of square root function ($\surd$) into hardware design is a success. The design is verified and able to be used for further implementation using an FPGA platform which is needed to further extracting signals' features based on SODS technique, especially algorithm ω. It has been found that module designed for $\surd$ utilised of 2432 logic elements and 16 DSP blocks.

Acknowledgements The authors would like to express their appreciation to the Ministry of Higher Education and Universiti Teknologi Malaysia for supporting and funding this study under Research University Grant No. Q.J130000.3001.01M13.

References

1. Barlow, J.: A systematic review of the benefits of home telecare for frail elderly people and those with long-term conditions. J. Telemed. Telecare **13**(4), 172–179 (2007)
2. Barranco, F., Tomasi, M., Díaz, J., et al.: Pipelined architecture for real-time cost-optimized extraction of visual primitives based on FPGAs. Dig. Signal Proc. **23**(2), 675–688 (2013)
3. Hu, W., Lin, C.C., Shyu, L.Y. et al.: An implementation of a real-time and parallel processing ECG features extraction algorithm in a field programmable gate array (FPGA). Comput Cardiol, pp. 38–41 (2011)

4. Fang, W.C., Huang, H.C., Tseng, S.Y.: Design of heart rate variability processor for portable 3-lead ECG monitoring system-on-chip. Expert Syst. Appl. **40**(5), 1491–1504 (2013)
5. Ballesteros, D.M., Marcela, D., Andrés, M.: FPGA compression of ECG signals by using modified convolution scheme of the discrete wavelet transform. Ingeniare Rev. Chil. Ing. **20**(1), 8–16 (2012)
6. Li, C., Balla-Arabé, Yang F.: Embedded multi-spectral image processing for real-time medical application. J. Syst. Archit. **64**, 26–36 (2016)
7. Chou, C., Fang, W., Huang, H.: A novel wireless biomedical monitoring system with dedicated FPGA-based ECG processor. In: IEEE 16th International Symposium on Consumer Electronics, pp. 2–5 (2012)
8. Chatterjee, H.K., Gupta, R., Mitra, M.: Real time p and t wave detection from ECG using FPGA. Proc. Technol. **4**, 840–844 (2012)
9. Chowdhury, S.R.: Field programmable gate array based fuzzy neural signal processing system for differential diagnosis of QRS complex tachycardia and tachyarrhythmia in noisy ECG signals. J. Med. Syst. **36**(2), 765–775 (2012)
10. Cvikl, M., Zemva, A.: FPGA-oriented HW/SW implementation of ECG beat detection and classification algorithm. Dig. Signal Proc. **20**(1), 238–248 (2010)
11. Sun, Y., Cheng, A.C.: Machine learning On-A-Chip: a high-performance low-power reusable neuron architecture for artificial neural networks in ECG classifications. Comput. Biol. Med. **42**(7), 751–757 (2012)
12. Lin, C.H., Lin, G.W.: FPGA implementation of fractal patterns classifier for multiple cardiac arrhythmias detection. J. Biomed. Sci. Eng. **5**(3), 120–132 (2012)
13. Bahoura, M., Ezzaidi, H.: FPGA-implementation of parallel and sequential architectures for adaptive noise cancelation. Circuits Syst. Signal Proc. **30**(6), 1521–1548 (2011)
14. Jindapetch, N., Chewae, S., Phukpattaranont, P.: FPGA implementations of an adaline adaptive filter for power-line noise cancellation in surface electromyography signals. Measurement **45**(3), 405–414 (2012)
15. Monmasson, E., Cirstea, M.N.: FPGA design methodology for industrial control systems—a review. IEEE Trans. Ind. Electron. **54**(4), 1824–1842 (2007)
16. Huang, C., Ye, S., Chen, H., et al.: A novel method for detection of the transition between atrial fibrillation and sinus rhythm. IEEE Trans. Biomed. Eng. **58**(4), 1113–1119 (2011)
17. Lee, J., Reyes, B.A., McManus, D.D., et al.: Atrial fibrillation detection using an iphone 4s. IEEE Trans. Biomed. Eng. **60**(1), 203–206 (2013)
18. Dash, S., Chon, K.H., Lu, S., et al.: Automatic real time detection of atrial fibrillation. Ann. Biomed. Eng. **37**(9), 1701–1709 (2009)
19. Babaeizadeh, S., Gregg, R.E., Helfenbein, E.D., et al.: Improvements in atrial fibrillation detection for real-time monitoring. J. Electrocardiol. **42**(6), 522–526 (2009)
20. Jiang, K., Huang, C., Ye, S., et al.: High accuracy in automatic detection of atrial fibrillation for holter monitoring. J. Zhejiang Univ. Sci. B **13**(9), 751–756 (2012)
21. Zidelmal, Z., Amirou, A., Ould-Abdeslam, D., et al.: ECG beat classification using a cost sensitive classifier. Comput. Methods Progr. Biomed. **111**(3), 570–577 (2013)
22. Daqrouq, K., Alkhateeb, A., Ajour, M.N., et al.: Neural network and wavelet average framing percentage energy for atrial fibrillation classification. Comput. Methods Progr. Biomed. **113**(3), 919–926 (2014)
23. Lee, J., Nam, Y., McManus, D.D., et al.: Time-varying coherence function for atrial fibrillation detection. IEEE Trans. Biomed. Eng. **60**(10), 2783–2793 (2013)
24. Oresko, J.J., Jin, Z., Cheng, J., et al.: A Wearable smartphone-based platform for real-time cardiovascular disease detection via electrocardiogram processing. IEEE Trans. Inf. Technol. Biomed. **14**(3), 734–740 (2010)
25. Abdul-Kadir, N.A., Safri, N.M., Othman, M.A.: Dynamic ECG features for atrial fibrillation recognition. Comput. Methods Progr. Biomed. **136**, 143–150 (2016)
26. Abdul-Kadir, N.A., Safri, N.M., Othman, M.A.: Classification of paroxysmal atrial fibrillation using second order system. J. Teknol. **67**(3), 57–64 (2014)

27. Othman, M.A., Safri, N.M., Ghani, I.A., et al.: A new semantic mining approach for detecting ventricular tachycardia and ventricular fibrillation. Biomed. Signal Proc. Control **8**(2), 222–227 (2013)
28. Othman, M.A., Safri, N.M.: Characterization of ventricular arrhythmias using a semantic mining algorithm. J. Mech. Med. Biol. **12**(03), 1250049 (2012)
29. Quiroz-Juarez, M.A., Vazquez-Medina, R., Ryzhii, E., et al.: Quasiperiodicity route to chaos in cardiac conduction model. Commun. Nonlinear Sci. Numer. Simulat. **42**, 370–378 (2017)
30. Kaplan, B.Z., Gabay, I., Sarafian, G., et al.: Biological applications of the "Filtered" Van der Pol oscillator. J. Frankin Inst. **345**(3), 226–232 (2008)
31. McSharry, P.E., Clifford, G.D., Tarassenko, L., et al.: A dynamical model for generating synthetic electrocardiogram signals. IEEE Trans. Biomed. Eng. **50**(3), 289–294 (2003)
32. Shekofteh, Y., Jafari, S., Sprott, J.C., et al.: A gaussian mixture model based cost function for parameter estimation of chaotic biological systems. Commun. Nonlinear Sci. Numer. Simulat. **20**(2), 469–481 (2015)
33. Abdul-Kadir, N.A., Safri, N.M., Othman, M.A.: ASIC design of natural frequency of ECG signal for atrial fibrillation detection module using high-level synthesis approach. J. Teknol. **74**(6), 103–108 (2015)
34. Gajski, D., Ramachandran, L.: Introduction to high-level synthesis. IEEE Des. Test Comput. **11**(4), 44–54 (1994)
35. Corvino, D., Epicoco, L., Ferrandi, F. et al.: Automatic VHDL restructuring for RTL synthesis optimization and testability improvement. In: Proceedings International Conference on Computer Design: VLSI in computers and processors, pp. 436–441 (1998)
36. Rose, C.W., Albarran, M.: Modeling and design description of hierarchical hardware/software systems. In: Proceedings of the 12th Design Automation Conference 421–430 (1975)
37. Sprott, J.C.: Some simple chaotic jerk functions. Am. J. Phys. **65**(6), 537–543 (1997)
38. Abdul-Kadir, N.A.: Human heart oscillatory behaviour for atrial fibrillation detection based on second order dynamic system. Ph.D. Thesis, Universiti Teknologi Malaysia (2017)

Part III
Cardiovascular Biomechanics

Mock Circulatory Loop for Cardiovascular Assist Device Testing

A. Z. M. Khudzari, M. R. A. Kadir, K. Osman and A. H. M. Hudzari

Abstract Medical device prototype developed for use on human has to be tested in various aspect; from numerical simulation, laboratory testing, testing on animals and finally human trial, increasing scope in term of complexity and number of participants. This chapter reviews one of the testing methods used by researchers in cardiovascular engineering in order to verify the performance of the prototype developed as per intended which is the in vitro test using the mock circulatory loop (MCL). MCL has many types and functions, depending on the medical device prototype intent of use. From simple hydraulic loop to a more complex loop combining hydraulic and computer interface. This chapter consists of condition of heart failure (HF), treatment modality, cardiovascular assist device types, MCL types, components in MCL, current literature and lastly, an example of an MCL is presented.

1 Introduction

1.1 Cardiovascular Disease

Heart failure has long been the leading factor of mortality in the world, accounting for almost 50% of deaths within the cardiovascular diseases (CVD) bracket. The World Heart Organization (WHO) reported that in 2004, CVD contributed to almost 30% of death of estimated worldwide death figures (58 million), with almost 32% of deaths being women, followed by men at almost 27%. The overall death reported worldwide

A. Z. M. Khudzari (✉) · K. Osman · A. H. M. Hudzari
School of Biomedical Engineering and Health Sciences, Faculty of Engineering, Universiti Teknologi Malaysia, 81310 Skudai, Johor, Malaysia
e-mail: zahran@biomedical.utm.my

K. Osman · A. H. M. Hudzari
IJN-UTM Cardiovascular Engineering Centre, Institute of Human Centered Engineering, Universiti Teknologi Malaysia, 81310 Skudai, Johor, Malaysia

M. R. A. Kadir
Faculty of Mechanical Engineering, Universiti Teknologi Malaysia, 81310 Skudai, Johor, Malaysia

D. E. O. Dewi et al. (eds.), *Cardiovascular Engineering*, Series in BioEngineering,
https://doi.org/10.1007/978-981-10-8405-8_8

due to cardiovascular disease was a little over 17 million and is projected to increase almost 30% to 23.4 million by 2030 [1], this further consolidated by Benjamin et al. [2] report stating the most common primary cause of death worldwide in 2013 was contributed by CVD. There are approximately 20% of the world's population developing heart failure [3, 4]. In the UK, it was reported around 650,000 patients suffering from definite heart failure (HF), and another 225,000 patients with possible HF [5]. In United States of America (USA), there are about 7 million people with HF [6], and the number is expected to raise up further. In Southeast Asia, around 9 million people are suffering from HF, out of approximately 648 million population, of which 6.7% of prevalence is contributed from Malaysia [7] i.e. 603,000 patients and this is translated into 2% of Malaysian population.

In some cases, heart failure onset is caused by an existing disease such as diabetes or hypertension, but in other cases the cause is unknown (idiopathic cardiomyopathy). Various aetiology studies have discovered that coronary artery disease (CAD), hypertension, valve dysfunction, cardiomyopathy (e.g. dilated, hypertrophic and alcoholic), cardiac arrhythmias/conduction disturbance, pericardial disease, or viral infection (rheumatic fever or pregnancy) are causing heart failure in general population [8–11]. Regardless of aetiology, heart failure treatment involves variety of modalities—lifestyle adjustment, pharmacological therapies, surgical therapy, heart assist device, and heart transplantation as the long term definitive treatment modality [12].

Hence, the demand in treating HF is progressively increased due to the increasing number of HF patients with the heart transplant being the gold standard treatment. Nevertheless, there is a significant gap between the available heart grafts and the number of registered HF candidates who are still on the waiting list [13]. There was a total of 4477 heart transplants performed in 2013 from 252 centres which is still inadequate to meet the demand of 15,000 patients currently on the waiting list for transplant. Number of donor increased consistently up until 2010, especially in Europe, and has remained stable over the last few years [7]. Other than that, some heart failure patients are contra-indicated with the transplant as well as refractory to drug treatment plus several other factors. In dealing with that restriction, one of the favourable alternative treatment is the utilization of mechanical heart assist device. Theoretically, the number of potential HF patients that would benefit from mechanical heart assist system implantation would be around 0.05–0.08% of the population [6] (USA: 150,000–250,000 patients, South East Asia: 324,000–540,000 patients) as illustrated in Fig. 1.

1.2 Cardiovascular Assist Device (Mechanical Circulatory Support)

The history of mechanical circulatory support (MCS) i.e. heart assist device can be traced back since the end of 1800s. Etienne-Jules Marey published a description and illustrations of an artificial heart back in 1881 [14]. In the middle of twentieth century,

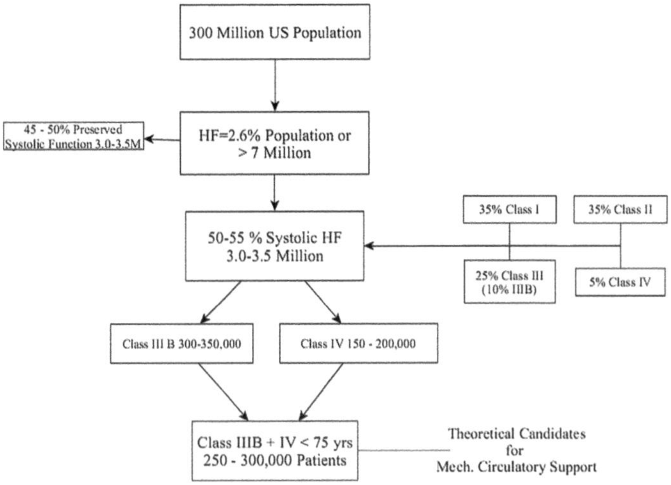

Fig. 1 Theoretical number of patients eligible for mechanical circulatory support in the USA [6]

the effort to create an independent artificial heart has intensified with the starting of a project by the National Heart Institute in the USA to develop artificial heart by awarding research contracts in the early 1960s, which spurred R&D activities leading to successful usage of mechanical heart assist device to support patient through waiting period for transplantation (as a bridge to transplantation—BTT device) in 1978. DeBakey group in particular, have used mechanical heart assist device since its conception in 1966 [15]. Since then, there are various types of MCS that were produced and implanted into patients.

MCS can be categorised by several classifications; (1) the position of the device in blood circulation track (either series or parallel), (2) the type of blood pump whether it is rotary or displacement, and (3) duration of devices being used (acute, medium, or long term) [16]. In the following subsection, several MCS devices were discussed that necessitate the development of MCL for its in vitro testing platform.

1.2.1 Left Ventricular Assist Device (LVAD)

LVAD is commonly being used for patients suffering end-stage HF. It is an electro-mechanical device used to partly substitute the function of a failing heart, assisting to circulate blood to the body. This device is implanted by bypassing the left ventricle chamber of the heart where the input cannula is connected to the apex of left ventricle [17]. Short term use of the LVAD is for bridge to transplantation (BTT) which is when the device is implanted as a temporary measure until the heart donor is found. Long term use however can either be bridge to recovery (BTR) or destination therapy (DT). BTR is when the LVAD is used in sustaining the normal cardiac output for patients in heart failure while promoting the cardiac recovery effect. DT is when the device is considered as an ultimate treatment if the patient is contra-indicated to transplant and refractory to drugs [18].

LVAD has been implemented as mechanical heart assist device for decades and has undergone several improvisations throughout generations. The first generation of LVAD implements the pulsatile displacement configuration in which the pump is made of a positive displacement diaphragm accompanied by pneumatic or electrical driveline to generate the pulsatile flow. Due to the excessive size and extra moving parts that leads to high rate of thrombosis, it has been replaced by continuous rotary configuration of blood pump or as known as second generation of LVAD.

The LVAD in this generation utilized an axial and centrifugal rotating impeller in a unidirectional blood movement along the axis of rotation. The drawback of this design configuration is that the miniature pump size requires high speed of rotation (especially for axial type impeller configuration) in order to drive the blood flow which consequently result in hemolysis (damage of red blood cells) due heat generation of the contact bearing during impeller rotation [18]. This has been addressed by applying the different bearing system on the same impeller configuration and implementation of levitation concept to suspend the impeller without a driving shaft or bearing contacting the blood thus reducing likelihood of blood damage. These improvements became the third generation LVAD.

LVAD as a viable HF treatment are constantly improved too continuously reduce the risk of complications such as hemodynamic response, occurrence haemolysis and thrombosis. Prior to in vivo testing, LVAD are usually tested through in vitro method to gauge its hemodynamic response. Mock Circulatory Loop (MCL) usually being setup to provide the testing platform in analysing the hemodynamic effect of the design.

1.2.2 Mechanical Heart Valves (MHV)

Other than LVAD, mechanical heart valves are also examples of cardiovascular devices that have been progressively investigated either numerically or experimentally. For those patients who suffered from valvular disease, MHV can be best used to replace the malfunctioned native heart valve. Nevertheless, the hemodynamic effect of fluid passing through the MHV still being rigorously investigated since the unusual hemodynamic flow may affect the cardiac output as well as durability of the device itself. Additionally, the anticoagulation prescription is usually being administered during post-surgery of MHV replacement. If the durability of MHV to withstand blood clotting mechanism is not properly examined, there would be high risk of the device to malfunction in a short period of time.

An early design of MHV used the ball-and-cage valve structure in which the ball was placed inside of the metal cage. For this design, in preventing the backflow of blood, the valve orifice was blocked by the ball during diastolic phase. During systolic phase, the ball was pushed downstream (within the area of metal cage) by high blood pressure and the orifice was opened to allow the flow of blood. Another design that was similar to ball-and-cage valve is caged-disc valve wherein the occluder was no longer a ball, but a disc. Due to some major reported drawbacks of these designs, a tilting disc and bileaflet configurations of MHV were introduced. For the tilting disc design, the orifice is totally occluded by a disc and the one-way flip movement of the disc was maintained by a retaining strut. The bileaflet designs are the most recent MHV as the valve is made of two semi-circular occluder [19]. This bileaflet design of MHV offers a better blood flow with much lower turbulence flow profile as well as reduces the possibility of backflow.

The most important consideration when designing the MHV is the hemodynamic effect of blood flow passing through it. The formation of vortices flow profile can lead to the production of bubble cavitation which may cause the damage of blood cell and most crucially affect the durability of the valve [20]. MCL have been developed specifically for studying the flow through MHV. For ease of visualization of the hemodynamic effect, the MCL with Particle Image Velocimetry (PIV) setup can be utilized.

1.2.3 Stents

Stents have been investigated for ages as the number of patients with blockage blood vessels related disease elevated annually. This device is a tiny tube that intended to be inserted into blockage blood vessels resulted from plaque formation to keep it open and restore the normal flow of blood or any other fluid depending on where it is placed. Despite computational study of stents, the effect of stent implantation still presents numerous uncertainties. Similar to the other devices that have been discussed in the previous subsections, the most vital effect of device implantation lies on hemodynamic of the blood flow specifically the interaction with blood cells. There are several types of stents have been developed which are Dual Therapy Stent (DTS); most recent type of stent, Drug Eluting Stent (DES), Bare Metal Stent (BMS), Bioresorbable Vascular Scaffold (degradable version of DES) and Bioengineered Stent. Generally, the difference of these types of stents is mainly on its materials and functionality.

In order to examine the hemodynamic effect of blood flow in stented region, an in vitro approach is crucial before progressing to clinical application. MCL development is useful for in vitro testing platform of stent devices as it offers the simulated human condition and the applicability of the experimental setup to provide a good interaction of blood-like fluid with the stent itself. The hemodynamic parameters outcome in the stented region can be obtained easily as the flexibility of MCL setup can be adjusted according to researcher's intention. In vitro investigation using MCL provides a good platform to study and improve not only stents but many other cardiovascular devices as stent developers commonly focused more on structural performance of stent as well as the effect of stent material degradation rather than bulk haemodynamic performance and fluid flow visualisation.

2 Mock Circulatory Loop (MCL) Recent Literature

In order to test the mechanical response of those devices, an essential experimental rig or likely to be called as Mock Circulatory Loop (MCL) is the most encouraging method being used by many researchers in which it is capable of simulating physiological condition of native heart. Moreover, this loop resembles the cardiovascular system of human heart where it can consist whether either systemic and pulmonary circulation, or only one of these subsystems. It also offers the user to emulate various heart disease condition that suitable for the testing of specific heart assist device [3]. The data obtained from this testing helps in refining and validating the design of tested devices.

The underlying reason in developing MCL is to provide the platform for testing procedure of cardiovascular assist device. Even though there were some differences of components involved in developing MCL by different researchers, but there were also several common criteria that must be met in MCL development:

1. MCL must consistently emulate the native heart physiology with varying degrees of heart conditions ranging from Normal to Heart Failure (HF).
2. It must be adherence to Frank-Starling mechanism i.e. cardiac output corresponds to preload of blood from venous return.
3. MCL must be flexible in providing an appropriate cannulation site to accommodate various kinds of cardiac assist devices.

2.1 Review on Human Mock Circulatory Loop (MCL)

MCL has been utilized about 70 years ago when the very first MCL ever developed in 1950s called pulse duplicator was successful in simulating appropriate haemodynamic activity of heart valve prostheses. This MCL was developed by McMillan et al. (1952) which later modified in 1955 intended to examine the flow past a heart valve [21, 22]. The main highlight about this design was that, the real human heart was explanted to the mechanical pulse duplicator as illustrated in Fig. 2.

The inspiration of their MCL design has attracted another research group to develop a mechanical pulse duplicator with the purpose of heart valve's flow visualization. In this design, a linkage system of cam and piston (refer to Fig. 4) were used wherein it allows the flexible simulation of pressure waveforms of ventricle [23]. The development of pulse duplicator was improved from time to time by many other researchers as it has undergone the several enhancements of varying degree of complexity, from steady flow circuit to more refined and precise ones [24–28].

The pulse duplicator design was further utilized by Cornhill et al. (1975) with three-element Windkessel model consist of proximal resistance, peripheral resistance and arterial compliance. In order to investigate the effectiveness of pulse duplicator system and Oxford aortic-pulmonary heart valve prostheses, the input impedance

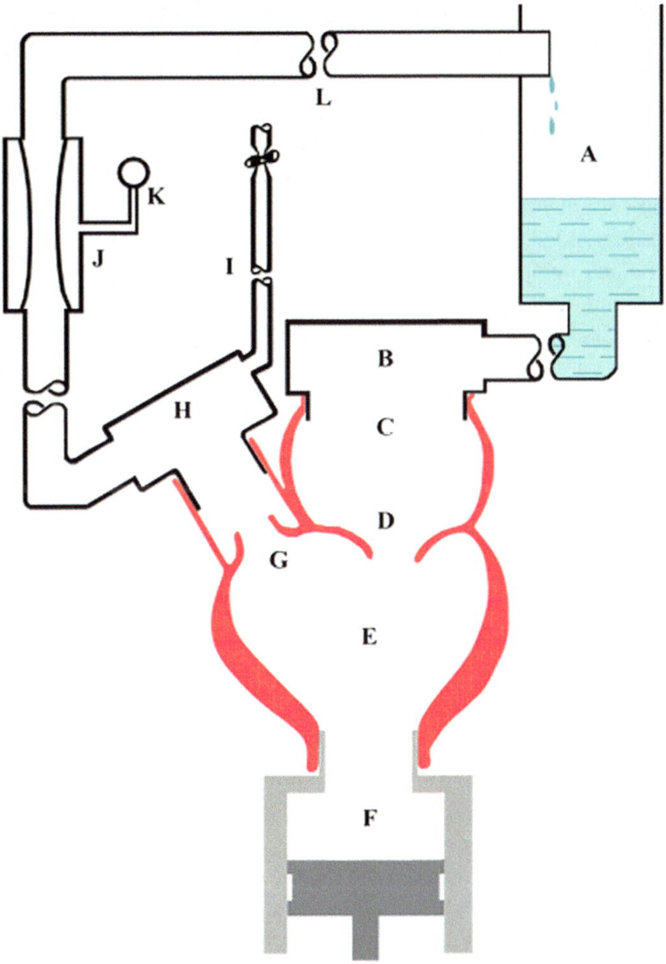

Fig. 2 The MCL developed by McMillan et al. [22] called pulse duplicator used for evaluation of heart valve visualization explanted of native human heart. **a** Reservoir cum left atrium, **b** atrial viewing chamber, **c** left atrium, **d** mitral valve, **e** ventricle, **f** piston pump, **g** aortic valve, **h** aortic valve viewing chamber, **i** aortic tube, **j** peripheral resistance simulating chamber, **k** peripheral resistance manometer, **l** site for rotameter [23]

analysis was applied which resulting in physiologically acceptable of flow and pressure waveforms for the characteristic of mechanical heart valves [24]. Recently, there was an example of pulse duplicator designed for heart valve testing wherein it is constructed to be in a closed incubator.

After the successful of heart valve investigation using MCL, an early model of Total Artificial Heart (TAH) was proposed to be tested on MCL as well. It was done in a rather crude MCL that was capable of simulating both systemic and pulmonary circulation of native heart. This MCL was constructed with two water columns distal to both ventricles (left and right) to provide an afterload simulation specifically 60 mmHg for aorta and 20 mmHg for pulmonary artery. Nevertheless, there was no compliance chamber to imitate aortic pressure waveforms. At the same time, there was no peripheral resistance element implemented even though the author mentioned about the increase of resistance with flow [29].

Another MCL developed for early TAH evaluation was done by Donovan [30] in which the design consisted of systemic and pulmonary circulation. An additional component was added where the pulmonary and systemic resistances were controlled by the bellow-operated valves which was set automatically. This component was added to simulate the baroreceptor response in systemic arteries and passive pulmonary resistance response in pulmonary arteries. This evaluation of Kwan-Gett TAH showed a comparable result to in vivo outcome from a calf [30]. Donovan's MCL became well known which attracted further attention to adapt their design for in vitro TAH examination.

In early 1970s, a research team studying biventricular assist device (BiVAD) adopted the MCL method to test their device. Rosenberg's team developed a MCL for BiVAD and TAH experiment started in 1971 and has been improved 10 years later. The design utilized analogue computer simulation for the MCL. The systemic and pulmonary circulation were made of capacitance, resistive and inertance elements together with reservoir preload chamber [31]. Another research team called Orime's group developed a similar looking MCL for testing Baylor TAH before further in vivo testing. This group work has shown results supporting the stability of the TAH system as well as good biocompatibility [32].

Cassot et al. [33] carried out an investigation on flow visualization in anatomically correct Left Ventricle (LV) using ultrasonic Doppler Velocimeter wherein it is capable to measure a bidirectional velocity of the flow. In this experiment, they have shown the successful results in term of Fourier and impedance analysis and good agreement with in vivo data collected from canine [33].

A deformable silicon modelled was then started to be implemented in MCL development as it offered better representation of flow dynamics of LV cardiac cycles. Gregory et al. [34] developed MCL for the placement of LVAD inflow cannula experiment in the dilated LV. Although MCL is widely being used for VADs or TAHs testing procedure, it also used as a teaching tool. Some interactive components were assembled in order to develop a simple yet comprehensible design of MCL (refer Fig. 3). In the design, LV was represented by syringe controlled by a motor to push the syringe back and forth. It also used a custom made mitral and aortic valves. The atrium and aorta were imitated by a glass cylinder and a silicon rubber respectively. To apply the Frank-Starling mechanism, an external chamber was placed around the syringe where it is made of surgical glove. A notable weakness in this design is that, there was a possibility of impedance mismatch from the length of silicon aorta [35].

In 2006, a complete MCL of systemic and pulmonary circulation was developed by Liu et al. [36]. It was constructed to test an axial rotary blood pump (RBP) in which it consisted three-element Windkessel model comprises of a venous compliance, an arterial compliance and a peripheral resistance element. A highly accurate result was obtained where the aortic pressure waveforms was in an excellent agreement with

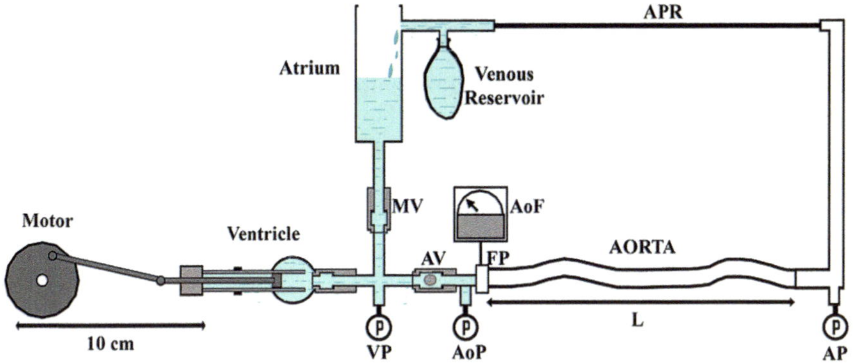

Fig. 3 Vertical and horizontal arrangement of Mock circulatory loop (MCL) intended for teaching purposes [35] (VP: Ventricular pressure, AoP: aortic pressure, AP: arterial pressure, MV: mitral valve, AV: aortic valve, APR: arterial resistance, AoF: aortic flowmeter)

physiological waveforms [36]. A year later, Wu's group adopted their MCL design for the determination the effectiveness of LVAD physiological control system [37].

Although the physical hydraulic simulation of MCLs have proven an excellent platform for medical device testing, it was quite expensive to develop. Numerical simulation was then introduced as an alternative as it offered a relatively low cost with flexible and accurate outcome but in return the physical devices such as LVAD and heart valve prostheses cannot be tested. To address this problem, a hybrid approach of MCL was established. Ferrari's group was one of the team that constructed the hybrid model of numerical-physical model of MCL [37–42].

Timms et al. [43] constructed a compact mock circulatory system with the dimension of 600 mm × 600 mm × 600 mm. It consisted of both systemic and pulmonary circulation applicable for VAD and BiVAD examination at the same time, it can simulate congenital heart disease and structural failure [43]. The design is then upgraded to its final version in 2011 [4]. In the final version of that particular MCL, various kinds of cardiovascular conditions were capable to be simulated from normal (resting) to both left and right HF and many other cardiovascular complications such as myocardial infarction (MI), structural failure like aortic/mitral valve regurgitation and atrial/ventricular septal defect. To make it more advance, Timms and his teammates developed the MCL based on a five-element Windkessel comprises of arterial compliance, characteristic resistance, inertial components, peripheral resistance and venous compliance element. The simulation of vasculature compliance was represented by using multi-chamber Windkessel wherein it was controlled by a solenoid valve across connecting pipes. The adherence of Frank-Starling we also incorporated in the design.

Fairly recently, Vilchez-Monge et al. [44] constructed a simple and inexpensive MCL using easily obtain materials. This MCL was intended to be used for preliminary VAD testing in which it only consist of systemic circulation of cardiovascular system where it emulated the physiological condition found on an average healthy adult person. This design contains a passive left atrium and pulsatile left ventricle accompanied with arterial and venous, compliances and resistance. The Graphic User Interface (GUI) was implemented to control the parameters that have a direct implication of MCL [44].

Generally, Mock Circulatory Loop (MCL) aims to mimic native heart function as accurate as possible, so that the devices can be tested realistically. Most researchers aim to reproduce cardiac function values such as cardiac output (CO), heart rate (HR), compliance, resistance etc.

2.2 *MCL Physiological Parameters*

In order to mimic the physiological flow characteristics of native human heart, there are several variables need to be simulated, e.g. flow rates, aortic pressure, heart chamber pressure fluctuations, both systemic and pulmonary vascular resistances, arterial and venous compliances. All these parameters must be set as accurate as possible in MCL development.

2.2.1 Cardiac Output (CO)

The most vital parameter need to be simulated correctly is the cardiac output (CO) as it represents specific heart conditions. The reduction of CO is among the characteristics of HF. Normally, the CO found on a healthy adult is 5.0 L/min. For patients suffering end-stage HF possibly due to dilated cardiomyopathy is usually experienced relatively depressed CO value of 3.0 L/min, while recovering condition at about 4.0 L/min.

In physical MCL design, the contractility level of LV chamber must be designed to be flexibly adjusted to get a desired CO values. In another words, increasing contractility of LV chamber must be able to induce increase in stroke volume thus the CO value itself. In addition, adjustable peripheral resistance element is also capable in simulating the CO fluctuation. Apart from that, by using cam-controlled piston illustrated in Fig. 4, it can be an alternative technique in controlling the amount of CO. The amount of CO ejection is dependent on the stroke length while the velocity of rotational determines the value of Heart Rate (HR) [23]. Other MCL configuration may choose to increase CO value by increasing piston stoke length, or by increasing pneumatic pressure on silicone chamber.

2.2.2 Heart Rate (HR)

The mechanism of HR in MCLs is often controlled by pressure activation mode. For instance, the rotational speed of piston is easily being increased or decreased in order to simulate the HR mechanism. The normal ratio of systole to diastole of HR

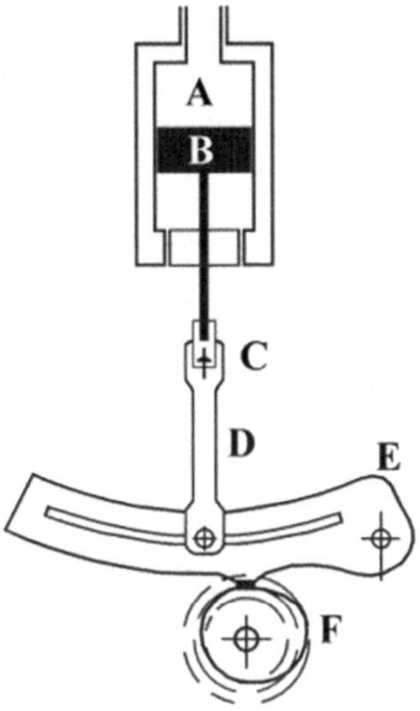

Fig. 4 Configuration of cam-controlled piston. **a** Pump cylinder, **b** piston, **c** universal joint, **d** piston drive shaft, **e** rocker-arm, **f** cam [23]

is selected to be 40/60 even though there are several researchers tend to use slightly different values. As mentioned by Levick [45], the normal physiological value of systolic to diastolic ratio is in the range around 34/66 to 39/61. The examples of HR value collected from many researchers can be found in Table 1.

2.2.3 Stroke Volume (SV)

Stroke Volume (SV) is the amount of blood volume ejected from left ventricle chamber during systolic phase. SV is much close related to the Ejection Fraction (EF) in which EF can be defined as the ratio between SV and end-diastolic blood volume (LVVed). The following equations represents the relation between SV with CO, HR, EF and LVVed:

$$SV = \frac{CO}{HR} \tag{2.1}$$

$$EF = \frac{SV}{LVVed} \tag{2.2}$$

LVVed is the volume of blood in LV right before it is ejected during systole or the volume of blood during end-diastolic period. Usually, the value of LLVed is 120 mL with 80 mL of normal SV. This is also rated 67% of EF value. concerning about EF, the range of normal physiological value of EF is in between 65 and 70% and patients suffering from HF have much lower EF value which is normally reduced to 25%.

Table 1 Summary of physiological parameters of MCL

Researchers	Heart condition	Physiological parameters of MCL			
		HR (bpm)/(ratio of systole to diastole)	Systemic vascular resistance, SVR (mmHg s/mL) (dyne s/cm^2)	Compliance value (mL/mmHg)	Coronary artery flowrate (mL/min)
Timms et al. [43]	Normal	**60** (all conditions)/(40/60) [43]	1.08 (1439)	2.0 (aortic)	N/A
				22.5 (systemic venous)	
				2.95 (pulmonary artery)	
				7.1 (pulmonary venous)	
	HF		0.95 (1266)	1.2 (aortic)	
				11.5 (systemic venous)	
				2.95 (pulmonary artery)	
				7.1 (pulmonary venous)	
Pantalos et al. [47]	Normal	**72** (all conditions)/(NA) (Koening et al. 2004; Litwak et al. 2005)	1.00 (1335)	1.3	N/A
	HF		1.43 (1902)		
Kolyva et al. (2010a)	HF	**60**/(40/60) (Kolyva et al. 2010)	1.22 (1624)	0.94	N/A
Ferrari et al. [38]	Recovering HF	N/A	0.98 (1300)	1.8 (total arterial)	N/A
				80 (systemic venous)	
Cornhill [24]	Normal	N/A	1.86 (2485)	N/A	N/A
Knierbein et al. [52]	Normal	N/A	N/A	1.33	N/A

(continued)

Table 1 (continued)

Researchers	Heart condition	Physiological parameters of MCL			
		HR (bpm)/(ratio of systole to diastole)	Systemic vascular resistance, SVR (mmHg s/mL) (dyne s/cm^2)	Compliance value (mL/mmHg)	Coronary artery flowrate (mL/min)
Donovan [30]	Normal	N/A	N/A	1.0 (systemic arterial)	N/A
				10 (systemic venous)	
				1 (pulmonary arterial)	
				5 (pulmonary venous)	
Rosenberg et at. [31]	Normal	N/A	N/A	1.55 (systemic)	N/A
				4.84 (pulmonary)	
Liu et al. [36]	Normal	N/A	N/A	2.0 (systemic arterial)	N/A
				50 (systemic venous)	
Koenig et al. (2004)	HF	N/A	N/A	N/A	67
	Normal				97
	Recovering				104
Geven et al. [54]	Mean (normal)	N/A	N/A	N/A	80

2.2.4 Pressure

Figure 5 illustrates the pressure distribution in native human heart. The main pressure that has to be considered is LV pressure where it can be achieved by adjusting its chamber's contractility. There are some techniques in adjusting the contractility of LV chamber for instance, the use of pneumatic system in which LV pressure can easily be controlled by altering the amplitude of air compressor. Other than that, a piston pump mechanism is also one of the technique in controlling the LV pressure with fixed SV displacement. In much advance MCL, a programmable linear actuator can also be used as it can generate pre-programmed LV sinusoidal waveform.

Other than LV pressure, aortic pressure is also important to be simulated correctly. It can be achieved by the addition of air chamber using Windkessel chamber principle. On the other hand, by controlling the spring constant parameter or altering the suitable

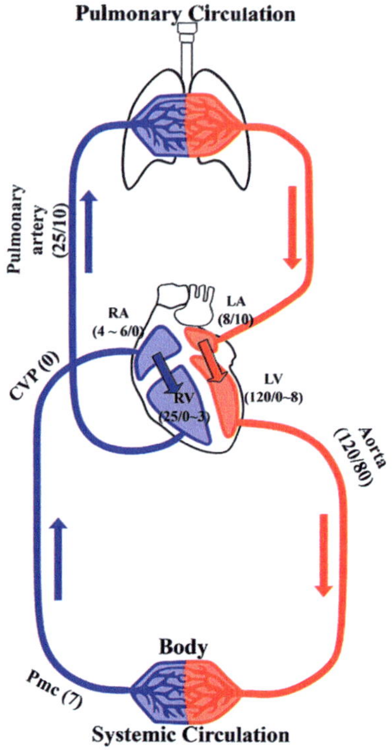

Fig. 5 The schematic diagram of pressure distribution in native human heart. LV: Left Ventricle, LA: Left Atrium, RV: Right Ventricle, RA: Right Atrium, CVP: Central Venous pressure, Pmc: Mean circulatory pressure (mmHg)

point of peripheral resistance are also the alternative technique in order to control the aortic pressure mechanism.

Some MCLs may use the pneumatic pump of the atrium chamber to simulate an atrial kick during diastolic phase [43]. In case of MCL consists of only systemic circulation in which the fluid is directly enters the left atrium, its pressure can be obtained by maintaining the height of water column in atrial chamber as it provides the potential energy according to the formula $P = \rho gh$ [ρ = fluid density, h = height of water column, g = standard gravity constant (9.81 m/s^2)].

Most of the MCLs use LA pressure as an indicator for the preload. Hence, preload is usually being assumed as the amount of volume or pressure during end-diastolic phase. That is slightly incorrect because, the normal value of LV end-diastolic pressure is around 8 mmHg while, but concerning the HF condition such as dilated heart, the pressure often increase beyond 15 mmHg. Thus, the LA pressure should be properly simulated with the wide range of value covering from normal heart condition to HF condition [43]. For the mean arterial pressure (MAP), most of the MCLs set it to be 95–100 mmHg for normal heart condition, while 55–60 mmHg for HF and 80 mmHg for Recovering heart condition.

2.2.5 Resistance

There are two main resistances that need to be considered in developing the MCLs which are systemic vascular resistance (SVR) and pulmonary vascular resistance (PVR). The SVR is capable to be calculated using the equation below:

$$CO = \frac{MAP}{SVR} \tag{2.3}$$

Considering the normal healthy male with MAP = 94 mmHg and CO = 5 L/min, SVR is calculated to be 18.8 mmHg min/L or 1.13 mmHg s/mL. As mentioned by Levick [45], the textbook values of SVR ranges from 1.0 to 1.2 mmHg s/mL. For PVR, it is about 0.18 mmHg s/mL [45].

There are many mechanisms can be used in regulating the peripheral resistance in MCL development. The simplest way is by implementing the simple screw clamp that can be operated manually for controlling the degree of peripheral resistance [25]. Other than that, there are also more complicated technique used by other researchers such as automatic bellows and valve [30], sintered aluminium oxide [24], the gate-type valve [46], open cell foam [47], computer controlled proportional control pinch [43, 48], and the adjustable tube tourniquet/clamp. The list of SVR values from other researchers can be seen in Table 1.

2.2.6 Compliance

Vascular compliance is defined as the ability of blood vessel wall to contract and expand in sustaining blood volume changes due to the pressure changes in the vessel. The normal value physiological value of systemic arterial compliance is about 2.0 mL/mmHg [45]. Due to eight times greater of distensibility, systemic venous has 24 times higher of compliance compared to systemic arterial compliance. Moreover, venous blood vessel capable to store up to 64% of blood volume compared to arterial blood vessel which is only at 20% of blood volume [49].

Arterial compliance can be computed using an equation below. Considering the aortic pressure of 120/80 mmHg (PP = 40 mmHg), and SV = 80 mL:

$$C = \frac{dV}{dP} = \frac{SV}{PP} = \frac{80}{(120-80)} = \frac{80}{40} = 2\left[\frac{\text{mL}}{\text{mmHg}}\right] \tag{2.4}$$

If the runaway blood is taken into account in which during cardiac cycle, since aorta is not a close chamber, around 16% of blood usually drains into peripheral arteries [50]:

$$C = \frac{dV}{dP} = \frac{67.2}{40} = 1.68\left[\frac{\text{mL}}{\text{mmHg}}\right] \tag{2.5}$$

Windkessel chamber is usually being used to simulate the arterial compliance in most MCLs. During systole, the air in the chamber is compressed while during diastole, the compressed air is expanding, and the pressure waveform follows the physiological natural decay of native heart vessel compliance [36, 43, 51].

In developing the MCL that uses Windkessel chamber to simulate vessel compliance, it must capable to store enough air volume precisely [36]. Boyle's Principle is one of the approach that can be used to calculate the expansion of air during cardiac cycle (assumed an adiabatic air expansion).

$$PV = constant \tag{2.6}$$

The following equation can be used to determine the air volume needed to attain normal heart condition with 120/80 mmHg of aortic pressure and 83 ml of SV:

$$P_1 V_1 = P_2 V_2 \tag{2.7}$$

It is assumed that no runaway fluid into peripheral artery circulation where (1) diastolic phase, (2) systolic phase, $P_1 = 840$ mmHg (80 mmHg + 760 mmHg), $P_2 = 880$ mmHg, $V_2 = V_1 - SV$:

$$P_1 V_1 = P_2(V_1 - SV)$$

$$V_1 = \frac{P_2 \times SV}{P_2 - P_1} = \frac{P_2 \times SV}{PP}$$

$$V_1 = \frac{880 \times 83}{40}$$

$$V_1 = 1826 \text{ mL} \simeq 1.83 \text{ L}$$

While for arterial compliance:

$$C = \frac{SV}{PP} = \frac{83}{40} \simeq 2.08 \left[\frac{\text{mL}}{\text{mmHg}}\right]$$

The range of the regulation of air volume in Windkessel chamber to be in between 1034 and 2068 mL so that it can reaches the compliance level of 1.06–2.13 mL/mmHg [52].

Although the utilization of Windkessel chamber offered the most compromise compliance simulation technique, there were some researchers tend to use other method i.e. some researchers fabricate a soft and thin elastomeric material from silicon or rubber [35], the attachment of silicon aorta to the Windkessel chamber to simulate larger range of aortic compliance [33, 53] and others. Furthermore, the use of Windkessel chamber offers the minimal cost consumption although normally venous compliance chamber is big and tall [36]. The list of compliance values used in other studies can be referred in Table 1.

2.2.7 Simulation of Left Coronary Artery in MCL

In native human heart, the left and right coronary arteries are originated from root of aortic arch specifically at coronary Ostia several millimetre distal to aortic valve. The role of coronary circulation is to supply the oxygen and nutrient needed by myocardium. The region covers by left coronary artery (LCA) includes left atrium and left ventricle. In most MCL constructed, LCA is normally being simulated rather than the right one due to its greater significance of covering the left heart side.

There are several factors that may affect the flow of coronary artery circulation such as myocardium contraction and fluctuation of the pressure in ascending aorta. On the other hands, aortic pressure is the driving force factor in governing the coronary flow rate in coronary perfusion [54].

In simulating the LCA in MCL development, one major problem faced by many researchers was the reflection of coronary flow value with that of physiological value. In most literature of in vivo studies, the coronary flow reverse is used as an indicator and the coronary normalized is expressed with heart mass which is not possible to be adapted directly in this study. Several comparable values of coronary artery flowrate in mL/min unit can be referred in Table 1.

The use of silicon or latex collapsible tube that can be pressurized simultaneously with LV chamber is capable in producing the unique biphasic coronary waveform. In order to introduce the pressure onto the soft tube, it can be achieved either by compressing the air for the LV sac, or directed fluid originates from LV chamber. As proposed by Downey and Krik [55] regarding the waterfall model, the compression of subepicardial part of myocardium vascular bed can be imitated by this simultaneous compression of latex collapsible tube during the systolic phase.

3 Mock Circulatory Loop (MCL) Development and Working Principle

To investigate the effect of a novel device for heart failure (HF) treatment, the Chronic Intermittent Mechanical Support (CIMS) balloon pump, a MCL was developed [56]. The incorporation of improved hardware and data acquisition system (DAQ) was implemented on the basic MCL developed at University of Nottingham. This MCL implements two-element Windkessel model which consists of a compliance chamber (represents the atrial chamber) and an element of total peripheral resistance. Figure 6 shows the final version of MCL developed.

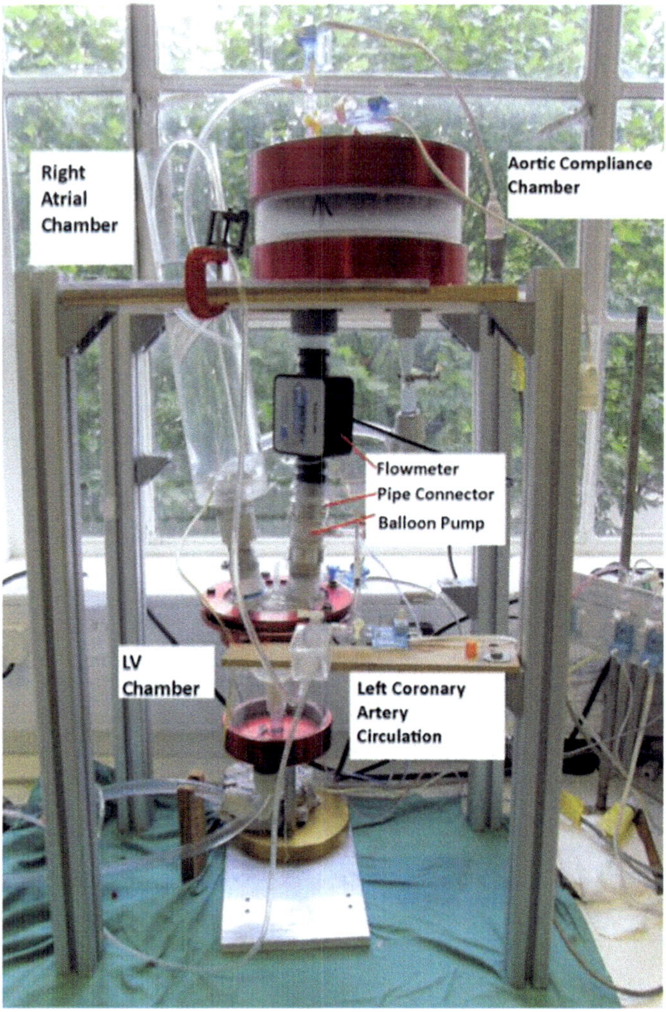

Fig. 6 The final version of MCL constructed [56]

3.1 Overview of MCL Construction

Figure 7 illustrates the schematic diagram of connection of MCL with other components such as DAQ system, flowmeter and IABP driver.

The pulsatile action of LV silicon sac to mimic the physiological pulsatile nature was promoted by the 3-way proportional pressure regulator in response from a square wave signal delivered to it. The change of amount of pressurized air in the LV chamber was translated from the fluctuation of the square wave amplitude.

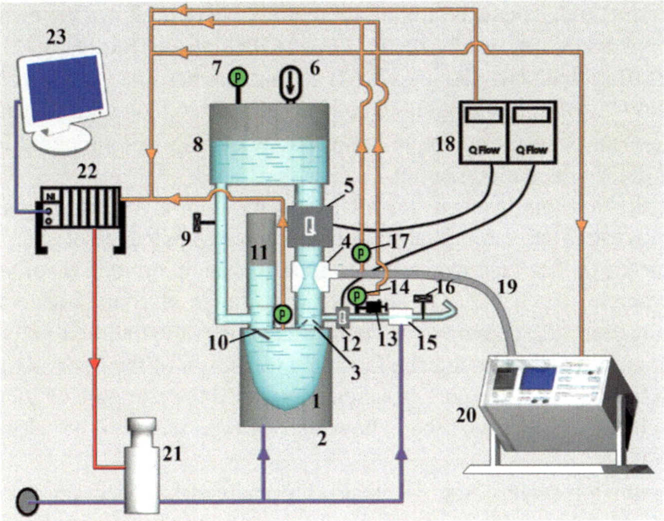

Fig. 7 The schematic diagram of the human MCL for investigation of the effect of CIMS balloon pump. 1 LV silicone sack, 2 LV chamber, 3 aortic valve, 4 CIMS balloon pump, 5 flow sensor, 6 air bellow, 7 aortic pressure transducer, 8 arterial compliance chamber, 9 resistor, 10 mitral valve, 11 left atrial chamber, 12 flow sensor, 13 compliance syringe, 14 left coronary artery pressure transducer, 15 systolic resistor, 16 resistor, 17 balloon pump pressure transducer, 18 flow meter, 19 helium gas driveline, 20 IABP console, 21 proportional pressure regulator, 22 DAQ system, 23 personal computer (PC)

During systole, the air is pressurized to fill up the LV chamber and being accurately cut off upon completing the diastolic phase. An oversize hole was created at the bottom of LV chamber to pass the conduit originates from apex of LV silicon sac and to allow the air to escape from LV chamber during diastole while facilitates the passive filling of LV silicon sac. Furthermore, the thick wall of LV silicon sac (about 2 mm) promotes an additional assistance on passive filling action that would naturally re-assume the original shape of LV sac during diastole. To simulate a various range of heart physiological condition, the LV sac was fabricated to sustain up to 250 mL of water corresponds to almost double the normal physiological value. It is capable to simulate the LV volume during dilated cardiomyopathy condition.

Several millimetres (15 mm to be exact) distal from aortic valve, two holes were created for the insertion of Tygon tubes, R3603, Saint-Gobain Inc., France (ID: 3.2 mm, OD: 4.8 mm) and being fixed in place by applying the silicone sealant, Loctite 595 Clear Silic RTV, Loctite Inc., USA (refer Fig. 3.50 in [56]). One tube was connected to the LCA circulation and the other one to pressure transducer intended to measure the pressure just distal to the aortic valve.

The CIMS balloon pump was connected in series distal to aortic valve (in-line with ascending aorta) using two-pipe connectors. The flow sensor was located 25 mm distal to CIMS balloon pump intended to measure the volumetric fluid movement. A non-resistive flow sensor was used in which it is capable to measure a bidirectional flowrate of the fluid. This is an important feature where it allows the measurement to be in a forward or backward direction. The IABP console [Bard H-8000 TransAct IABP (C. R. Bard Inc., New Jersey, USA)] was used to activate the counter-pulsatile mechanism of CIMS balloon pump to be in synchronize phase with LV sac contraction. This console transferred the helium gas to the CIMS balloon pump in order to inflate and deflate the balloon pump.

One highlighted component in this MCL design was the air chamber for simulating the physiological arterial compliance or also known as Windkessel chamber (refer Fig. 3.52 in [56]). The chamber was designed to sustain up to 3.1L of volumetric capacity wherein it is capable to simulate wide range of compliance values. For easy visualization, the chamber was fabricated using a cylindrical body made of transparent acrylic with lids attached to top and bottom of the body. At the top of the body, a handheld bellow port was created to assist the pumping of additional air inside the chamber. Moreover, the bellow also allowed the air to be released to the surrounding.

At the same top surface, there was also a port for the connection of two 3-way stopcock connectors used to connect a pressure transducer to measure the air trapped inside the chamber and further assist of bleeding air to the surrounding. One pressure sensor was devoted to DAQ system for data recording and displaying whereas the other one was dedicated for IABP console input. 25 mm (ID) of PVC tube was used as inflow water while 12 mm (ID) tube for outflow water. It was due to the reduction of size as well as the existence of branching of flow past the aorta. The outflow tube was connected to the left atrial chamber and the flow resistor was place in the middle of it, so that the total peripheral resistance (TPR) can be manually adjusted.

Another Windkessel element was implemented in the left atrial chamber of the MCL model. The chamber was made of acrylic tube with the dimensions of 76 mm(OD) $\times$ 3 mm $\times$ 335 mm and capable to fill up to 1.23 L of water corresponds to the potential pressure around 23 mmHg. A conduit was inserted at the bottom of chamber allowing pressure transducer to measure the left atrial pressure (LAP). LAP is the pressure on top of mitral valve in which during calibration of LAP, the pressure difference between mitral valve and the pressure point must be taken into consideration.

3.2 Development of Left Coronary Artery (LCA) Circulation

Similar to systemic circulation of the MCL, two-element Windkessel model was implemented wherein it composed of coronary compliance and total coronary artery resistance. The circulation was fabricated by Tygon® Tube with 3.2 mm ID. An ultrasound flow sensor was used to measure the flowrate of the fluid inside of the Tygon® tube and 1 mL syringe was implemented into the circulation to simulate the LCA compliance in which the appropriate air volume was maintained which was around 0.5–0.6 mL.

In the native human heart, the fluctuation of LCA flow characteristic is mainly affected by a crucial factor which is the early systolic depression resulted from coronary arteries myocardium compression. Adapting the similar method used by Geven and Gaillard, the simulation of early systolic compression was done by implementing a 45 mm × 30 mm × 30 mm dimension of watertight small acrylic box. That small acrylic box contains the deformable silicon tube connected in-line with LCA circulation in which the tube was compressed simultaneously with the compression of LV silicon sac. Hence, the physiological waveform of LV compression was similar to that of deformable silicon tobe compression.

3.3 MCL Data Acquisition System (Hardware and Software)

In every experimental work, the outcome results must be recorded appropriately. The suitable data acquisition tools must be selected as such it has the ability to record all the parameters intended to be measured. In this MCL, a chassis (NI cDAQ-9172, NI, Austin, Texas, USA) of DAQ hardware was used as an interconnection terminal for the receiving input and delivered output between MCL model components and DAQ software processing unit. The chassis was capable to hold up to 8 different input and output I/O connections.

The flowmeter (TS410, Transonic System Inc., Ithaca, NY, USA) was used to measure the instantaneous flowrate at the LCA circulation and ascending aorta and the flowmeter was connected to one of input cartridge of DAQ hardware chassis and hence recording and displaying the simultaneous mean and instantaneous flowrates. Another type of transducer that has been used in this MCL was disposable pressure transducer for measuring the fluid pressure.

The data received by DAQ chassis were then transmitted to the DAQ software for data processing and display. A graphical programming language software called Labview (National Instrument, Austin, Texas, USA) was used as it was capable to process the input and output data from and to DAQ hardware. Various processing method was applied in order to get the best output results such as sample scaling, sample averaging and signal filtration. All these processing stages was applied with the suitable threshold value for medical range signal processing application.

4 Conclusions

In summary, Mock Circulatory Loop (MCL) has been utilized over decades and has become a favourable platform for in vitro testing procedure. Although there were some difference of configurations and the use of components, all models that have been developed by many researchers were directed towards the primary goal of MCL development which is to reproduce various cardiac parameters as close as possible to the actual physiological cardiac values. With the suitable combination of hardware and software tools, MCLs have undergone a design evolution from a simple circulatory circuit to much more complex and flexible configuration models. Thus, the successful testing of cardiovascular devices marks the impactful and practicability of MCL as an in vitro testing platform prior to the clinical testing procedure.

Acknowledgments The authors would like to thank all team members for helpful discussion as well as their contributions in reviewing and editing this review topic.

References

1. Mathers, C.: The Global Burden of Disease: 2004 Update. World Health Organization (2008)
2. Benjamin, E.J., et al.: Heart disease and stroke statistics—2017 update: a report from the American Heart Association. Circulation **135**(10), e146–e603 (2017)
3. Lloyd-Jones, D.M., et al.: Lifetime risk for developing congestive heart failure: the Framingham Heart Study. Circulation **106**(24), 3068–3072 (2002)
4. Timms, D.L., et al.: A compact mock circulation loop for the in vitro testing of cardiovascular devices. Artif. Organs **35**(4), 384–391 (2011)
5. Petersen, S., Rayner, M., Wolstenholme, J.: Coronary Heart Disease Statistics: Heart Failure Supplement, 2002 edn. University of Oxford (2002)
6. Miller, L.W.: Left ventricular assist devices are underutilized response to miller. Circulation **123**(14), 1552–1558 (2011)
7. Savarese, G., Lund, L.H.: Global public health burden of heart failure. Cardiac Failure Review **3**(1), 7–11 (2017)
8. Cowie, M., et al.: Survival of patients with a new diagnosis of heart failure: a population based study. Heart **83**(5), 505–510 (2000)
9. McDonagh, T.A., et al.: Symptomatic and asymptomatic left-ventricular systolic dysfunction in an urban population. The Lancet **350**(9081), 829–833 (1997)
10. McKee, P.A., et al.: The natural history of congestive heart failure: the Framingham study. N. Engl. J. Med. **285**(26), 1441–1446 (1971)
11. McMurray, J.J., Stewart, S.: Epidemiology, aetiology, and prognosis of heart failure. Heart **83**(5), 596–602 (2000)
12. Birks, E.J., et al.: Left ventricular assist device and drug therapy for the reversal of heart failure. N. Engl. J. Med. **355**(18), 1873–1884 (2006)
13. Dorent, R., et al.: Assessment of potential heart donors: a statement from the French heart transplant community. Arch. Cardiovasc. Dis. (2017)
14. Ratner, B.D., et al.: Biomaterials Science: An Introduction to Materials in Medicine. Elsevier, Amsterdam (2004)
15. DeBakey, M.E.: Left ventricular bypass pump for cardiac assistance: clinical experience. Am. J. Cardiol. **27**(1), 3–11 (1971)
16. Reul, H.M., Akdis, M.: Blood pumps for circulatory support. Perfusion **15**(4), 295–311 (2000)
17. Lim, K.M., et al.: Computational analysis of the effect of the type of LVAD flow on coronary perfusion and ventricular afterload. J. Physiol. Sci. **59**(4), 307–316 (2009)
18. Doost, S.N., Zhong, L., Morsi, Y.S.: Ventricular assist devices: current state and challenges. J. Med. Dev. **11**(4), 040801–040801-11 (2017)
19. Dasi, L.P., et al.: Fluid mechanics of artificial heart valves. Clin. Exp. Pharmacol. Physiol. **36**(2), 225–237 (2009)
20. Li, C.-P., et al.: Role of vortices in cavitation formation in the flow at the closure of a bileaflet mitral mechanical heart valve. J. Artif. Organs **15**(1), 57–64 (2012)
21. McMillan, I.: Aortic stenosis: a post-mortem cinephotographic study of valve action. Br. Heart J. **17**(1), 56 (1955)

22. McMillan, I., Daley, R., Matthews, M.: The movement of aortic and pulmonary valves studied post mortem by colour cinematography. Br. Heart J. **14**(1), 42 (1952)
23. Davila, J.C., et al.: A simple mechanical pulse duplicator for cinematography of cardiac valves in action. Ann. Surg. **143**(4), 544 (1956)
24. Cornhill, J.: An aortic-left ventricular pulse duplicator used in testing prosthetic aortic heart valves. J. Thorac. Cardiovasc. Surg. **73**(4), 550–558 (1977)
25. Duran, C., Gunning, A., McMillan, T.: A simple versatile pulse duplicator. Thorax **19**(6), 503 (1964)
26. Hildebrand, D.K., et al.: Design and hydrodynamic evaluation of a novel pulsatile bioreactor for biologically active heart valves. Ann. Biomed. Eng. **32**(8), 1039–1049 (2004)
27. Morsi, Y.S.: In vitro comparison of steady and pulsatile flow characteristics of jellyfish heart valve. J. Artif. Organs **3**(2), 143–148 (2000)
28. Raftery, E., Dayem, M., Melrose, D.: Mechanical performance of Hammersmith mitral valve prosthesis. Br. Heart J. **30**(5), 666 (1968)
29. Kolff, W.J.: Mock circulation to test pumps designed for permanent replacement of damaged hearts. Clevel. Clin. Q. **26**, 223 (1959)
30. Donovan, F.M.: Design of a hydraulic analog of the circulatory system for evaluating artificial hearts. Biomater. Med. Devices Artif. Organs **3**(4), 439–449 (1975)
31. Rosenberg, G., et al.: Design and evaluation of the Pennsylvania State University mock circulatory system. ASAIO J. **4**(2), 41–49 (1981)
32. Orime, Y., et al.: In vitro and in vivo validation tests for total artificial heart. Artif. Organs **18**(1), 54–72 (1994)
33. Cassot, F., et al.: New versatile physical model fitting the systemic circulation accurately. Med. Biol. Eng. Comput. **23**(6), 511 (1985)
34. Gregory, S., et al.: A naturally shaped silicone ventricle evaluated in a mock circulation loop: a preliminary study. J. Med. Eng. Technol. **33**(3), 185–191 (2009)
35. Zannoli, R., Corazza, I., Branzi, A.: Mechanical simulator of the cardiovascular system. Physica Medica: Eur. J. Med. Phys. **25**(2), 94–100 (2009)
36. Liu, Y., et al.: Construction of an artificial heart pump performance test system. Cardiovasc. Eng. **6**(4), 151–158 (2006)
37. Wu, Y., et al.: Modeling, estimation, and control of human circulatory system with a left ventricular assist device. IEEE Trans. Control Syst. Technol. **15**(4), 754–767 (2007)
38. Ferrari, G., et al.: A hybrid mock circulatory system: testing a prototype under physiologic and pathological conditions. ASAIO J. **48**(5), 487–494 (2002)
39. Ferrari, G., et al.: A hybrid (numerical-physical) model of the left ventricle. Int. J. Artif. Organs **24**(7), 456–462 (2001)
40. Ferrari, G., et al.: Development of hybrid (numerical-physical) models of the cardiovascular system: numerical-electrical and numerical hydraulic applications. Biocyber. Biomed. Eng. **25**(4), 3–15 (2005)
41. Ferrari, G., et al.: Mock circulatory system for in vitro reproduction of the left ventricle, the arterial tree and their interaction with a left ventricular assist device. J. Med. Eng. Technol. **18**(3), 87–95 (1994)
42. Kozarski, M., et al.: A new hybrid electro-numerical model of the left ventricle. Comput. Biol. Med. **38**(9), 979–989 (2008)
43. Timms, D., et al.: A complete mock circulation loop for the evaluation of left, right, and biventricular assist devices. Artif. Organs **29**(7), 564–572 (2005)
44. Vilchez-Monge, M., Truque-Barrantes, A., Ortiz-Leon, G.: Design and construction of a hydro-pneumatic mock circulation loop that emulates the systemic circuit of the circulatory system. In: 36th Central American and Panama Convention (CONCAPAN XXXVI). IEEE (2016)
45. Levick, J.: Overview of cardiovascular system. In: An Introduction to Cardiovascular Physiology, pp. 1–13 (2003)
46. Sharp, M.K., Dharmalingham, R.: Development of a hydraulic model of the human systemic circulation. ASAIO J. (American Society for Artificial Internal Organs: 1992), **45**(6), 535–540 (1999)

47. Pantalos, G.M., et al.: Characterization of an adult mock circulation for testing cardiac support devices. ASAIO J. **50**(1), 37–46 (2004)
48. Timms, D., et al.: Evaluation of left ventricular assist device performance and hydraulic force in a complete mock circulation loop. Artif. Organs **29**(7), 573–580 (2005)
49. Guyton, A.C., Hall, J.: Insulin, glucagon, and diabetes mellitus. Textbook Med. Physiol. **8**, 855–867 (2006)
50. Berne, R.M., Levy, M.N.: Principles of Physiology. Mosby Incorporated (2000)
51. Donovan, F.: Design of a hydraulic analog of the circulatory system for evaluating artificial hearts. Biomater. Med. Devices Artif. Organs **3**(4), 439–449 (1975)
52. Knierbein, B., et al.: Compact mock loops of the systemic and pulmonary circulation for blood pump testing. Int. J. Artif. Organs **15**(1), 40–48 (1992)
53. Kolyva, C., et al.: A mock circulatory system with physiological distribution of terminal resistance and compliance: application for testing the intra-aortic balloon pump. Artif. Organs **36**(3) (2012)
54. Geven, M.C., et al.: A physiologically representative in vitro model of the coronary circulation. Physiol. Meas. **25**(4), 891 (2004)
55. Downey, J.M., Kirk, E.S.: Inhibition of coronary blood flow by a vascular waterfall mechanism. Circ. Res. **36**(6), 753–760 (1975)
56. Khudzari, A.Z.B.M.: The development and investigation of a novel pulsatile heart assist device (2012)

Computational Fluid Dynamics Application in Reducing Complications of Patent Ductus Arteriosus Stenting

M. I. Kori, K. Osman, A. Z. M. Khudzari and I. Taib

Abstract In some cases, especially in neonates, ductus arteriosus needs to remain patent for multiple medical purposes. In order to achieve this, current practice involves inserting stent in the ductus arteriosus. This condition is called patent ductus arteriosus (PDA). For this process, stents such as coronary stent are commonly used due to unavailability of customized stent for PDA in neonates. The usage of coronary stent however, opens the possibility of acute stent thrombosis and other complications. Therefore, there is a high need of special and customized stents to be used for PDA in neonates. This customized stent has to be able to sustain the hemodynamic effects of the flow inside the PDA. The stent has to be able to support the ductus wall compression and contraction due to arterial compliance. What is more important is that the stent must properly fit into various morphologies of the ductus. There are several different morphologies of PDA identified and the stents must be able to sustain the various shapes and tortuosity. In addition, the stent has to be tested for biocompatibility and practicality. Therefore, the customized design of the PDA stents can be derived from the concept of coronary stents and in compliance with all the mentioned characteristics. However, further analysis has to be completed ensure proper compatibility with neonates. In conclusion, the biggest challenge is to customize a stent that fits all the PDA morphologies.

M. I. Kori · K. Osman (✉) · A. Z. M. Khudzari
Faculty of Biosciences and Medical Engineering, Universiti Teknologi Malaysia, 81310 Skudai, Johor, Malaysia
e-mail: zahran@biomedical.utm.my

K. Osman · A. Z. M. Khudzari
IJN-UTM Cardiovascular Engineering Centre, Institute of Human Centered Engineering, Universiti Teknologi Malaysia, 81310 Skudai, Johor, Malaysia

I. Taib
Faculty Mechanical and Manufacturing Engineering, Universiti Tun Hussein Onn Malaysia, 86400 Parit Raja, Johor, Malaysia

D. E. O. Dewi et al. (eds.), *Cardiovascular Engineering*, Series in BioEngineering,
https://doi.org/10.1007/978-981-10-8405-8_9

1 Introduction

The implantation of tubular wire mesh called stent in ductus arteriosus has been selected as an alternative palliative treatment for neonates diagnosed with cyanotic congenital heart disease. This procedure or treatment has been selected as a temporary treatment before any permanent surgical shunts take place. By maintaining a patent ductus arteriosus (PDA) using stent, the neonates will be able to gain sufficient weight before undergoing surgical repair.

Implantation of stent however has caused various complications on several patients after the procedure. Between 2001 and 2003, approximately 8.9% of all patients whose undergo PDA stenting procedure were deemed unsuccessful due to complications as reported by Institut Jantung Negara (IJN) [1]. One of the most common complication is in-stent restenosis. This complication may require the patient to undergoes subsequent stent redilation to reopen the stenosed vessel. Furthermore, the necessity of reintervention is highly dependent on the stent design and ductal morphology of the patient.

Researchers have found out that the stent strut configurations and PDA morphologies contribute to the formation of in-stent restenosis. Thus, understanding and understanding of the effect of PDA morphologies and the stent strut configuration are important. This will lead to improvement on the current stent design in order to reduce the risk of in-stent restenosis formations. Various methods of analysis have been employed in order to understand the flow behaviour in blood vessels and one of them is via numerical analysis.

Numerical analysis such as computational fluid dynamics (CFD) have been implemented to study the effect of stent strut designs and blood vessel morphologies to the flow behaviour inside blood vessel and stented vessel. Using this method, the performance of the stent to mitigate in-stent restenosis can be predicted and in addition, better stent design can be derived and customized with better hemodynamic performance. Solving fluid-structure interaction (FSI) problems on the hand may help determine the stent structural strength due to compliance effect of the artery and pulsatile blood flow. Experimental analysis using particle image velocimetry are used to validate the numerical results.

2 Patent Ductus Arteriosus Stenting

Ductus arteriosus is a vascular structure that connects the aorta to the roof of the pulmonary artery [2, 3]. Ductus arteriosus shunts the blood away from the lungs, by redirecting blood from the pulmonary artery. Changes of blood flow happen after birth, with the increase of pulmonary blood flow and the closure of ductus arteriosus. These changes are dynamic depending on the pressure gradient between the pulmonary and systemic circulation [4]. However, there are also chances in which the vessel remain patent, called patent ductus arteriosus (PDA) which is abnormal in neonates, 72 h after birth [5]. This condition is unfavourable especially in normal patient, as it may cause further complication which leads to oxygen-rich blood to recirculate to lungs instead of the body [6], thus causing heart failure [7].

On the other hand, for new-born with cyanotic congenital heart disease, maintaining the patency of ductus arteriosus has been selected as an alternative treatment before permanent surgical repair in order to maintain sufficient blood flow throughout the body. This can be achieved by implanting stent, which acts as a scaffold to maintain the patency of the vessel. However, various considerations need to be made before stent implantation in PDA can be performed, including the morphology of the PDA itself.

2.1 Stenting of Patent Ductus Arteriosus

Patent ductus arteriosus stenting is a less invasive palliative treatment introduced for neonates with cyanotic congenital heart disease with duct-dependent pulmonary circulation [3, 8–12]. Patency of ductus arteriosus via stenting maintains the blood flow and acts as a temporary shunt [8]. This procedure will implant a device called stent to maintain the patency of ductus arteriosus temporarily for 6–12 months before the patients undergo surgical repair for palliative over conduit surgery or first stage cavopulmonary anastomosis [8, 13]. The patient will gain sufficient weight before any surgical repair takes place [10]. Stenting of PDA able to reduce the risk of complications on premature infants with cyanotic congenital heart disease that will lead to morbidity and mortality [14] (Fig. 1).

There are several indications for PDA stenting to be selected for treatment. The first indication is to increase the pulmonary blood flow to the artery [15]. Second indication is the need to increase the pulmonary atresia with intact ventricular septum (PAIVS) and tricuspid atresia. Third indication is Ebstein's Anomaly in order to prolong the infusion of prostaglandin while the last indicator is simple transposition of great arteries (TGA) with inflated left ventricle [15] (Fig. 2).

Odemis et al. [16] had reported 13 patients who were implanted with coronary stents. All patients were less than two months old and the average weight was 3.1 kg. All thirteen patients had the coronary stent implanted in them for the treatment of various complications. The maximum diameter of the stents used were 4 mm. Various types of congenital heart diseases were reported for all patients. They were pulmonary atresia (PA) with intact ventricular septum and ventricular septal defect. A few other were diagnosed to have critical pulmonary stenosis (CPS).

Not all patients diagnosed with congenital heart disease are suitable for stent implantation. Alwi et al. has list out exclusion criteria which indicates whether the patients are not suitable for stenting. The criteria are complex PDA morphology, insufficient constricted PDA, pre-existing branch pulmonary artery stenosis and insufficient body weight, which is less than 2.5 kg [10, 17, 18]. Based on the exclusion criteria of PDA from stenting, understanding of the PDA morphologies are required, since the complexity of the stenting procedure are dependent on the mor-

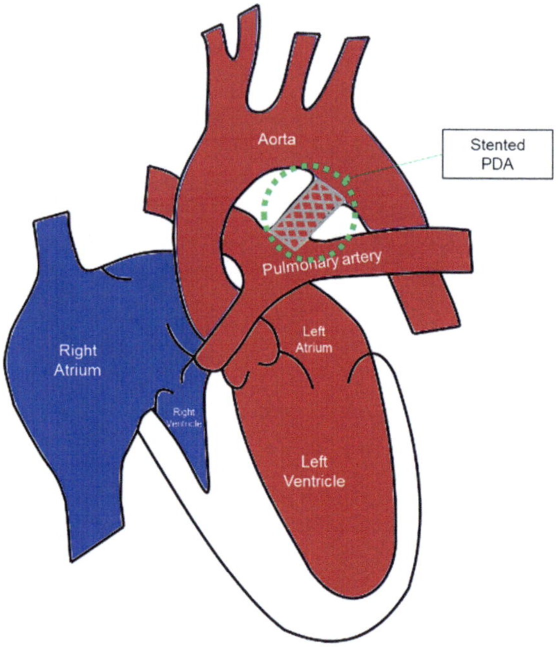

Fig. 1 Illustration of ductus arteriosus stenting procedure for neonate

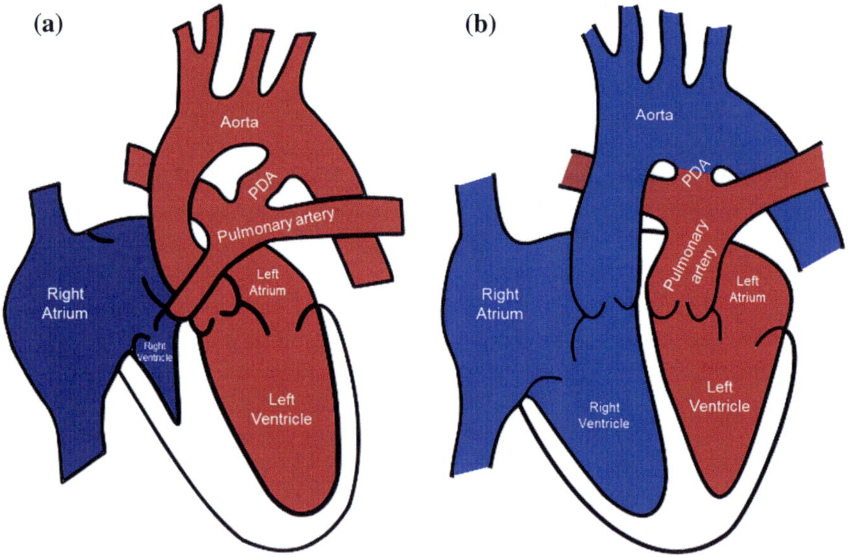

Fig. 2 Morphology of cyanotic congenital heart disease. **a** PAIVS, **b** transposition of great artery which may require PDA stenting

phology itself. Several studies on the classification of PDA morphologies are done and discussed in the next section.

2.2 Morphologies of PDA

Ductus arteriosus may persist in various configurations and sizes [2]. Researchers have characterised the ductal morphologies based on the origin of the aorta, size and shape, length, tortuosity and insertion onto the pulmonary artery [8]. Understanding of the PDA morphologies is important especially in predicting the technical difficulty in stenting, risk of stenosis and the necessity of reintervention [9]. In addition, identification of the morphology can be used to determine the resistance of blood flow through the PDA [2]. In this section, we will discuss mainly on the shape of the PDA, origin from the aorta and the insertion onto the pulmonary artery.

Krichenko et al. [19] had categorised PDA morphologies based on the shape of the PDA into 5 main variations as discussed in Table 1 and illustrated in Fig. 3. The most frequent shapes observed is Type A, with a conical insertion into the pulmonary artery identified [19]. If the PDA exists as an isolated lesion, majority of the PDA are short, and straight conical shape [10].

Alwi and Mood on the hand identified the PDA for patient with cyanotic congenital heart disease tends to be longer, tubular in shape with having more than one curve. Isolated PDA is characterized by the tortuosity and acute bends [17] (Fig. 4).

Schneider et al. has classified the PDA based on the origin from aorta. The arterial ducts are found to be either originated from the descending aorta (Fig. 5a), distal aortic arch (Fig. 5b) and from the aortic arc (Fig. 5c) [3]. Alwi and Mood on the hand found out that the origin of PDA either start at the proximal descending aorta, proximal of the underside of the arch opposite the origin of the left subclavian artery or left common pulmonary artery [17]. As for the insertion point at pulmonary artery, Alwi et al. found out that the insertion of PDA to the pulmonary artery is at the origin or

Table 1 PDA shape discussed by Krichenko et al.

Type	Shape	Descriptions
A	Conical shape	The ampulla diameter is quite larger adjacent to proximally descending aorta and constriction near the left pulmonary artery end
B	Window shape	Very large PDA insertion at pulmonary artery end and short in length
C	Tubular shape	PDA is present without constriction
D	Complex shape	PDA body is similar to fusiform aneurysm shape which has multiple constriction
E	Elongated shape	Stenosis remote from the anterior edge of trachea

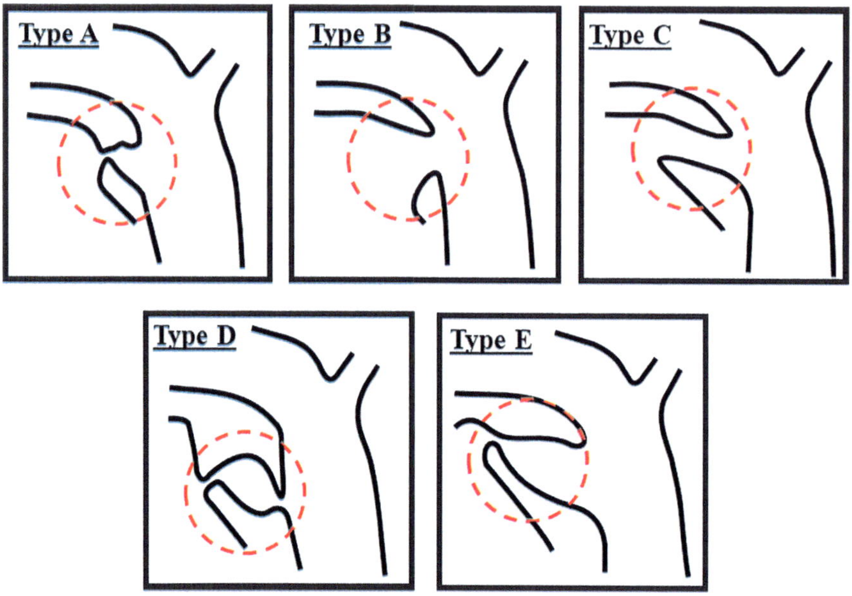

Fig. 3 Lateral angiogram images of variation of PDA configuration as discussed by Krichenko et al. Circled is the location of PDA

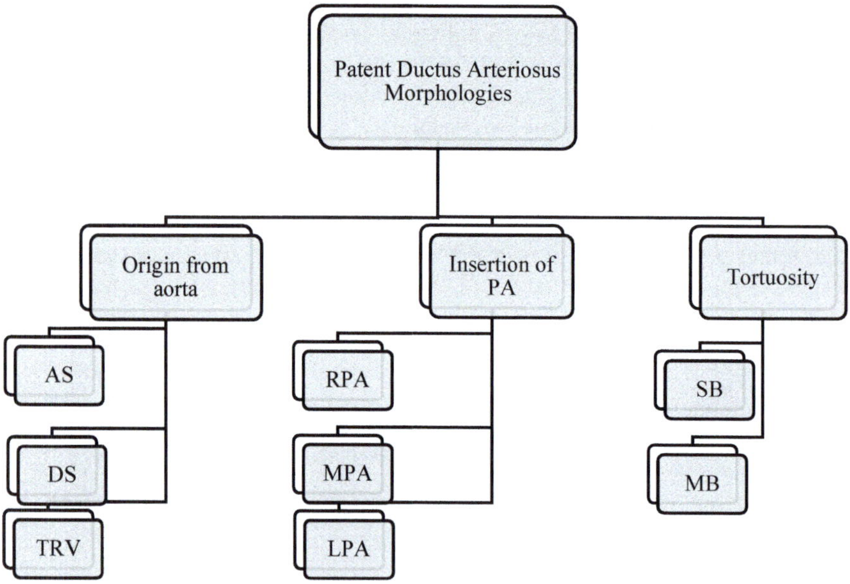

Fig. 4 Various PDA stenting morphologies as discussed by Alwi et al. RAC: left aortic arc; LRC: left aortic arc; AS: ascending aorta; DS: descending aorta; TRV: transverse aorta; SC: subclavian artery; RPA: right pulmonary artery; LPA: left pulmonary artery

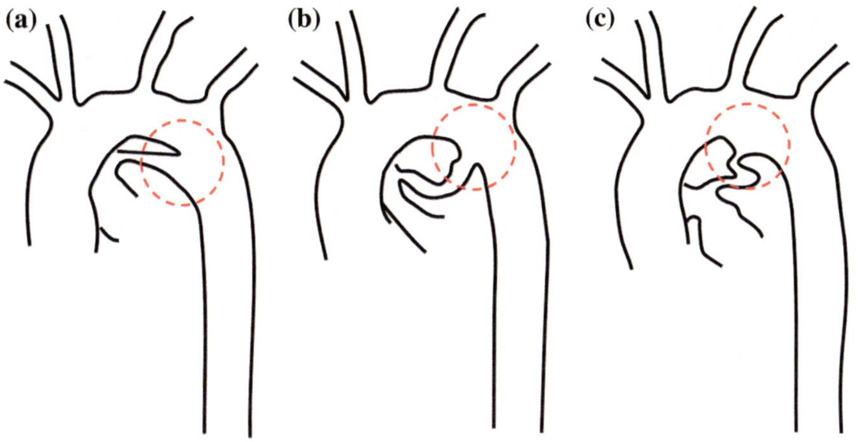

Fig. 5 Different origin of PDA as described by Schneider et al. Circled is the origin of PDA from the aorta

distal of the left pulmonary artery and also at the dome of the main pulmonary artery. The PDA is observed to be straight or tortuous [17].

Various PDA morphologies were identified, however, not all morphologies are suitable for PDA stenting. Considerations need to be made whether these morphologies are suitable for PDA stenting or otherwise to avoid further complications post-implantation of stent. Detailed evaluation of PDA morphologies are needed and done using angiography [8]. Alwi identified that the most suitable PDA morphologies originated proximally of the descending aorta and normal straight course with constriction as it insert to main pulmonary artery, while long and tortuous PDA, with significant constrictions is not suitable for stenting [8]. Boshoff et al. indicates stenting can be performed on tortuous PDA, however, it might be very challenging and cannot be done safely [9]. Branch pulmonary artery stenosis is likely to occur [8].

2.3 Complications After PDA Stenting

Implantation of stent in PDA however introduced several complications which are commonly associated with stenting in other vessels such as acute stent thrombosis, vascular injury, cardiac failure, PDA spasm, stent migration and haemorrhage [17, 20]. In-stent restenosis is common complication and become hindrance in maintaining the patency of ductus arteriosus post stenting [14]. Restenosis is a process of vessel narrowing down after stent implantation in the region of implanted stent. Murphy et al. had identified four main stages of restenosis, which are thrombosis formation, arterial inflammation, neointimal hyperplasia and arterial remodelling [21].

Table 2 Summary on the complications after PDA stenting [1]

Authors/year	Types of stent	Total number of stents	Total patients	Total successful procedure	Total unsuccessful procedure	Complications
Schneider et al. [3]	Coronary stent	32	21	21	–	Arterial damage, early re-intervention
Gibbs et al. [22]	Coronary stent	–	19	17	2 died (tortuosity)	Spasm, stent thrombosis
Alwi et al. [13]	Coronary stent	56	61	56	5 (tortuosity)	Stent embolization, cardiac failure
Gewillig et al. [12]		13	10	8	2	Ductal occlusion
Alwi [20]	Coronary stent	37	37	37	–	Acute stent thrombosis

Various researchers and medical practitioners had highlighted the impact of PDA stenting to in-stent restenosis [12, 14, 22]. The efficacy of the PDA stenting gradually lessen after six months due to ductal tissue prolapse through the stent struts and intrastent endothelial hyperplasia [14]. Due to in-stent restenosis, the patient may require further interventions which includes stent redilation and restenting [14]. Medical practitioner even did not recommend PDA stenting due to its difficulty and potential complications of the procedure [22]. However, with the advancement in paediatric cardiac surgery, the inferior durability of stented PDA may not be an issue [23].

In order to mitigate this issue, stent coated with antiproliferative drug is shown to be able to reduce the risk of restenosis by up to 85% if used in stenting procedure [24]. However, late thrombosis which leads to in-stent restenosis is known to be a long-term issue once the drugs finished, similar to bare metal stents [25]. This indicates that further refinement on the stent, especially on the design are required [24]. One way to predict the performance of the stent after implantation, various method can be applied including design performance evaluation in reducing the risk of in-stent restenosis via computational fluid dynamics (CFD) modelling of stented vessel (Table 2).

3 Hemodynamic Performance of Stents

Implanting stent in PDA has induced blood flow alterations leads to the development of in-stent restenosis. Due to stent implantation, maintaining unobstructed ductus arteriosus has become a challenge due to this complication. Researchers had found out that the formation of restenosis near to the stented region are due to the stent design and the ductal morphologies. Various studies had been performed and concluded that the divergence of flow, local to the stented region are shown to be causing excessive growth of cell proliferation [17]. To identify the effect of the stent implantation in PDA to the formation of in-stent restenosis, hemodynamic performance of stent strut designs are required. Evaluation of the stent hemodynamic performance after implantation in PDA are needed to predict the efficacy in reducing the risk of in-stent restenosis.

3.1 Evaluating PDA Stent Hemodynamic Performance

Making sure the modifications of stent design maybe able to reduce the risk of in-stent restenosis, various methods have been implemented and used to evaluate the hemodynamic performance of the stents. The hemodynamic performance of the stent can be analysed by numerical modelling with experimental analysis as a validation.

3.1.1 Evaluation of Stent Design via Numerical Modelling

Numerical modelling of stented PDA has been selected as one of the method to evaluate the stent hemodynamic performance due to its lower cost as compared to experimental-based laboratory studies. Numerical modelling also able to predict accurately the mechanism of in-stent restenosis after stent implantation via computer simulations. By using computational fluid dynamics (CFD) analysis, researchers rank commercial stents used in PDA based on its hemodynamic performance in reducing the risk of in-stent restenosis to occur [24]. Others also evaluates various design parameters via numerical modelling to improve the blood flow characteristic in the vicinity of the stent wall. He et al. concluded that the strut design should align with the flow of the blood and flow circulation region need to be reduced [26].

To predict the stent hemodynamic performance after implantation, various hemodynamic parameters need to be evaluated. Parameters such as wall shear stress, wall shear stress gradient, and oscillatory shear index have been used to predict the performance of stent after stent implantation [21]. Other parameter, such as flow separation parameter is small and can be neglected in stent performance evaluation.

Wall shear stress (WSS) have been commonly used to quantitatively predict the formation of in-stent restenosis and the stent performance. WSS is a tangential force exerted on the arterial wall by the blood flow. WSS is governed by Eq. 3.1, where the scalar dot product of the unit normal to vector n to a surface and the fluid viscous stress tensor on the surface with direction of the viscous stress acting on the surface [21].

$$WSS, \boldsymbol{\tau}_w = \boldsymbol{n} \cdot \boldsymbol{\tau}_{ij} \tag{3.1}$$

In term of time-dependent flow, time-averaged wall shear stress (TAWSS) is commonly used and defined as shown in Eq. 3.2, where T s the duration of one cardiac cycle and dt s a small increment of time.

$$TAWSS = \int_0^T |\boldsymbol{\tau}_{\boldsymbol{w}}| dt \tag{3.2}$$

The normal average of WSS over a cardiac cycle is between 0.5 and 2 Pa. Endothelial cells subjected to WSS larger than 1 Pa tend to elongate and aligned to the flow. At WSS less than 0.5, the flow exhibit low and oscillatory WSS and tend to be circular in shape with no preferred alignment pattern, which indicates increase of uptake of blood borne particles to the artery. This is due to increase in residence time and permeability at the site of low WSS. Low WSS able to promote platelet activation and plaque build-up and more susceptible to formation of neointimal hyperplasia.

Another parameter that is taken into consideration in evaluating the hemodynamic performance of stent is wall shear stress gradient (WSSG). Arterial wall with WSSG of higher than 200 N/m^3 is susceptible to neointimal hyperplasia, which increased the permeability of arteries [21, 27]. Growth of endothelial cells is predicted with WSSG value of more than 5000 N/m^3 [21]. WSSG lower or equal than 5000 N/m^3 is taken as indication of good stent hemodynamic performance [1]. WSSG is governed as shown in Eq. 3.3. m is the WSS direction tangential to the endothelial cell surface and n s the tangential to the endothelial surface and normal to m

$$WSSG = \left[\left(\frac{\partial \tau_{w,m}}{\partial m} \right)^2 + \left(\frac{\partial \tau_{w,n}}{\partial n} \right)^2 \right]^{1/2} \tag{3.3}$$

Next parameter is oscillatory shear index (OSI), which described the cyclic variation which measure the deviation of WSS vector from its predominant axial direction of flow over a cardiac cycle. It is governed by the equation:

$$OSI = \frac{1}{2} \left(1 - \frac{\left| \int_0^T \boldsymbol{\tau}_{\boldsymbol{w}} dt \right|}{\int_0^T |\boldsymbol{\tau}_{\boldsymbol{w}}| dt} \right) \tag{3.4}$$

OSI takes value from 0 to 0.5, scaled from fully predominant flow in primary direction to WSS vector that is not aligned to the flow. Threshold value of 0.2 is considered before causing degenerative changes in arterial wall endothelium [28]. A high OSI value indicates disturbed flow region [29], with greater risk of arterial narrowing due to atherosclerosis.

Relative residence time is a parameter which predicts the residence time for stagnation flow to occur at the arterial wall. RRT is governed by:

$$RRT = \frac{1}{[1 - 2(OSI)]TAWSS} \tag{3.5}$$

Table 3 Parameters used in predicting in-stent restenosis

Author/year	Hemodynamic parameters
LaDisa et al. [27]	$WSS \leq 5$ dyn/cm^2
Balossino et al. [33]	$WSS \leq 5$ dyn/cm^2
Jimenez et al. [34]	WSS ≤ 5 dyn/cm^2
Pant et al. [40]	$WSS \leq 5$ dyn/cm^2
Duraiswamy et al. [35]	$WSS \leq 5$ dyn/cm^2 $WSSG \leq 200$ dyn/cm^3
Gundert et al. [31]	$WSS \leq 4$ dyn/cm^2 $OSI \geq 0.1$
Taib et al. [24]	$WSS \leq 5$ dyn/cm^2 $WSSG \leq 200$ dyn/cm^3
Taib [1]	$WSS < 5$ dyn/cm^2 $OSI > 0.2$ $RRT \leq 10$ Pa^{-1}
Chiastra et al. [41]	$WSS < 0.4$ dyn/cm^3 $OSI < 0.2$ $RRT > 4.17$ Pa^{-1}

Decrease in TAWSS managed to increase RRT value otherwise increase in OSI causes the increase of RRT.

These parameters were employed by researchers to evaluate the stent performance of the stent. Table 3 summarised some parameters used in the prediction of in-stent restenosis growth through computational analysis.

In the case of CFD, the stent and the blood vessel are assumed to be rigid. However, physiologically, the arterial wall and the stent are not rigid due to the elasticity of the blood vessel and the stent. One way to integrate the elastic property of the stent is by using fluid-structure interaction (FSI) modelling. From FSI, various structural strength due to the hemodynamic behaviour can be determine, and one of the method used is by evaluating the von Mises stress exerted on to the stent strut. Von Mises stress is the most common way to determine whether the structure will fail or fracture at a given load [1]. By using FSI modelling, the stent structure integrity due to the pulsatile blood flow and the compliance effect of the artery can be predicted by analysing the stress distributions of the stent strut configurations. Further modification on the stent strut are needed if failure due to the compliance effect of the arterial wall and pulsatile blood flow. On the other hand, the modifications on strut should not alter the hemodynamic behaviour which may induce in-stent restenosis.

3.1.2 In Vitro Analysis

One way to check the accuracy of hemodynamic characteristic obtained from numerical modelling, validation via experimental analysis. Particle image velocimetry (PIV)

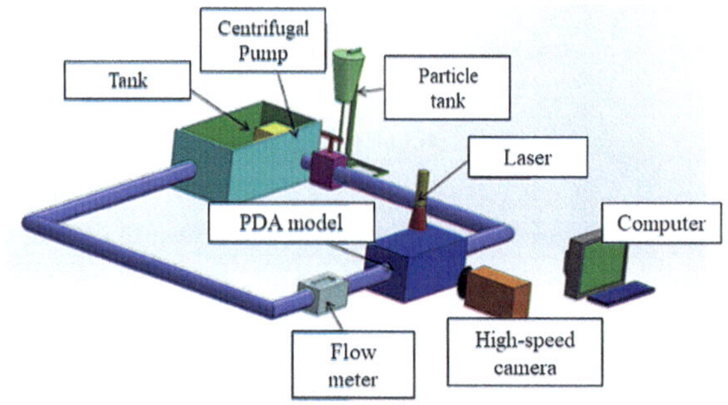

Fig. 6 Schematic diagram of an PIV experimental setup for PDA analysis

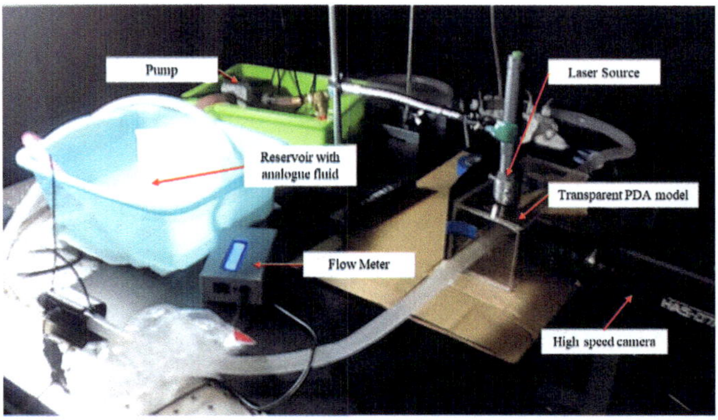

Fig. 7 Simple equipment setup for PIV experiment

is an optical visualization method that is able to qualitatively and quantitatively determine the velocity field in a plane. PIV requires a high-speed camera which is used to track the particles such as spherical aluminium particles, seeded in the flow field and a laser light that can create a laser sheet to determine the plane of analysis [30]. The model of the blood vessel is made using optically transparent material. The model of the blood vessel is attached to the mock loop which supply analogue fluid similar to blood (Figs. 6, 7 and 8).

PIV has the following advantages and disadvantages, which are shown in Table 4.

3.2 *PDA Stent Design Considerations*

Stent strut design plays a significant role in the development of in-stent restenosis in PDA as the strut caused diversion of blood flow which leads to excessive growth of cell proliferation with the embedded stent into fibrotic tissue [17]. Several consider-

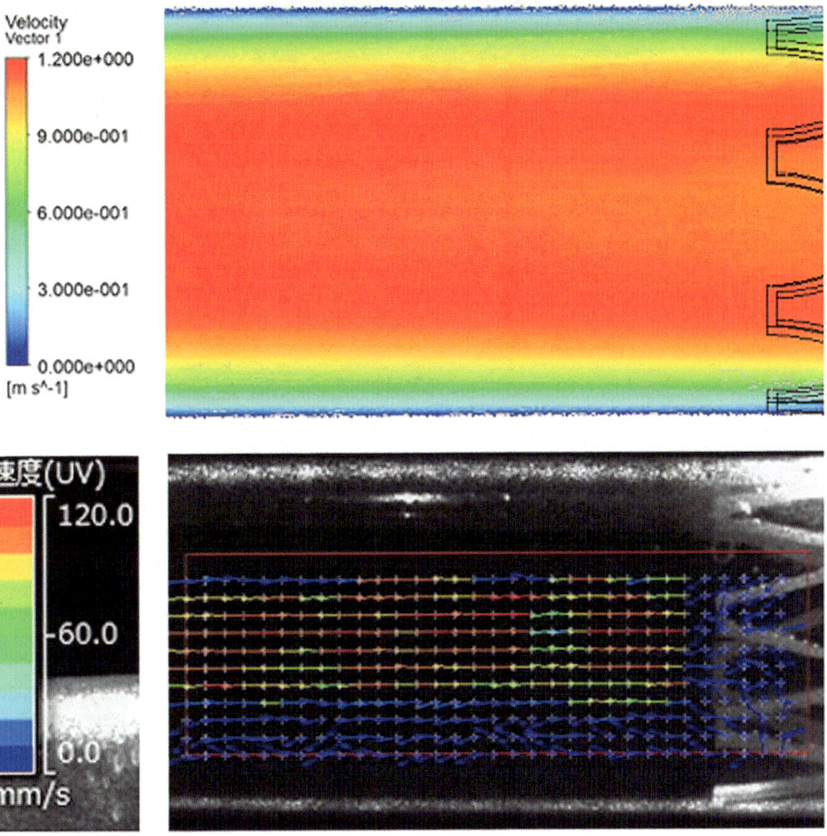

Fig. 8 Comparison of velocity contour of CFD modelling (top) with PIV results (bottom) in a simplified model of PDA

Table 4 Advantages and disadvantages of PIV [30]

Advantages	Disadvantages
Non-intrusive technique with negligible distortion due to added particles. Optical measurement avoids the usage of the intrusive flow measuring probes	Particles with higher density than fluid will not follow perfectly the flow of the fluid. May require tweaking in Reynolds number
Real-time or later-time measurement of fluid flow due to high speed data processing	PIV in general will not be able to measure components towards or away of the camera. Parallax error can occur. Solvable by using stereoscopic PIV
Subpixel displacement values allow high degree of accuracy	Accuracy of the flow is affected due to the averaging of resultant velocity components

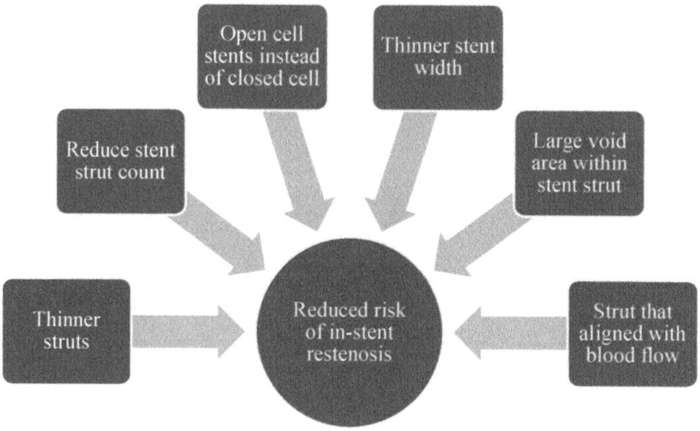

Fig. 9 Design consideration in reducing the risk of in-stent restenosis

ations on the stent strut configuration has been identified to be able to reduce the risk of in-stent restenosis. Modification on the design parameters are found to be able to reduce the risk of restenosis by reducing the deposition of platelets in the vicinity of the stent struts.

Several modifications on the stent strut configurations have been proposed to reduce the risk of in-stent restenosis as flow alterations of the local hemodynamic near to the stent strut configurations leads to the development of restenosis [24, 31]. Blood flow divergence create deposition of platelet which accelerated the progression of late thrombosis [25, 31, 32]. Using CFD analysis, researchers found out that by reducing the stent strut count will reduce the region affected by low WSS, which reduce the risk of formation of in-stent restenosis [27, 33]. Thinner strut also has been seen as a way to reduce the flow recirculation, which in turn reduce the area affected by low WSS [1, 27, 34, 35]. In addition, thinner strut promotes endothelialization process [27]. Open-cell stents also show good hemodynamic performance than a closed-cell stents in reducing the risk of restenosis [24, 35]. The stent strut configurations should be positioned so that it is aligned with the blood flow to reduce flow recirculation [1, 26]. Large void area within the stent strut configuration also shows favourable outcomes in reducing the risk of in-stent restenosis [1] (Fig. 9).

Severe branch pulmonary artery stenosis after PDA stenting is likely to occur and common in many forms of cyanotic congenital heart disease. Surgical reconstruction repair is likely to be problematic and unsatisfactory with a metallic stent embedded in dense fibrotic tissue [10]. Researchers recommended the usage of biodegradable stent as an alternative to current metallic stent. Biodegradable metal such as magnesium and biodegradable or bioabsorbable polymer like polylactic acid have been used in fabrication of stents and showed favourable results. However, drawbacks on the mechanical performance of polymer-based stents is found, in which the mechanical performance of the stent is not as strong as metallic stent counterpart [10]. Therefore,

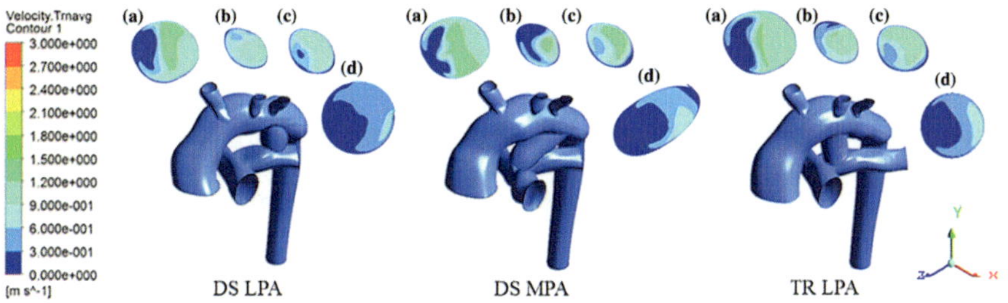

Fig. 10 Velocity contour of different morphologies in PDA. Dark blue region indicates region of flow recirculation which is susceptible to the formation of in-stent restenosis [38]

further modifications on the polymer mechanical properties and the stent design should be considered especially in polymer-based stents.

Thinner strut able to reduce the formation of in-stent restenosis and improved deliverability [36]. However, improvement of the stent materials is required to accommodate thinner strut configurations. Alloys such as cobalt chromium able to improve the stent structure by increasing the strength and radio-opacity as compared to the stainless steel. More complicated and novel design of stent able to be achieved which previously not possible with lower strength material [36].

The design of the stent should also consider the ductal morphologies of the PDAs. Since there is no specific study on the effect of PDA morphologies to the formation of restenosis at stented PDA, researchers have concluded that flow disturbances especially at flow divider region [37], similar to flow characteristic in PDA morphologies [38], has caused higher risk of in-stent restenosis. Neointimal build up is found to be localized at the lateral wall of stented vessel due to complex geometry at the region of flow divider [39]. These studies conclude that complex morphologies in relation with the location of stent implantation has substantial contribution to the formation of in-stent restenosis and need to be put in consideration in designing stent suitable for PDA applications (Fig. 10).

4 Conclusions

In summary, the implantation of stent in patent ductus arteriosus has been selected as a treatment for patient with cyanotic congenital heart disease. However, the implantation of stent in PDA has caused complication such as in-stent restenosis and requires further treatment. Researchers have concluded that the morphology of the vessel and stent design contribute to the formation of restenosis. One way to predict the formation of in-stent restenosis after stent implantation is by analysing the hemodynamic characteristic in stented PDA using CFD. By implementing CFD analysis, the effect of the design of the stent strut configurations and understanding of the complex PDA morphologies to the hemodynamic behaviour in stented vessel can be predicted and it is important in developing a new customized stent suitable for PDA stenting application.

Acknowledgments The support of the University of Teknologi Malaysia, under the Ministry of Education Malaysia and IJN-UTM Cardiovascular Centre grant, led by Dr. Kahar Osman under grant number 4F240, 01G77, 01G17 and 4L638 are gratefully acknowledged.

References

1. Taib, I.: Improvement of Haemodynamic Stent Strut Configuration for Patent Ductus Arteriosus. Universiti Teknologi Malaysia (2016)
2. Schneider, D.J., Moore, J.W.: Patent ductus arteriosus. Circulation **114**, 1873–1882 (2006). https://doi.org/10.1161/CIRCULATIONAHA.105.592063
3. Schneider, M., Zartner, P., Sidiropoulos, A., et al.: Stent implantation of the arterial duct in newborns with duct-dependent circulation. Eur. Heart J. **19**, 1401–1409 (1998)
4. Crossley, K.J., Allison, B.J., Polglase, G.R., et al.: Dynamic changes in the direction of blood flow through the ductus arteriosus at birth. J. Physiol. **587**, 4695–4704 (2009). https://doi.org/10.1113/jphysiol.2009.174870
5. LeRoy, S.: Patent Ductus Arteriosus. CS Mott Children's Hospital: University of Michigan Health System (2012). http://www.mottchildren.org/medical-services/ped-heart/conditions/patent-ductus-arteriosus. Accessed 17 Aug 2014
6. Bakhshali, M.A., Mafi, M., Daneshvar, S.: Mathematical modelling of the patent ductus arteriosus (PDA). Math. Comput. Model Dyn. Syst. **19**, 238–249 (2013). https://doi.org/10.1080/13873954.2012.727187
7. Murphy, P.J.: The fetal circulation. Contin. Educ. Anaesth. Crit. Care Pain **5**, 107–112 (2005). https://doi.org/10.1093/bjaceaccp/mki030
8. Alwi, M.: Stenting the ductus arteriosus: case selection, technique and possible complications. Ann. Pediatr. Cardiol. **1**, 38–45 (2008). https://doi.org/10.4103/0974-2069.41054
9. Boshoff, D.E., Michel-behnke, I., Schranz, D.: Stenting the neonatal arterial duct. Expert Rev. Cardiovasc. Ther. **5**, 893–901 (2007)
10. Alwi, M.: Stenting the patent ductus arteriosus in duct-dependent pulmonary circulation: techniques, complications and follow-up issues. Future Cardiol. **8**, 237–250 (2012). https://doi.org/10.2217/fca.12.4
11. Vida, V.L., Speggiorin, S., Maschietto, N., et al.: Cardiac operations after patent ductus arteriosus stenting in duct-dependent pulmonary circulation. Ann. Thorac. Surg. **90**, 605–609 (2010). https://doi.org/10.1016/j.athoracsur.2010.04.007
12. Gewillig, M., Boshoff, D.E., Dens, J., et al.: Stenting the neonatal arterial duct in duct-dependent pulmonary circulation: new techniques, better results. J. Am. Coll. Cardiol. **43**, 107–112 (2004). https://doi.org/10.1016/j.jacc.2003.08.029
13. Alwi, M., Choo, K.K., Latiff, H.A., et al.: Initial results and medium-term follow-up of stent implantation of patent ductus arteriosus in duct-dependent pulmonary circulation. J. Am. Coll. Cardiol. **44**, 438–445 (2004). https://doi.org/10.1016/j.jacc.2004.03.066
14. Matter, M., Almarsafawey, H., Hafez, M., et al.: Patent ductus arteriosus stenting in complex congenital heart disease: early and midterm results for a single-center experience at children hospital, Mansoura. Egypt. Pediatr. Cardiol. **34**, 1100–1106 (2013). https://doi.org/10.1007/s00246-012-0608-x
15. Feltes, T.F., Bacha, E., Beekman, R.H., et al.: Indications for cardiac catheterization and intervention in pediatric cardiac disease: a scientific statement from the American Heart Association. Circulation **123**, 2607–2652 (2011). https://doi.org/10.1161/CIR.0b013e31821b1f10

16. Odemis, E., Haydin, S., Guzeltas, A., et al.: Stent implantation in the arterial duct of the newborn with duct-dependent pulmonary circulation: single centre experience from Turkey. Eur. J. Cardio-thorac. Surg. **42**, 57–60 (2012). https://doi.org/10.1093/ejcts/ezr258
17. Alwi, M., Mood, M.C.: Stenting of lesions in patent ductus arteriosus with duct-dependent pulmonary blood flow. Interv. Cardiol. Clin. **2**, 93–113 (2013). https://doi.org/10.1016/j.iccl.2012.09.011
18. Alwi, M.: Patent ductus arteriosus stenting—problems, complications and technical consideration. Congenit. Cardiol. Today **3**, 1–16 (2005)
19. Krichenko, A., Benson, L.N., Burrows, P., et al.: Angiographic classification of the isolated, persistently patent ductus arteriosus and implications for percutaneous catheter occlusion. Am. J. Cardiol. **63**, 877–880 (1989). https://doi.org/10.1016/0002-9149(89)90064-7
20. Alwi, M.: Patent Ductus arteriosus stenting: problems, complications and technical consideration. Neonatol. Today **6** (2011)
21. Murphy, J., Boyle, F.: Predicting neointimal hyperplasia in stented arteries using time-dependant computational fluid dynamics: a review. Comput. Biol. Med. **40**, 408–418 (2010). https://doi.org/10.1016/j.compbiomed.2010.02.005
22. Gibbs, J.L., Uzun, O., Blackburn, M.E., et al.: Fate of the stented arterial duct. Circulation **99**, 2621–2625 (1999). https://doi.org/10.1161/01.CIR.99.20.2621
23. Alwi, M.: Core curriculum management algorithm in pulmonary atresia with intact ventricular septum. Catheter. Cardiovasc. Interv. **67**, 679–686 (2006). https://doi.org/10.1002/cci.20672
24. Taib, I., Kadir, M.R.A., Azis, M.H.S.A., et al.: Analysis of hemodynamic differences for stenting patent ductus arteriosus. J. Med. Imaging Heal Inform. **3**, 555–560 (2013). https://doi.org/10.1166/jmihi.2013.1197
25. Wernick, M.H., Jeremias, A., Carrozza, J.P.: Drug-eluting stents and stent thrombosis: a cause for concern? Coron. Artery Dis. **17**, 661–665 (2006). https://doi.org/10.1097/MCA.0b013e32801122b1
26. He, Y., Duraiswamy, N., Frank, A.O., Moore, J.E.: Blood flow in stented arteries: a parametric comparison of strut design patterns in three dimensions. J. Biomech. Eng. **127**, 637–647 (2005). https://doi.org/10.1115/1.1934122
27. LaDisa, J.F., Olson, L.E, Guler, I., et al.: Stent design properties and deployment ratio influence indexes of wall shear stress: a three-dimensional computational fluid dynamics investigation within a normal artery. J. Appl. Physiol. **97**, 424–30; discussion 416 (2004). https://doi.org/10.1152/japplphysiol.01329.2003
28. Glor, F.P., Ariff, B., Hughes, A.D., et al.: Image-based carotid flow reconstruction: a comparison between MRI and ultrasound. Physiol. Meas. **25**, 1495–1509 (2004). https://doi.org/10.1088/0967-3334/25/6/014
29. Caballero, A. D., Laín, S.: A review on computational fluid dynamics modelling in human thoracic aorta. Cardiovasc. Eng. Technol. **4**, 103–130 (2013). https://doi.org/10.1007/s13239-013-0146-6
30. Rathakrishnan, E.: Instrumentation, Measurements, and Experiments in Fluids. CRC Press, Taylor & Francis Group (2016)
31. Gundert, T.J., Dholakia, R.J., McMahon, D., LaDisa, J.F.: Computational fluid dynamics evaluation of equivalency in hemodynamic alterations between driver, integrity, and similar stents implanted into an idealized coronary artery. J. Med. Device **7**, 011004 (2013). https://doi.org/10.1115/1.4023413
32. Alwi, M., Choo, K.K., Radzi, N.A.M., et al.: Concomitant stenting of the patent ductus arteriosus and radiofrequency valvotomy in pulmonary atresia with intact ventricular septum and intermediate right ventricle: early in-hospital and medium-term outcomes. J. Thorac. Cardiovasc. Surg. **141**, 1355–1361 (2011). https://doi.org/10.1016/j.jtcvs.2010.08.085
33. Balossino, R., Gervaso, F., Migliavacca, F., Dubini, G.: Effects of different stent designs on local hemodynamics in stented arteries. J. Biomech. **41**, 1053–1061 (2008). https://doi.org/10.1016/j.jbiomech.2007.12.005
34. Jiménez, J.M., Davies, P.F.: Hemodynamically driven stent strut design. Ann. Biomed. Eng. **37**, 1483–1494 (2009). https://doi.org/10.1007/s10439-009-9719-9

35. Duraiswamy, N., Schoephoerster, R.T., Moore Jr., J.E.: Comparison of near-wall hemodynamic parameters in stented artery models. J. Biomech. Eng. **131**, 1–22 (2010). https://doi.org/10.1115/1.3118764.Comparison
36. O'Brien, B., Carroll, W.: The evolution of cardiovascular stent materials and surfaces in response to clinical drivers: a review. Acta Biomater. **5**, 945–958 (2009). https://doi.org/10.1016/j.actbio.2008.11.012
37. Nakazawa, G., Yazdani, S.K., Finn, A.V., et al.: Pathological findings at bifurcation lesions. The impact of flow distribution on atherosclerosis and arterial healing after stent implantation. J. Am. Coll. Cardiol. **55**, 1679–1687 (2010). https://doi.org/10.1016/j.jacc.2010.01.021
38. Kori, M.I., Osman, K., Taib, I.: Hemodynamic prediction in patent ductus arteriosus morphologies. APRN J. Eng. Appl. Sci. **12**, 3156–3160 (2017)
39. Koskinas, K.C., Chatzizisis, Y.S., Antoniadis, A.P., Giannoglou, G.D.: Role of endothelial shear stress in stent restenosis and thrombosis: Pathophysiologic mechanisms and implications for clinical translation. J. Am. Coll. Cardiol. **59**, 1337–1349 (2012). https://doi.org/10.1016/j.jacc.2011.10.903
40. Pant, S., Bressloff, N.W., Forrester, A.I.J., Curzen, N.: The influence of strut-connectors in stented vessels: a comparison of pulsatile flow through five coronary stents. Ann. Biomed. Eng. **38**, 1893–1907 (2010). https://doi.org/10.1007/s10439-010-9962-0
41. Chiastra, C., Gallo, D., Tasso, P., et al.: Healthy and diseased coronary bifurcation geometries influence near-wall and intravascular flow: a computational exploration of the hemodynamic risk. J. Biomech. **58**, 79–88 (2017). https://doi.org/10.1016/j.jbiomech.2017.04.016

Part IV
Cardiovascular Biomaterials

Biopolymers as Potential Carrier for Effervescent Reaction Based Drug Delivery System in Gastrointestinal Condition

I. I. Muhamad, S. Selvakumaran, M. A. M. Sis, N. Pa'e and M. H. Salehudin

Abstract Biopolymers are naturally occurring materials formed in nature during the life cycles of organisms. Biopolymers include the polysaccharides, carbohydrates and protein such as cellulose, starch, wool, silk, gelatine and collagen. In recent years, biopolymer-based hydrogels become important area of research in pharmaceutical aspects because of their promising properties in drug delivery system. These properties include low toxicity, biodegradability, stability and renewable nature. Numerous studies have been carried out in order to develop carrier from biopolymers with better controlled release properties. This is important to ensure precisely desired concentration of drug or essential nutrient absorption into the blood or tissue could be achieved. Among other different approaches, floating system is one of the most convenient, economical, and effective drug delivery system. Floating delivery system could potentially achieve longer retention time of carrier with capsulated bioactive drug or functional nutrients in the gastrointestinal tract. The floating behaviour of carrier could be induced by effervescent reactions. Effervescent reaction occurs between acidic gastric content and pore forming agent such as carbonates or bicarbonates salts incorporated into the carrier. This chapter discusses some of the use of biopolymers in drug delivery systems for effervescent reaction in gastrointestinal tract.

Keywords Biopolymers · Carrier · Effervescent reaction · Drug delivery system · Gastrointestine

---

I. I. Muhamad (✉) · M. A. M. Sis · N. Pa'e · M. H. Salehudin
Department of Bioprocess and Polymer Engineering, School of Chemical and Energy Engineering, Faculty of Engineering, Universiti Teknologi Malaysia, 81310 Johor Bahru, Johor, Malaysia
e-mail: idaidayu@utm.my

I. I. Muhamad
Cardiac Biomaterials Cluster, IJN-UTM Cardiovascular Engineering Center, Level 2, Block B, Building V01/FBME, UTM, 81310 Johor Bahru, Johor, Malaysia

S. Selvakumaran
Department of Biotechnology, School of Science and Engineering, Manipal University, Nilai, Negeri Sembilan, Malaysia

D. E. O. Dewi et al. (eds.), *Cardiovascular Engineering*, Series in BioEngineering,
https://doi.org/10.1007/978-981-10-8405-8_10

1 Introduction

Drug delivery system (DDS) is the process of delivering a pharmaceutical compound to achieve a therapeutic effect in humans or animals [1]. Drug delivery system becomes one of the most promising areas, in which pharmaceutical scientists are keen to develop various methods to deliver drugs at a controlled rate and slow delivery in order to enhance human health care. To date, great effort has been taken in developing potential drug vehicle or carrier in drug delivery system to increase the efficiency of drug delivery in human body which is safe to consume and comply with patient convenience [2].

Biopolymers are polymeric biomolecules produced by living organism. Biopolymers have been accepted and widely used as carriers for drug delivery due to their non-toxic properties and biocompatibility [3]. Recently, many researchers start to focus on the utilization of biopolymer based hydrogels for drug delivery application. Hydrogel is a three dimensional, hydrophilic, cross linked polymeric network that capable to imbibe huge amount of water or biological fluid. The nature of hydrogel such as biocompatible, non-toxic and biodegradable, release of entrapped molecules in a controlled manner, tissue-like and soft physical properties. These allow high permeability to undersized molecules, and low interfacial tension hence make hydrogel an excellent carrier for drugs, protein, peptides and other biological compound [4–7]. Hydrogels also have the ability to protect the drug from hostile environments such as the presence of enzyme and various pH conditions [8, 9]. The remarkable physiochemical and biological features of hydrogels have led them a promising vehicle for drug delivery system. Besides that, hydrogels are also widely explored in diverse biomedical fields, including biosensor, tissue engineering, wound dressing, biomolecule separation, and the contact lenses [10, 11].

Recently, a great deal of research focused on floating hydrogels as drug carrier for controlled drug release. Floating hydrogels are derived from the floating drug delivery system (FDDS). FDDS is a bulk density, lower than the gastric fluid and remains float over the stomach for an extended period without affecting the gastric emptying rate. While the system is floating on the gastric fluid, the drugs are released slowly at desired rate and time from the system [12]. Tablets, capsules, powders, granules, laminated films, hollow microspheres and hydrogels are examples of developed buoyant systems [13]. Floating hydrogels can be prepared from natural or synthetic polymers including chitosan, alginates, carrageenan, polyester and acrylic acid.

Floating hydrogels maximize the amount of drugs reaching their absorption site in solution and hence ready for absorption [14]. In addition, floating hydrogel results in dissolution of drug in the gastric fluid. This would then make it available for absorption in the small intestine after emptying the stomach content. It is expected that the drug will be fully absorbed from floating hydrogels if it remains in a solution form even at alkaline pH of the intestinal system [15, 16]. Moreover, the formation of porosity within hydrogels enables drug loading into the gel matrix and consequently release drug at a pre designed rate.

Table 1 Various drug delivery systems

Systems	Description
High density system	Remaining in the stomach for prolonged periods by sedimenting to the folds of the stomach
Bio-adhesive system	Enabling the localized retention system in the stomach
Swelling and expanding system	Preventing transit from the gastric sphincter
Floating system	Low density system which provide adequate buoyancy over the gastric contents

2 Drug Delivery Systems

The ideal drug delivery system is important in treating various diseases by providing a stable, economical delivery system and effective therapeutic effect to human or animal [17, 18]. A prolonged residence time of the delivery system will lead to a complete drug release and thus improve the drug bioavailability. There are several approaches to prolong the residence time of dosage form or drug carrier in the upper part of the gastrointestinal tract including bio-adhesive system [19], swelling and expanding systems [20], high density systems, and floating system. Table 1 shows various systems that can be used to prolong the residence time in the GIT [14].

2.1 Drug Delivery System in Gastrointestinal Tract

The muscular tube that link between mouth and anus is called gastrointestinal tract or the other name is gastric mucosa tract [21]. While eating, the food will enter the mouth and pass through the gastrointestinal tract and will be eliminated by digestion, secretion, absorption, and excretion process [22]. There are three main parts in gastrointestinal tract which are stomach, small intestine and large intestine [23]. All this part plays different fundamental function in the body.

In the stomach, the food consumed will be temporarily stored and undergo digestion process [22]. Then, digested food slowly transfer into duodenum or in familiar term is small intestine [24]. The stomach is an important site for enzyme production and hormone secretion [25]. Anatomically the stomach has three main regions there are fundus, body and pyrolic antrum [22, 25]. The various parts of the stomach are shown in Fig. 1.

The fundus is the dome like projection of the stomach which is the part for food storage. The body part of the stomach contains parietal cell which secretes hydrochloric acid, HCl whereas the pyrolic antrum functioning as mixing and grinding and responsible for hormonal phase of gastric acid secretion [25].

In biomedical, nutraceutical and pharmaceutical research, the gastrointestinal tract is the part that playing foremost important roles in drug delivery system [22]. By studying the characteristics either physically or mechanically of gastrointestinal tract,

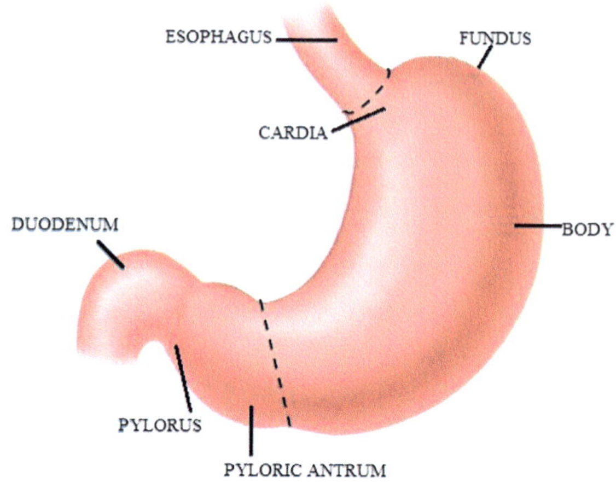

Fig. 1 The parts of the stomach

the great system of drug delivery can be created [26]. The transitions of an orally consumed drug or functional food through the gastrointestinal tract as well as drug release profile exert an excellent influence on the efficacy of medicine and functional food.

In these cases, many researchers design more advance drug delivery system to optimize the absorption of drug into a specific tissues or organs. Floating biopolymer based hydrogel is one of effort to prolong the gastric retention time [26]. The longer the retention time of hydrogel will achieve desired concentration of drug absorbed in blood or tissues [22]. The more concentrated of drug or functional food is absorbed will give higher therapeutic efficacy [27].

2.2 *Floating Systems*

The promise of increased bioavailability in gastric retentive drug delivery system has led to a significant level of effort devoted to enhance of such encapsulated drug for many years [28]. This improvement of bioavailability is expected be particularly significant to replace the conventional dosage which is wasting of drug due to incomplete drug release [29]. Within this context, the phrase 'gastric retentive dosage form' is intended to indicate the capsulated drug that can retain longer in the upper of gastrointestinal tract [16].

Davis, in 1968 is the first person described floating system [22]. The floating system is the low density system that has capabilities to float over gastric contents due to the buoyancy force that can remain in the stomach to prolong retention time [26]. Figure 2 shows the localization of floating microcapsule and high density system in stomach. In this system, the drug will release slowly into the gastric fluid and then it will be actively absorbed and transported by stomach wall [22]. The desired rate

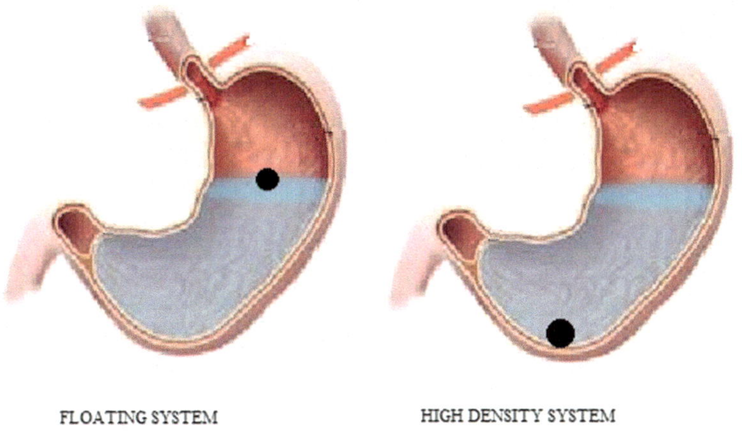

Fig. 2 Schematic localization of floating microcapsule and high density system in body of stomach

of drug release will increase gastro-retention time and reduces fluctuation in plasma drug concentration [24].

Increase of the retention time period of drug in gastrointestinal tract will be affected by several factors, but the gastric emptying time in individuals is particularly influenced by the size of the orally managed object [22]. Therefore, many studies for prolonging drug using gastro retention and floating system have been found to be promising because that can increase the gastro retention without depending on the dose size of orally uptake.

Many factors are affecting the gastric retention include density, size, shape of dosage form, single or multiple unit formulation and coloric content in feeding of natural of meal. Gastrointestinal tract is a function of amount dosage form buoyancy that hooked on the density [30]. The floating drug delivery system is classified base on the use of two formulation variables [16]. There are effervescent and non-effervescent systems. The effervescent floating drug delivery refers to the swellable matrix with cross-linked with swellable biopolymers, for example methyl cellulose and chitosan. These matrixes were formulated with effervescent compounds such as sodium bicarbonate ($NaHCO_3$), calcium carbonate ($CaCO_3$), tartaric acid and citric acid which can produce carbon dioxide gas when they were in contact with acidic gastric juice in gastrointestinal tract [16, 22, 24]. The CO_2 gas will be entrapped in swollen hydrocolloids and then give buoyancy effect to the matrix gel. Figure 3 shows the different layers of matrix gel and the mechanism of floatation by CO_2 generation.

There are various mechanisms to retain the oral dosage forms in order to increase the gastro-retention time in the stomach. These attempts could be done by introduce gas generating system and swelling or expanding, high density system and modified shape systems [22, 24]. Other mechanism is bio adhesive or muco-adhesive system which is helping the stomach to reside the dosage form during fasting or emptying

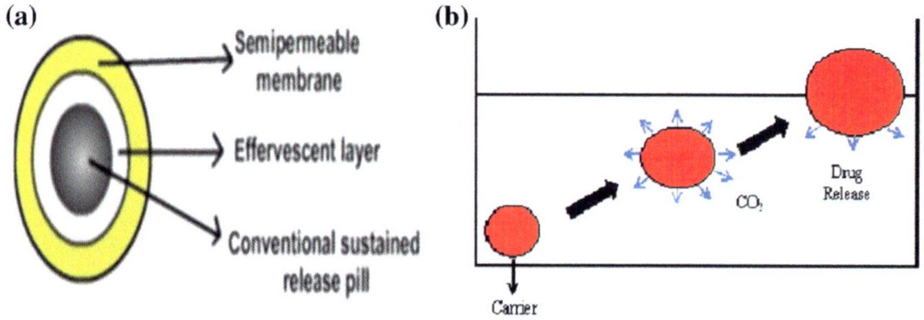

Fig. 3 **a** Three different layer of matrix gel; **b** mechanism of floatation via CO_2 generation

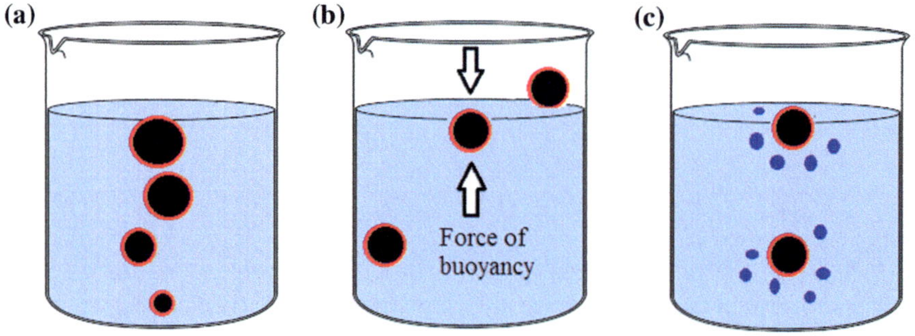

Fig. 4 Different mechanisms of floating system. **a** Swelled dosage forms; **b** density concept system; **c** gas generating system

the stomach [22]. By using these floating mechanisms, the drug release will be under control at desired rate. Figure 4 shows the different mechanisms of floating system.

According to Chandel et al. [24], longer retention time of drug in stomach will give many advantages in enhancing therapeutic effect of dosage, desirable absorption plasma drug concentration, and complete drug release. The floating micro particulate drug delivery system improves patient compliance by decreasing dosing uptake frequency, increase gastro retention time, hence site specific drug target is achieved and results in optimum pharmacotherapy due to continuous drug release [26].

3 Effervescent Reaction Using Pore Forming Agents

3.1 Calcium Carbonate as Pore Forming Agent

Calcium carbonate is one of the popular pore forming agents that is widely used in various applications. Calcium carbonate is a chemical compound with the chemical formula $CaCO_3$. It is formed by carbon, oxygen, and calcium. $CaCO_3$ has extensively

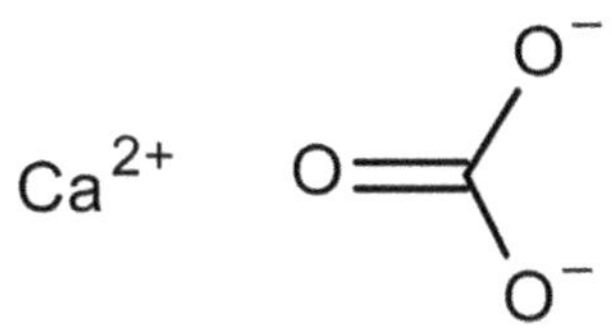

Fig. 5 Chemical structure of calcium carbonate

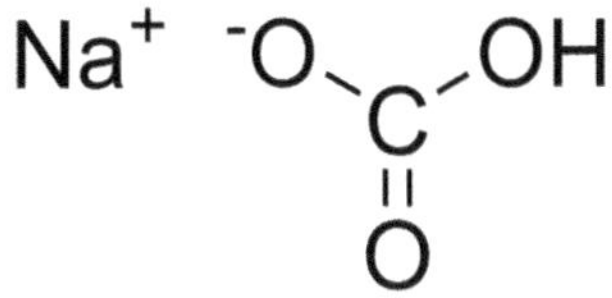

Fig. 6 Chemical structure of sodium hydrogen carbonate

been employed in industrial materials, health and dietary, environmental and food industry applications.

In medical application, $CaCO_3$ is used as an inexpensive dietary calcium supplement or gastric antacid. It is also used as a phosphate binder for the treatment of hyperphosphatemia (chronic renal failure) and as filler for tablets in pharmaceutical industry [31]. In food industry, $CaCO_3$ acts as food additive and is used as acidity regulator, anticaking agents and stabilizer, dietary calcium, soy milk and almond milks [32], and as firming agents in many canned or bottled vegetable products. The chemical structure of calcium carbonate is displayed in Fig. 5.

In drug delivery system, $CaCO_3$ has broad utilization due to its high availability, low cost, safety, biocompatibility, pH sensitivity, and slow biodegradability. Calcium carbonate is used as pore forming agents for developing a floating carrier in drug delivery system. $CaCO_3$ in carrier will react with gastric fluid and release CO_2 and permeate pore on hydrogels matrix causing the hydrogel to become less dense than the fluid, making it to remain afloat for a prolong time, and release drug in a controlled manner [33, 34].

Abou El Ela et al. [35] synthesized floating beads using calcium carbonate as gas forming agents to make a sustained release of Ketorolac tromethamine in the stomach. While Yellanki and Neralla [33] utilized calcium carbonate as gas forming agents to develop floating alginate beads for stomach specific drug delivery. It was found that calcium carbonate amount gives significant effect on size analysis, morphology test, drug entrapment efficiency, buoyancy and in vitro release. There are several research papers on calcium carbonate as pore forming agents for floating system [34, 36].

3.2 *Sodium Hydrogen Carbonate as Pore Forming Agent*

Sodium hydrogen carbonate (Sodium bicarbonate), commonly known as baking soda, is a chemical compound with the formula $NaHCO_3$. $NaHCO_3$ has regularly been used as a gas forming agent to develop floating carrier [37–41]. Figure 6 represent the chemical structure of $NaHCO_3$.

Chen et al. [42] developed floating tablets using $NaHCO_3$ as gas forming agents to examine the floating properties and swelling characteristics. It was found that by adding $NaHCO_3$ makes the tablets reached surfaces within 1 min and sustained buoyancy for more than 24 h. The buoyancy effect shortens the floating lag time further because gas bubbles are formed after reacting with acidic medium. They concluded that addition of $NaHCO_3$ could improve the floating capabilities (floating lag time and floating duration). There are also some research papers on potassium carbonates and calcium carbonates as pore forming agents in developing floating carriers [33, 43].

Choi et al. [44] prepared floating alginate beads to investigate the effect of two different CO_2 gas forming agents ($CaCO_3$ and $NaHCO_3$). Riboflavin was used as a model drug. The results showed that the drug release rate was increased consistently with the addition of $CaCO_3$ and $NaHCO_3$. They concluded that $CaCO_3$ is an excellent CO_2 gas forming agent compared to $NaHCO_3$ in alginate bead preparations.

4 Biopolymers as Hydrogel for Drug Delivery System

4.1 Starch

Starch is originated from variety of crops such as potato, wheat, rice and corn. The source is abundant, and readily available at low cost [45]. Starch in chemical term is a polysaccharide. It consists of a number of monosaccharides or glucose molecules joined together with a -D-(1–4) and/or a -D-(1–6) linkages. Main structural component of the starch includes amylose and amylopectin [46].

Figure 7 shows the structure of starch. However the relative amount of these components are varied as a function of the starch source (e.g., corn, potato, tapioca, wheat) and effects on the molecular order and crystallinity of the polysaccharide [47]. Amylose is linear or slightly branched, while amylopectin is highly branched. About 70% of the mass of starch granule is considered as amorphous and about 30% as crystalline and less than 1% lipids and protein from plant. Starch can be converted into hydrogel in a thermally three-step assisted, hydration-plasticization of the polymeric network [48]. Initially, adsorption of water in hydrophilic starch granules cause it to swell.

Next is gelatinization process that will occur after starch is dissolved by heating, resulting in leaching of the amylose component, irreversible physical changes, and the destruction of the granule structure. Finally, retrogradation step will continue where the starch hydrogel network is created upon cooling and aging, resulting in partial recrystallization and reorganization of the polysaccharide structure. Here, the amount of amylase and gelatinization temperature are the two main process parameters affecting the gel formation [49, 50].

Synthesis of hydrogel is achievable via various pathways. There are also studies that has been conducted on hydrogel synthesis. Preparation of the polysaccharide-

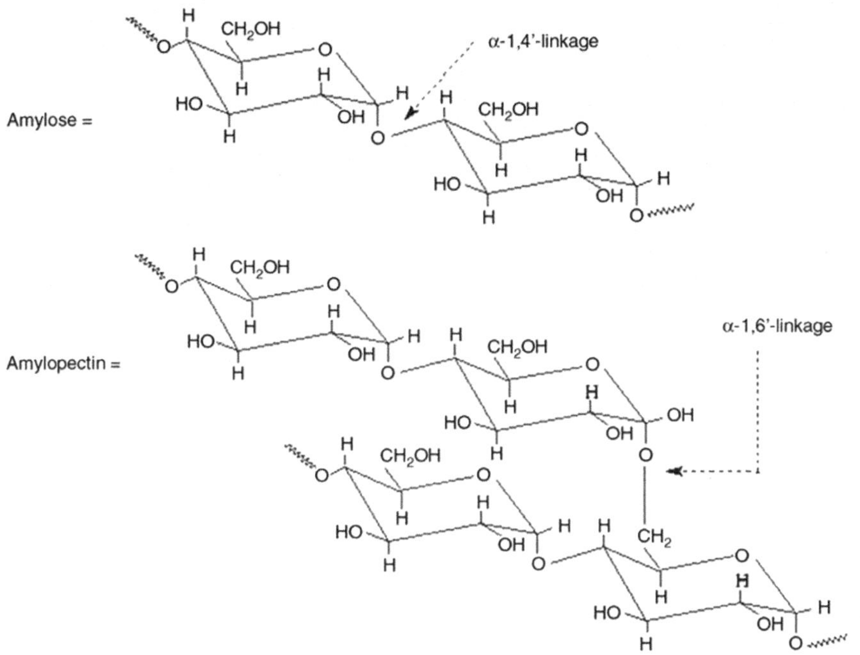

Fig. 7 Structure of amylopectin and amylose in starch

based hydrogels via reaction can be classified into two main groups: (1) graft copolymerization of vinyl monomers on polysaccharide in the presence of a crosslinker; and (2) direct crosslinking of polysaccharide. For example, previous study reported on the preparation of starch-based hydrogel through graft copolymerization of acrylic acid onto maize starch [51].

Differ from graft copolymerization, direct crosslinking of polysaccharides are using polyfunctional compounds (e.g., glycerol, glyoxal, and epichlorohydrine) or polyvinylic compounds (e.g., divinyl sulphone, DVS) [52]. An example of direct crosslinking was reported where the preparation of superabsorbent hydrogels derived from cellulose and crosslinked with citric acid (CA) [53].

4.2 *Carrageenan*

Carrageenan is extracted from certain types of red algae, a seaweed where is abound in coast of North America and Europe. It could be derived from *Kappaphycus alvarezii* a type of seaweed that was found in Tawau, Sabah which had been reported as the first food-grade natural κ-carrageenan in Malaysia [54]. In accordance with easy to get and abundant in Malaysia, κ-carrageenan is a new alternative for researchers to formulate hydrogel.

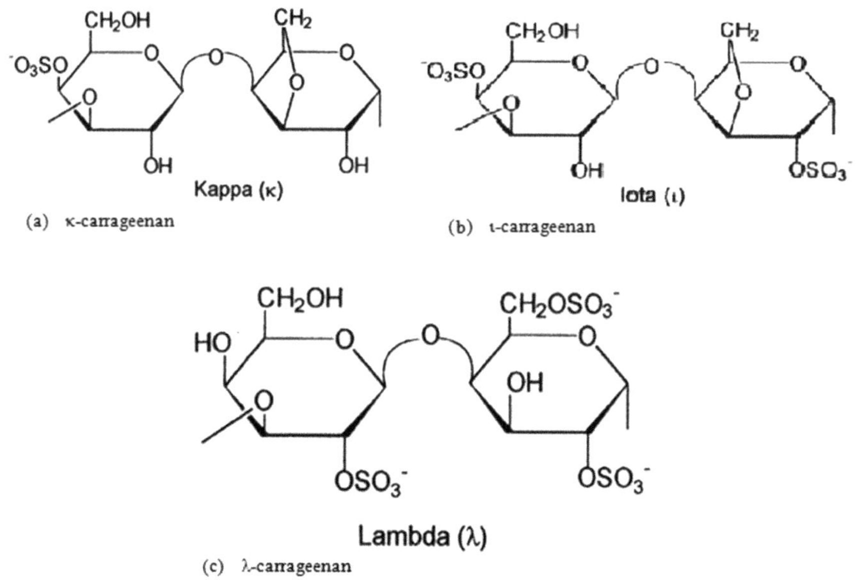

Fig. 8 Principle carrageenan structure of **a** κ-, **b** ι- and **c** λ-carrageenan

The carrageenan extracted from different species of marine red algae of class *Rhodophyceae sp.* is water-soluble, anionic and linear polysaccharide. Water-solubility of carrageenan is referred to sulphate group in its structure [55]. Several kinds of carrageenan exist such as kappa (κ), iota (ι), and lambda (λ) but the more important types are κ-carrageenan and ι-carrageenan because they are capable of forming thermo-reversible gels [55]. The different type of carrageenan is according to the position of sulphate group in its structure [56]. The arrangement of sulphate group in carrageenan structure would influence the gel formation [57].

Figure 8 shows the structures of ι-carrageenan, κ-carrageenan and λ-carrageenan. The idealized structures of ι-carrageenan and κ-carrageenan are the backbone structure of alternating β-1,3 and α-1,4-linked galactose residue (D-enantiomer) [58]. The structure of κ-carrageenan and ι-carrageenan are almost the same but differs in the presence of one additional sulphate ester group on the repeating disaccharide unit for ι-carrageenan.

Gel formation of carrageenan depends on the type of carrageenan. The κ-carrageenan and ι-carrageenan can form gel to produce hydrogel. Whereas λ-carrageenan could not form gel, but it is used as thickener which can form high viscosity solution [58]. Table 2 shows the different types of carrageenan.

By referring to Table 2, the behaviours of κ-carrageenan and ι-carrageenan almost similar. Imeson [59] stated κ-carrageenan gel is hard, strong and brittle gels during it was frozen, whereas ι-carrageenan gel is weak gels although it was frozen. Most ι- and κ-carrageenan solution will set into gel structure at ambient temperatures.

Table 2 Different properties of carrageenan types

Carrageenan types	Elasticity	Appearance	Gel formation	Stability	Viscosity
Iota (ι)	Elastic	Clear	Can form gel	Stable	Viscous
Kappa (κ)	More elastic	Clear	Can form gel	More stable	Viscous
Lambda (λ)	Non-elastic	Unclear	No gel formation	Unstable	High viscosity

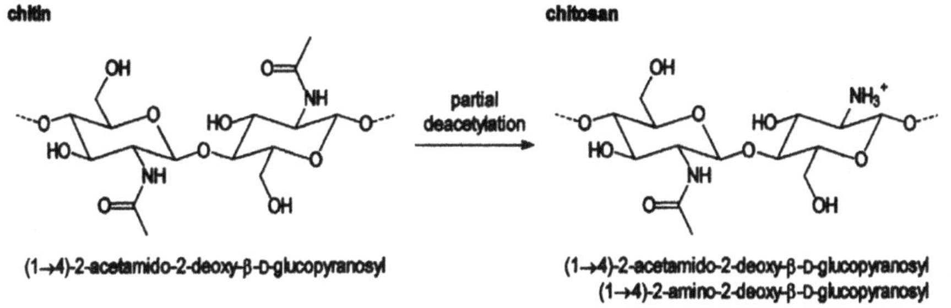

Fig. 9 Structure of chitin and chitosan

4.3 *Chitosan*

Chitosan is a biopolymer that has unique characteristics which is cationic, abundant, low toxicity, biodegradable, and non-immunogenic. Chitosan has the reinforcement capability which can enhance starch-based material. In other word, it shows that highly-acetylated chitosan can act both as an emulsifier and as a flocculant.

Chitosan–thioglycolic acid conjugates appear to be very promising new excipients for liquid or semisolid formulations based on their in situ gelling properties. As a result, they should stabilize themselves once applied on the site of drug delivery [60]. The relative presence of the two monomeric building units (N-acetyl-glucosamine and D-glucosamine) is crucial to whether chitosan is predominantly an ampholyte or predominantly a polyelectrolyte at acidic pH-values. The chemical composition is not only crucial to its surface activity properties, but also to whether and why chitosan can undergo a sol–gel transition [61]. Obtained from crustaceans such as crabs, prawns, shrimps, lobsters and cell wall of some fungi such as aspergillus and mucor.

Chitosan has a very good gelling properties to form hydrogel that suitable for biomedical applications, e.g., in tissue engineering and drug delivery, as well as the chitosan's surface activity and its role in emulsion formation, stabilization and destabilization. The chemical structure of chitin and chitosan are described in Fig. 9.

Chitosan is a polycationic polymer with a specific structure and properties. It contains more than 5000 glucosamine units and is obtained commercially from shrimp

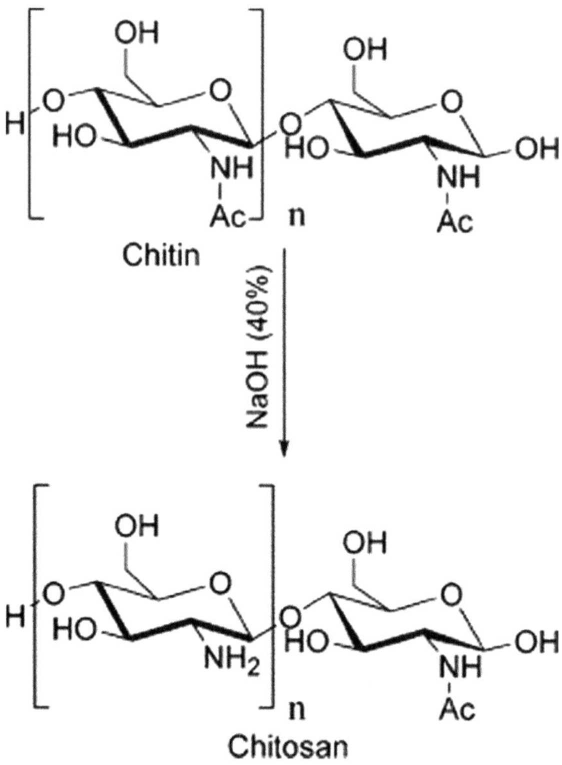

Fig. 10 Preparation of chitosan from chitin

and crab shell chitin (a N-acetylglucosamine polymer) by alkaline deacetylation2-4 (NaOH, 40–50%) as in Fig. 10.

4.4 Cellulose and Its Derivative

4.4.1 Microcrystalline Cellulose

Microcrystalline cellulose has many advantages in the formulation and production of solid dosage forms, however, it has some limitations such as sensitivity to lubricants, moderately low bulk density, loss of compatibility after wet granulation and moderate flowability [62].

In order to improve the function of the microcrystalline cellulose, silicification is applied. Therefore, some properties such as compressibility, enhanced density, compatibility, low moisture content, larger particle size, flowability, low moisture content, and lubricity will improve. By co-drying, a suspension of colloidal silicon dioxide like dried finished product contains 2% colloidal silicon dioxide and microcrystalline cellulose particles, silicified microcrystalline cellulose (SMCC) are manufactured.

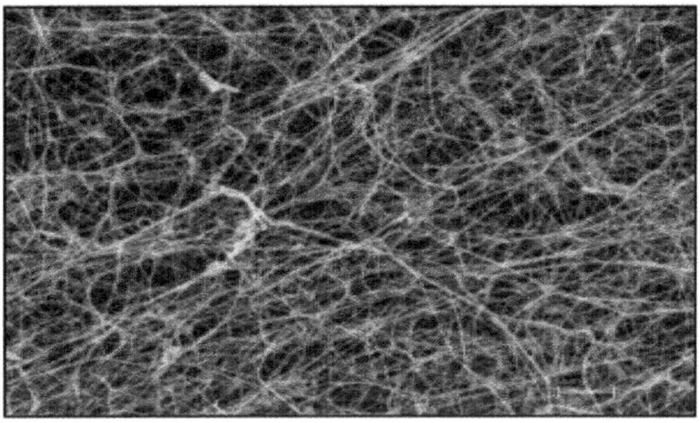

Fig. 11 SEM micrograph of the bacterial cellulose

The silicon dioxide remains on the surface of microcrystalline cellulose. Rather than the usual kinds of microcrystalline cellulose, silicified microcrystalline cellulose displays higher bulk density [63].

4.4.2 Bacterial Cellulose

Bacterial cellulose as hydrogel is a form of cellulose from bacteria. It is produced by bacteria from many genera such as *Acetobacter, Achromobacter, Agrobacterium and Sarcina*. Cellulose produced by microbes is uniquely different compared to cellulose acquired from plant with many points of interest that can be commercially valuable. Bacterial cellulose is well known for its purity. It is devoid of lignin and hemicellulose, extremely hydrophilic, and has excellent shape and strength retention. This is due to longer fiber length of bacterial cellulose. These characteristics of bacterial cellulose open many rooms for new applications in various fields. Figure 11 shows SEM image of bacterial cellulose.

Acetobacter xylinum (*A. xylinum*) which is acetic acid-producing bacterium has high capability to produce cellulose by converting carbon source like glucose to cellulose. *A. xylinum* produces cellulose from glucan chains. The chains extrude into the fermentation medium from *A. xylinum* pores. Each single cell of the bacteria continuously secretes nanofibres that scattered randomly and woven into fibrous layers of segmented nanofibres. These processes repeated until a bundle of microfibrils gathered and form bacterial cellulose. Figure 12 demonstrates formation of microfibrils by *A. xylinum*. The cellulose synthesis will continue until a limited condition such as when not enough carbon sources exists, change of pH [64] or when the bacterial cellulose fills the discs in fermentation using rotary discs reactor [65].

Bacterial cellulose pellicle is extremely hydrophilic, absorbing 60–700 times its weight in water. The nanofibers presence in a structure of multiple cellulose layer in bacterial cellulose which capable to hold an extensive amount of liquid between the fibrous layers to form a hydrogel [66]. Modification of bacterial cellulose had been done to enhance properties of native BC and impart some additional properties for certain specific application. Microfibrils of BC become denser with time and produce

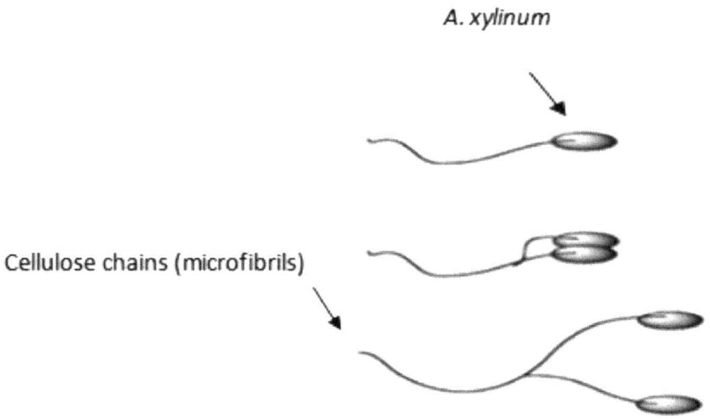

Fig. 12 Secretion of microfibrils by *Acetobacter* cells

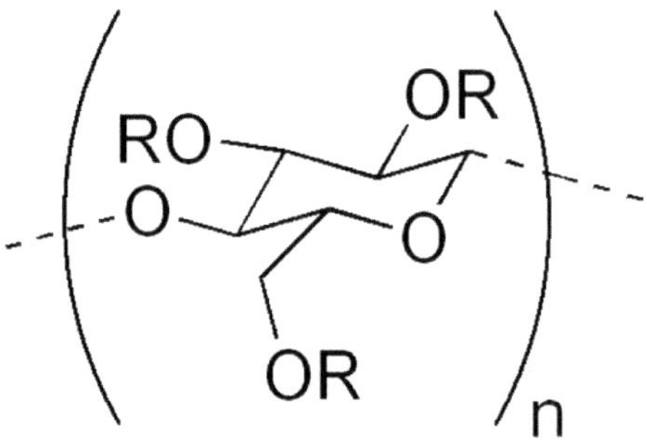

Fig. 13 Molecular structure of carboxymethyl cellulose ($R = H$ or CH_2CO_2H)

a web shaped structure [67, 68] that can trap various materials added to the medium [69]. The encaged materials become part of the bacterial cellulose fibril network, resulting in bacterial cellulose composites. A variety of additives materials had been used as a filler resulted to development of many new composites materials design for application on different fields. By adding certain substrates as filler, it is possible to change the properties of the cellulose.

4.4.3 Carboxymethyl Cellulose (CMC)

Figure 13 shows the molecular structure of carboxymethyl cellulose which is often used as its sodium salt, sodium carboxymethyl cellulose. It is a soluble polyanionic polysaccharide derivative of cellulose that has been employed as an emulsifying agent in the cosmetic and pharmaceutical industry.

This polymer has many crucial properties suitable for many applications such as stabilizer, thickener, film-former and binder [70]. One of its applications in biomedicine after a surgical procedure is to prevent epidural scare and soft tissue adhesions. Other applications including the therapeutic application of the superox-

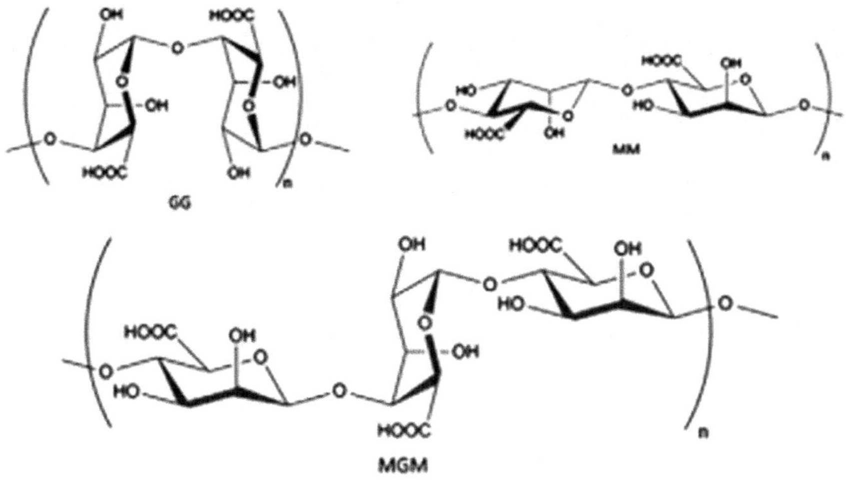

Fig. 14 Arrangement of b-D-mannuronic and a-L-guluronic acids in alginate structure

ide dismutase enzyme (SOD), as hydrogels of CMC carrying the enzyme for its controlled release and also as water absorbent in treating edemas [71].

Therapeutic use of SOD enzyme is limited by its fast clearance from the bloodstream and inactivation by its own reaction product, i.e. hydrogen peroxide. Meanwhile NaCMC is being used in the preparation of the semi-interpenetrating network of polymer, glutaraldehyde as a crosslinker [72].

4.5 Alginates

Various algae species are used in the alginate production that involve extraction of polysaccharide. Naturally, it consists of b-D-mannuronic and a-L-guluronic acids and structure sequential distribution arrangement could be either in blockwise fashion as homopolymer blocks (MM, GG) or alternating blocks of M and G with different M/G ratios as illustrated in Fig. 14 [73, 74]. It is widely available in the market molecular weights from tens to hundreds of kilo daltons [75].

The advantages of using alginate for producing hydrogel are because of its simplicity, GRAS (generally recognized as safe) status, non-toxicity, biocompatibility and low cost [73, 75, 76]. However, several aspects could be considered such as the process is difficult to scale-up and the hydrogel is very porous. Different approaches have been performed previously in reducing the porous effect such as mixing or coating the alginate hydrogel with other polymer compounds or incorporate different additives to modify the alginate structure [73, 76]. This characteristic allows alginate to be widely applied in agriculture, food and pharmaceutical industries [74].

In the alginate transformation, the presence of divalent ions such as Ca^{2+} is crucial to bind to guluronic residues and subsequently cross-linking the alginate structure. This is due to negative charge presence in the alginate which attributed to the structure of carboxylic acid groups [74, 75]. The size of alginate hydrogel formed depends on the methodology used in producing the hydrogel. For example, the hydrogel size approximately of ten microns could be produced using spray technology, however the extrusion method produces up to millimetre size of hydrogel [75].

5 Biocompatibility of Polymeric Hydrogels

Polymeric hydrogels are widely used as drug vehicle in drug delivery application. Therefore, it is essential for hydrogels as drug carrier to be biocompatible and non-toxic in order to be safely used. Biocompatibility is also known as the ability of materials to perform with an appropriate host response in specific application without any side effect [77].

Normally, polymeric hydrogels are biocompatible, biodegradable, and show excellent biological activity. However, low mechanical strength and high rates of degradation of natural polymers often results in their use in combination with nanoparticles, or in subsequent crosslinking reactions to reduce the degradation rates. Moreover, unreacted monomers, initiators, and crosslinker can leach out during application and cause toxicity to human body [78, 79].

It is necessary to carry out a number of tests on hydrogels to evaluate their biological tolerance and biocompatibility. The test may involve in vivo analysis using animals such as mice and in vitro analysis using cell line. In vivo test is normally expensive and involves regulatory concerns. Due to this in vitro testing are frequently performed to characterize newly developed biomaterial. In vitro tests for biocompatibility analysis involves cytotoxicity tests of the aspect of the material in the presence of live host cells and can usually be done in two ways in direct method and indirect method (MTT assay).

Li et al. [80] investigated the cytotoxicity and biocompatibility evaluation of N,O-carboxymethyl chitosan/oxidized alginate hydrogel for drug delivery application using NH3T3 cells line. The results clearly showed the NH3T3 cell spread and grow on the cell culture with the same morphology as those on normal cell culture after 24, 48, and 72 h of seeding. Thus, this indicates that the developed hydrogels are a safe, non-toxic carrier which is suitable for several drug delivery application.

Popa et al. [81] evaluated the in vitro and in vivo biocompatibility of kappa carrageenan based hydrogels using mouse fibroblast L929 cell line. In vitro cytotoxicity test was conducted using indirect method (MTS assay). They found that kappa carrageenan does not induce any harmful effect on cell line, thus showing that kappa carrageenan has low cytotoxic level and can be used safely for various biomedical applications, especially fast cell or bioactive molecules delivery.

The L929 (mouse fibroblast) cell line was used to investigate the cytotoxicity studies of bacterial cellulose/acrylic acid hydrogel in two different electron-beam radiations at 35 and 50 kGy. It was found that the percentage of viable cells after indirect and direct contact at 24 and 48 h were more than 85% for both hydrogels [82]. It has been suggested that any wound dressing that results in cell viability greater than 85% can be considered non-toxic for use as a wound dressing [83].

In addition, nanoparticles are another important constituent in nanocomposite hydrogels as it provides an improvement in physical application and also expansion of hydrogels application especially in drug delivery field. Thus, evaluating nanocomposite hydrogels safety is essential since the biocompatibility of the nanoparticles depends on a number of characteristics specific to the material including size, chemical composition, shape, and surface interactions. Toxicity analysis of nanocomposite hydrogels is very important as it can harm human by sticking to the cells or due to cellular uptake of the particles.

Meenach et al. [84] conducted biocompatibility analysis of magnetic hydrogel nanocomposites based on poly (N-isopropylacrylamide) and iron using NIH 3T3 murine fibroblasts cell line. The hydrogel and magnetic hydrogel nanocomposites involved in their study exhibited minimal cytotoxic effects on murine fibroblasts indicating the safety of their use in implant applications.

Moreover, there are many ongoing researches on the toxicity of the MgO nanoparticles towards cells and organs. The cytotoxicity analysis of MgO nanoparticles on human umbilical vein endothelial cells (HUVECs) was conducted by Ge et al. [85]. The result showed that MgO NPs below 200 g/mL exhibited non-toxicity while the growth rate of cell was lower if MgO NPs was higher than 500 g/mL.

6 Conclusion

Oral controlled-release drug delivery system provides drug release at a predetermined, predictable, and controlled rate and has drawn considerable attention. The literature survey has shown reports on developed floating system by various techniques including selected approaches clearly exhibited that residence time can be prolonged over long period. Although floating drug delivery system is necessary in gastrointestinal tract, there are few challenges arises in developing floating carrier in order to achieve optimum drug entrapment and drug released. To avoid this limitation, extensive research on modification of biopolymer based hydrogel need to be carried out in designing of sustained drug delivery system that can highly control the drug release, so the drug could be supplied continuously to its absorption site in gastrointestinal tract controlled manner.

Acknowledgements The authors would like to thank the Ministry of Science, Technology and Innovation of Malaysia, Ministry of Higher Education of Malaysia and Research Management Centre of Universiti Teknologi Malaysia (UTM) for the support of this project.

References

1. Tiwari, G., Tiwari, R., Sriwastawa, B., Bhati, L., Pandey, S., Pandey, P., Bannerjee, K.S.: Drug delivery system: an updated review. Int. J. Pharm. Investig. **2**(1), 2–11 (2012)
2. Saboktakin, M.Z.: Hydrogels as potential nano-scale drug delivery systems. Biopolymers, 575–596 (2010)
3. Sharma, A.K.: Biopolymers in drug delivery. Biopolym. Res. **1**(1) (2017)
4. Muhamad, I.I., Fen, L.S., Hui, H.Ng., Mustapha, N.A.: Genipin-cross-linked kappa-carrageenan/carboxymethyl cellulose beads and effects on beta-carotene release. Carbohydr. Polym. **83**, 1207–1212 (2011)
5. Peppas, N.A.: Hydrogels in Medicine and Pharmacy, vols. I–III. CRC, Boca Raton, FL (1987)
6. Ratner, B.D., Williams, D.F.: Biocompatibility of Clinical Implant Materials. CRC, Boca Raton, FL (1981)
7. Ratner, B.D., Hoffman, A.S., Andrade, J.D.: Hydrogels for Medical and Related Applications, p. 1. American Chemical Society, Washington, DC (1976)
8. Nho, Y.C., Park, S.E., Kim, H.I., Hwang, T.S.: Oral delivery of insulin using pH sensitive hydrogels based on polyvinyl alcohol grafted with acrylic acid/methacrylic acid by radiation. Nucl. Instrum. Methods Phys. Res. B **236**, 283–288 (2005)
9. Qiu, Y., Park, K.: Environment-sensitive hydrogels for drug delivery. Adv. Drug Deliv. Rev. **53**, 321–339 (2001)
10. Slaughter, B.V., Khurshid, S.S., Fisher, O.Z., Khademhosseini, A., Peppas, N.A.: Hydrogels in regenerative medicine. Adv. Mater. **21**, 3307–3329 (2009)
11. Yaszemski, M., Trntolo, D., Lewandrowski, K.U., Hasirci, V., Altobelli, D., Wise, D. (eds.): Tissue Engineering and Novel Delivery Systems. CRC Press (2004)
12. Singh, B.M., Kim, K.H.: Floating drug delivery systems: an approach to oral controlled drug delivery via gastric retention. J. Controlled Release **63**, 235–259 (2000)
13. Bhardwaj, V., Nirmala, Harikumar, S.L.: Floating drug delivery system: a review. Pharmacophore **4**(1), 26–38 (2013)
14. Dolas, T.R., Hosmani, A., Bhandari, A., Kumar, B., Somvanshi, S.: Novel sustained release gastroretentive drug delivery system: a review. Int. J. Pharm. Res. Dev. **2**(11), 26–41 (2011)
15. Chordiya, M.A., Senthilkumaran, K., Gangurde, H.H., Tamizharasi, S.: Floating drug delivery system: a versatile approach for gastric retention. Int. J. Front. Res. **1**(3), 96–112 (2011)
16. Mayavanshi, A.V., Gajjar, S.S.: Floating drug delivery systems to increase gastric retention of drugs: a review. Res. J. Pharm. Technol. **1**(4), 345–348 (2008)
17. Bajpai, S.K., Sonkusley, J.: Hydrogels for oral drug delivery of peptides: synthesis and characterization. J. Appl. Polym. Sci. **83**, 1717–1729 (2002)
18. Graham, N.B., Mc-Neil, M.E.: Hydrogels for controlled drug delivery. Biomaterials **5**(1), 27–36 (1984)
19. Tangri, T., Khurana, S., Madhav, S.N.V.: Mucoadhesive drug delivery: mechanism and methods of evaluation. Int. J. Pharma Bio Sci. **2**(1) (2011)
20. Omidian, H., Park, K.: Swelling agents and devices in oral drug delivery. J. Drug Deliv. Sci. Technol. **18**(2), 83–93 (2008)
21. Murata, Y., Kofuji, K., Kawashima, S.: Preparation of floating alginate beads for drug delivery to gastric mucosa. J. Biomater. Sci. Polym. Ed. **14**(6), 581–588 (2003)
22. Borase, C.B.: Floating systems for oral controlled release drug delivery. Int. J. Appl. Pharm. **4**(2), 1–13 (2012)
23. Mor, J.: Progress in floating drug delivery systems. A review. Int. J. Pharma Profe. Res. **4**(2), 441–446 (2011)
24. Chandel, A., Chauhan, K., Parashar, B., Kumar, H., Arora, S.: Floating drug delivery systems: a better approach. Int. Curr. Pharma. J. **1**(5), 110–118 (2012)
25. Ellis, H.A.: Basic Science: Anatomy of the Stomach: Surgery, vol. 29, pp. 541–543. Elsevier Ltd. (2011)
26. Dhole, A., Gaikwad, P., Bankar, V., Pawar, S.: A renew on floating multiparticulate drug delivery system: a novel approach to gastric retention. Int. J. Pharm. Sci. Rev. Res. **6**(2), 205–211 (2011)

27. Patil, C., Baklilwal, S., Rane, B., Gurathi, N., Pawar, S.: Floating microspheres: a promising approach for gastric retention. Ins. J. Pharma. Res. Dev. **12**(2), 26–38 (2011)
28. Cardinal, J.R.: Gastric Retentive Drug Delivery Systems, 1st edn. Wiley, Sharon, MA (2011)
29. Gadad, A.P., Patil, M.B., Naduvinamani, S.N., Matiholimath, V.S., Dandagi, P.M, Kulkarni, A.R.: Sodium alginate polymeric floating beads for the delivery of cefpodoxime proxetil. J. Appl. Polym. Sci. **114**(3), 1921–1926 (2009)
30. Goyal, M., Prajapati, R., Purohit, K., Mehta, S.: Floating drug delivery system: a review. J. Curr. Pharma. Res. **5**(1), 7–18 (2011)
31. Lieberman, H.A., Lachman, L., Schwartz, J.B.: Pharmaceutical Dosage Forms: Tablets, p. 153. Dekker, New York (1990)
32. Zhao, Y., Martin, B.R., Weaver, C.M.: Calcium bioavailability of calcium carbonate fortified soymilk is equivalent to cow's milk in young women. J. Nutr. **135**(10), 2379–2382 (2005)
33. Yellanki, S.K., Nerella, N.K.: Stomach specific drug delivery of riboflavin using floating alginate beads. Int. J. Pharm. Pharm. Sci. **2**(2), 160–163 (2010)
34. Sriamornsak, P., Sungthongjeen, S., Puttipipatkhachorn, S.: Use of pectin as a carrier for intragastric floating drug delivery: carbonate salt contained beads. Carbohydr. Polym. **67**(3), 436–445 (2007)
35. Abou El Ela, A.E.S.F., Hassan, M.A., El- Maraghy, D.: Ketorolac tromethamine floating beads for oral application: characterization and in vitro/in vivo evaluation. Saudi Pharm. J. **22**, 349–359 (2014)
36. Pahwa, R., Bhagwan, S., Kumar, V., Kohli, K.: Available online through role of natural polymers in the development of floating drug delivery systems. J. Pharm. Res. **3**(6), 1312–1318 (2010)
37. Park, Y.J., Liang, J., Yang, Z., Yang, V.C.: Controlled release of clot-dissolving tissue-type plasminogen activator from a poly (L-glutamic acid) semi-interpenetrating polymer network hydrogel. J. Controlled Release **75**, 37–44 (2001)
38. Baumgarter, S., Kristl, J., Vrecer, F., Vodopivec, P., Zorko, B.: Optimisation of floating matrix tablets and evaluation of their gastric residence time. Int. J. Pharm. **195**, 125–135 (2000)
39. Chen, J., Park, K.: Synthesis of fast-swelling, superporous sucrose hydrogels. Carbohydr. Polym. **41**, 259–268 (2000)
40. Mandel, K.G., Daggy, B.P., Brodie, D.A., Jacoby, H.I.: Alginate-raft formulations in the treatment of heartburn and acid reflux. Aliment. Pharmacol. Ther. **14**, 669–690 (2000)
41. Zaniboni, H.C., Fell, J.T., Collett, J.H.: Production and characterisation of enteric beads. Int. J. Pharm. **125**, 151–155 (1995)
42. Chen, Y.-C., Hoa, H.O., Lee, T.Y., Sheu, M.S.: Physical characterizations and sustained release profiling of gastroretentive drug delivery systems with improved floating and swelling capabilities. Int. J. Pharm. **441**, 162–169 (2013)
43. Krishnan, V., Sasikumar, S., Prabhu, F., Vijayaraghavan, R.: Effect of pore forming agents on the physical characteristics and release kinetics of levofloxacin hemihydrate from floating alginate drug delivery system—an in vitro study. Trends Biomater. Artif. Organs **24**(3), 139–145 (2010)
44. Choi, B.Y., Park, H.J., Hwang, S.J., Park, J.B.: Preparation of alginate beads for floating drug delivery system: effects of CO_2 gas-forming agents. Int. J. Pharm. **239**(1–2), 81–91 (2002)
45. Chang, P.R., Jian, R., Yu, J., Ma, X.: Starch-based composites reinforced with novel chitin nanoparticles. Carbohydr. Polym. **80**, 420–425 (2010)
46. Tester, R.F., Karkalas, J., Qi, X.: Starch—composition, fine structure and architecture. J. Cereal Sci. **39**, 151–165 (2004)
47. Ellis, R.P., Cochrane, M.P., Dale, M.F.B., Duffus, C.M., Lynn, A., Morrison, I.M., Prentice, R.D.M., Swanston, J.S., Tiller, S.A.: Starch production and industrial use. J. Sci. Food Agric. **77**, 289–311 (1998)
48. Ismail, H., Irani, M., Ahmad, Z.: Starch-based hydrogels: present status and applications. Int. J. Polym. Mater. Polym. Biomater. **62**, 411–420 (2013)
49. White, R.J., Budarin, V.L., Clark, J.H.: Tuneable mesoporous materials from α-D-polysaccharides. ChemSusChem **1**, 408–411 (2008)
50. García-González, C.A., Alnaief, M., Smirnova, I.: Polysaccharide-based aerogels—promising biodegradable carriers for drug delivery systems. Carbohydr. Polym. **86**, 1425–1438 (2011)

51. Athawale, V.D., Lele, V.: Graft copolymerization onto starch. II. Grafting of acrylic acid and preparation of it's hydrogels. Carbohydr. Polym. **35**, 21–27 (1998)
52. Zohuriaan-Mehr, M.J., Kabiri, K.: Superabsorbent polymer materials: a review. Iran. Polym. J. (Engl. Ed.) **17**, 451–477 (2008)
53. Demitri, C., Del Sole, R., Scalera, F., Sannino, A., Vasapollo, G., Maffezzoli, A., Ambrosio, L., Nicolais, L.: Novel superabsorbent cellulose-based hydrogels crosslinked with citric acid. J. Appl. Polym. Sci. **110**, 2453–2460 (2008)
54. Chan, S.W., Mirhosseini, H., Taip, F.S., Ling, T.C., Tan, C.P.: Comparative study on the physicochemical properties of κ-carrageenan extracted from *Kappaphycus alvarezii* (doty) doty ex Silva in Tawau, Sabah, Malaysia and commercial κ-carrageenans. Food Hydrocolloids **30**(2), 581–588 (2013)
55. Relleve, L., Yoshii, F., Dela Rosa, A.M., Kume, T.: Radiation-modified hydrogel based on poly (n-vinyl-2-pyrrolidone) and carrageenan. Radiat. Phys. Chem. **273**(1), 63–68 (1999)
56. Michel, G., Nyval-Collen, P., Barbeyron, T., Czjzek, M., Helbert, W.: Bioconversion of red seaweed galactants: a focus on bacterial agarases and carrageenanes. Appl. Microbiol. Biotechnol. **7**(1), 23–33 (2006)
57. Hui, N.H.: Effect of cross-linking on the properties of kappa-carrageenan/sodium carboxymethyl cellulose hydrogel. Degree thesis, Faculty of Chemical and Natural Resources Engineering, Universiti Tekbologi Malaysia (2009)
58. Flashaw, R., Bixler, H.J., Johndro, K.: Structure and performance of commercial kappa-2 carrageenan extracts I. Structure analysis. Food Hydrocolloids **15**(1), 441–452 (2001)
59. Imeson, A.: Thickening and Gelling Agents for Food, 2nd edn, pp. 45–83. Blackie Academic & Professional, London (1997)
60. Hornof, M.D., Kast, C.E., Bernkop-Schnürch, A.: In vitro evaluation of the viscoelastic properties of chitosan–thioglycolic acid conjugates. Eur. J. Pharm. Biopharm. **55**, 185–190 (2003)
61. Nilsen-Nygaard, J., Strand, S., Vårum, K., Draget, K., Nordgård, C.: Chitosan: gels and interfacial properties. Polymers **7**, 552 (2015)
62. Tuğcu-Demiröz, F., Acartürk, F., Erdoğan, D.: Development of long-acting bioadhesive vaginal gels of oxybutynin: formulation, in vitro and in vivo evaluations. Int. J. Pharm. **457**(1), 25–39 (2013)
63. Luukkonen, P., Schæfer, T., Hellén, L., Juppo, A.M., Yliruusi, J.: Rheological characterization of microcrystalline cellulose and silicified microcrystalline cellulose wet masses using a mixer torque rheometer. Int. J. Pharm. **188**(2), 181–192 (1999)
64. Zahan, K.A., Pa'e, N., Muhamad, I.I.: Process parameter for fermentation in rotary discs reactor for optimum bacterial cellulose production using response surface methodology. BioResources **9**(2), 1858–1872 (2014)
65. Pa'e, N., Zahan, K.A., Muhamad, I.I.: Production of biopolymer from *Acetobacter xylinum* using different fermentation methods. Int. J. Eng. Technol. (IJET-IJENS) **11**(5), 90–98 (2011)
66. Gao, X., Shi, Z., Lau, A., Liu, C., Yang, G., Silberschmidt, V.V.: Effect of microstructure on anomalous strain-rate-dependent behavior of bacterial cellulose hydrogel. Mater. Sci. Eng. C **62**, 130–136 (2016)
67. Horii, F., Yamamoto, H., Hirai, A.: Microstructural analysis of microfibrils of bacterial cellulose. Macromol. Symp. **120**, 197–205 (1997)
68. Tang, W., Jia, S., Jia, Y., Yang, H.: The influence of fermentation conditions and post-treatment methods on porosity of bacterial cellulose membrane. World J. Microbiol. Biotechnol. **26**, 125–131 (2010)
69. Ul-Islam, M., Khan, T., Park, J.K.: Water holding and release properties of bacterial cellulose obtained by in situ and ex situ modification. Carbohydr. Polym. **88**(2), 596–603 (2012)
70. Aziz, M.A., Cabral, J.D., Brooks, H.J.L., Moratti, S.C., Hanton, L.R.: Antimicrobial properties of a chitosan dextran-based hydrogel for surgical use. Antimicrob. Agents Chemother. **56**(1), 280–287 (2012)
71. Sannino, A., Madaghiele, M., Conversano, F., Mele, G., Maffezzoli, A., Netti, P.A.: Cellulose derivative–hyaluronic acid-based microporous hydrogels cross-linked through divinyl sulfone (DVS) to modulate equilibrium sorption capacity and network stability. Biomacromolecules **5**(1), 92–96 (2004)

72. Ganguly, K., Chaturvedi, K., More, U.A., Nadagouda, M.N., Aminabhavi, T.M.: Polysaccharide-based micro/nanohydrogels for delivering macromolecular therapeutics. J. Controlled Release **193**, 162–173 (2014)
73. Burgain, J., Gaiani, C., Linder, M., Scher, J.: Encapsulation of probiotic living cells: from laboratory scale to industrial applications. J. Food Eng. **104**, 467–483 (2011)
74. Dragan, E.S.: Design and applications of interpenetrating polymer network hydrogels. A review. Chem. Eng. J. **243**, 572–590 (2014)
75. Cook, M.T., Tzortzis, G., Charalampopoulos, D., Khutoryanskiy, V.V.: Microencapsulation of probiotics for gastrointestinal delivery. J. Controlled Release **162**, 56–67 (2012)
76. Krasaekoopt, W., Bhandari, B., Deeth, H.: Evaluation of encapsulation techniques of probiotics for yoghurt. Int. Dairy J. **13**, 3–13 (2003)
77. Williams, D.H.: Definitions in Biomaterials: Proceedings of a Consensus Conference of the European Society for Biomaterials, Chester, England (1986)
78. Shin, H., Temenoff, J.S., Mikos, A.G.: In vitro cytotoxicity of unsaturated oligo [poly (ethylene glycol) fumarate] macromers and their cross-linked hydrogels. Biomacromolecules **4**, 552–560 (2003)
79. Peppas, N.A., Bures, P., Leobandung, W., Ichikawa, H.: Hydrogels in pharmaceutical formulations. Eur. J. Pharm. Biopharm. **50**, 27–46 (2000)
80. Li, X., Kong, X., Zhang, Z., Nan, K., Li, L., Wang, X., Chen, H.: Cytotoxicity and biocompatibility evaluation of N, O-carboxymethyl chitosan/oxidized alginate hydrogel for drug delivery application. Int. J. Biol. Macromol. **50**, 1299–1305 (2012)
81. Popa, E.G., Carvalho, P.P., Dias, A.F., Santos, T.C., Santo, V.E., Marques, A.P., Viegas, C.A., Dias, I.R., Gomes, M.E., Reis, R.L.: Evaluation of the in vitro and in vivo biocompatibility of carrageenan based hydrogels. J. Biomed. Mater. Res. Part A **102**(11), 4087–4097 (2014)
82. Mohamad, N., Amin, M.C.I.M., Pandey, M., Ahmad, N., Rajabb, N.F.: Bacterial cellulose/acrylic acid hydrogel synthesized via electron beam irradiation: accelerated burn wound healing in an animal model. Carbohydr. Polym. **114**, 312–320 (2014)
83. Lin, W., Lien, C., Yeh, H., Yu, C., Hsu, S.: Bacterial cellulose and bacterial cellulose-chitosan membranes for wound dressing applications. Carbohydr. Polym. **94**, 603–611 (2013)
84. Meenach, S.A., Anderson, A.A., Suthar, M., Anderson, K.W., Hilt, J.Z.: Biocompatibility analysis of magnetic hydrogel nanocomposites based on poly(N-isopropylacrylamide) and iron oxide. J. Biomed. Mater. Res. Part A, 903–909 (2008)
85. Ge, S., Wang, G., Shen, Y., Zhang, Q., Jia, D., Wang, H., Dong, Q., Yin, T.: Cytotoxic effects of MgO nanoparticles on human umbilical vein endothelial cells in vitro. NanoBiotechnology **5**, 36–40 (2011)

3D Bioprinting of a Tissue Engineered Human Heart

Khalida Fakhruddin, Saiful Izwan Abd Razak, Nadirul Hasraf Mat Nayan and Mohammed Rafiq Abdul Kadir

Abstract Cardiovascular disease remains the leading cause of death in the world. Due to disadvantages of using the medical device to recover the function of the heart, researchers are getting interested to discover new intervention for cardiovascular diseases. Bioprinting is one facet of tissue engineering which has been emerging in recent years to build a three dimensional constructs containing biological cells and show a promising potential to overcome the challenges of current intervention of cardiovascular diseases. In this chapter, authors attempt to review and discuss on bioprinting of tissue engineered human heart. Explanation will be based on introduction of bioprinting, the materials it used, the potential of cell sources and previous work done.

Keywords Tissue engineering · 3D bioprinting · Cardiovascular

1 Introduction

Cardiovascular diseases (CVD) are the number one cause of death globally. According to World Health Organization, about 17.7 million people died from CVD in 2015 which represents 31% of total deaths globally [1]. Generally, CVD can be defined as a group of disorders of the heart and blood vessel. These include coronary artery diseases such as stroke, angina and myocardial infarction, hypertensive heart

K. Fakhruddin · S. I. A. Razak (✉) · M. R. A. Kadir
School of Biomedical Engineering and Health Sciences, Faculty of Engineering, Universiti Teknologi Malaysia, 81310 Skudai, Johor, Malaysia
e-mail: saifulizwan@utm.my

K. Fakhruddin
e-mail: dyah@utm.my

S. I. A. Razak
Centre for Advanced Composite Materials, Universiti Teknologi Malaysia, 81300 Skudai, Johor, Malaysia

N. H. M. Nayan
Faculty of Engineering Technology,
Universiti Tun Hussein Onn Malaysia, KM1 Jalan Panchor, 84600 Pagoh, Johor, Malaysia

D. E. O. Dewi et al. (eds.), *Cardiovascular Engineering*, Series in BioEngineering,
https://doi.org/10.1007/978-981-10-8405-8_11

disease, heart failure, congenital heart disease, cardiomyopathy, rheumatic heart disease, heart arrhythmia, valvular heart disease, carditis, aortic aneurysms, peripheral artery disease, thromboembolic disease, and venous thrombosis [1–3].

There are numerous of excellent surgical technique and pharmacological therapies that can slow down the progression to end-stage disease and prolong life, however, it still a challenge to prevent or reverse the progression of CVD [4]. Currently, the gold standard of end-stage CVD therapy is heart transplantation. However the problem of organ shortage and possibilities of immune-rejection make heart transplantation a challenging therapy.

Uses of medical device in treating CVD are current intervention for CVD. Insertion of stents, pacemaker, prosthetic valves, and patches are some of the examples of medical device that use in treating CVD. Although it can temporarily recover the function, however, it cant regenerate the damaged myocardium tissue as cardiomyocytes are limited in regenerative abilities. This also resulting a rise of problematic issue in congenital heart disease patients where the repair strategy must adapt the child's growth. Prosthetic heart valve is made from materials such as carbon and titanium which makes it more durable. However, these type of valves will provide a surface where blood clots can form easily and eventually cause heart attack or stroke [5, 6]. In order to prevent the development of blood clot, the patient will be on lifelong blood-thinning medication such as warfarin.

The intervention should be able to naturally restore the optimized and adaptive of biomechanical and biochemical of patient's original tissue in order to become an ideal intervention for CVD [7]. Cell-based therapy, tissue engineering and biofabrication or bioprinting are the current approaches to achieve the ideal intervention [8–10].

2 Three Dimensional Bioprinting

Tissue engineering and regenerative medicine is an interdisciplinary field which combines engineering and medicine. It involve use of combination of cell, engineering, and material to generate artificial tissues and organs for repairing and replacing the damaged tissues [11]. Bioprinting or also called as biofabrication is one facet of tissue engineering which has been gained more and more attention and interest in these recent years. Generally, bioprinting has capability to build a three dimensional constructs containing biological cells. It has a great potentials to fabricate human cell based tissue models, for drug development and disease modeling applications and also able to create functional tissues for regenerative medicine [12, 13].

Three-dimensional (3D) Bioprinting is made by advancement of 3D printing technology, cell biology and materials science [14]. Bioprinting differs from conventional 3D printing techniques where 3D printing usually print temporary cell-free scaffolds. Whereas bioprinting involve the process of building a 3D construct using computer-controlled devices by depositing the biomaterials and cells precisely into geometries that create anatomically correct biological structures. It prints by outputting layer-upon-layer of living cells in hydrogels or viscous fluid, or cell-seeded micro-carriers which is term as bioink [15].

Bioprinting requires a different technical approach that is compatible with depositing living cells which require extra complexities, such as the choice of materials, cell types, growth and differentiation factors, and technical challenges related to the sensitivities of living cells and the construction of tissues [14]. However, apart from the complexities, bioprinting has several advantages include scalability, high-resolution cell deposition, accurate control of cell distribution, and cost-effectiveness which makes the development and applications of bioprinting greatly increased during the last five years [16].

2.1 *Types of Bioprinter*

Bioprinting is present in various biologically applied printing systems, such as inkjet, micro-extrusion and laser-assisted bioprinting. These printing systems are varying in their method of deposition and patterning of biological materials. Among these three type of bioprinter, micro-extrusion bioprinter is the most commonly used technique to print cell and biomaterial scaffolds. Smith et al. [17] were the first to report the use of microextrusion bioprinter to deposit a living cell. Micro-extrusion bioprinter use pneumatic (air pressure) or mechanical (piston or screw) dispensing systems to extrude continuous beads of material and/or cells from the syringe or the nozzle to construct a 3D structure [18–21]. The amount of dispensed cells and biomaterials can be adjusted by varying the nozzle moving speed, nozzle size, and pneumatic or mechanical pressure [22].

Xu et al. [23] were the first researchers who use an inkjet printer for the biological application where they modified the commercially available 2D ink-base printer. Inkjet bioprinter heads are based on micro-electrochemical system device (MEMS) [24]. It used either thermal or acoustic force system to eject the drops of biological material liquid in a controllable size [23, 25]. Thermal inkjet printers electrically heat the print head to produce air-pressure pulses that force droplets from the nozzle, whereas acoustic printers use pulses formed by piezoelectric or ultrasound pressure [14, 26].

Laser assisted bioprinter is less commonly used than inkjet and microextrusion bioprinter because of high cost, cumbersome and complexities of the laser assisted bioprinter. However, laser-assisted bioprinter does not have the problems of nozzle clogging with cells or materials, which are major disadvantage of other bioprinting technologies as it is a nozzle-free technique [13]. Laser assisted bioprinter is based on laser direct-write and laser-induced forward transfer principle [27, 28]. It is a non-contacting method of material deposition that utilizes laser energy absorption to propel a cell-suspended hydrogel droplet to a receiving surface [29]. Table 1 shows the method of deposition and advantages and disadvantage of all the three type of bioprinter.

Table 1 Advantages and disadvantages of micro-extrusion, inkjet and laser-assisted bioprinter

Type of bioprinter	Method of deposition	Advantages	Disadvantages	References
Micro-extrusion 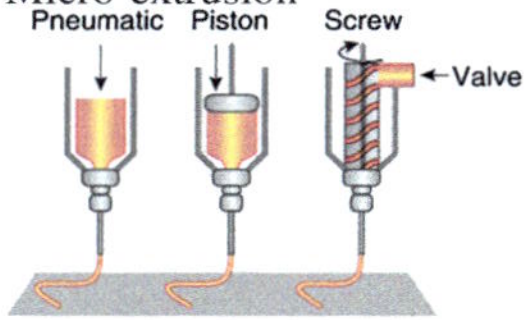	– Use pneumatic or mechanical (piston or screw) dispensing systems to extrude continuous beads of material and/or cells	– Create construct in short period of time – Capable to print with high viscous materials and cell spheroids through the nozzle – Can use multiple composition with dual extruder – Good mechanical properties	– Shear stress on nozzle tip wall – Low cell survival rates after printing (40–86%), survival rate decreases with increased extrusion pressure [7, 64] – Lower resolution (5 μm to mm)	[13, 14, 30–34]
Inkjet 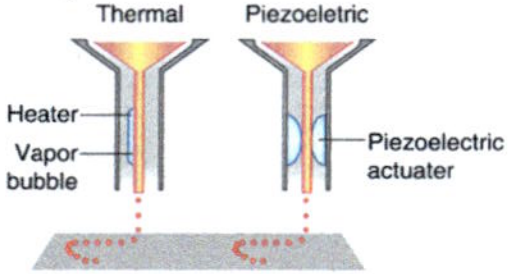	– Thermal inkjet printers electrically heat the print head to produce air-pressure pulses that force droplets from the nozzle – Acoustic printers use pulses formed by piezoelectric or ultrasound pressure	– Low cost – Versatile – High fabrication speed – Generate high resolution structure (20–100 μm) – Suitable to generate thin layers or patterned structures	– Limited to low viscous material, limited material choices – Thermal and mechanical stress – High viscous material will clog the nozzle	[13, 14, 30, 35]
Laser-assisted 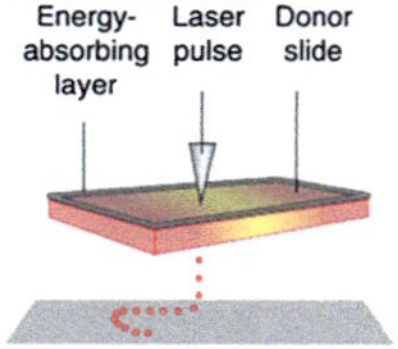	– Use lasers focused on an absorbing substrate to generate pressures that propel cell-containing materials onto a collector substrate	– Nozzle free technique, no nozzle clogging problem – High resolution (microscale) – High accuracy – Wide range of biomaterial viscosity (1–300 mPa/s)	– Cell-unfriendly – Complex as it involves laser/light – High cost	[13, 14, 30, 31, 36]

2.2 Bioprinting Process

Generally, bioprinting process can be divided into three sequential technological steps which are pre-processing, processing (actual printing), and post-processing (Fig. 1) [37]. Prior to these three processes, the desired cell that will be used in printing is must first been isolated and expanded or proliferated [16]. Pre-processing is a process of designing scaffold, tissue or organ using imaging and computer-aided design techniques. Then the designed is printed through a bioprinter. The post-processing of 3D bioprinted construct involves the process of tissue remodelling and maturation in a specially designed chamber bioreactor, which accelerates tissue maturation [31]. These bioprinted construct or called scaffolds could then be applied in therapeutic field as a testing platform for drug screening and discovery, or as an in vitro model system for disease or directly use for transplantation purposes [16].

Three-dimensional bioprinting method can precisely constructing a 3D bioprinted construct by finely tuning and defining material geometries, localization of biomolecular cues, and mechanical properties that can resemble endogenous tissues which have immobilized controlled concentrations of adhesive ligands, growth factors, or other signalling molecules to mimic cellular architectures in vivo [38, 39]. Because of this, 3D bioprinting is being applied to regenerative medicine. It also has already been used for the generation and transplantation of several tissues, including multi-layered skin, bone, vascular grafts, tracheal splints, heart tissue and cartilaginous structures.

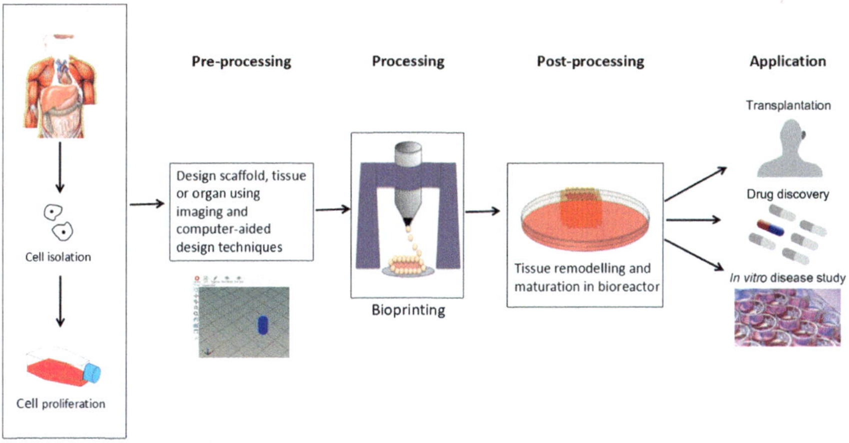

Fig. 1 Bioprinting process

3 Suitable Materials for Bioprinting of Tissue Engineered Human Heart

Successful implementation of bioprinting is depend on the integration with compatible biomaterials that are responsible for supporting the cellular components during and after printing process, and also compatible with the bioprinting devices. However until now, there is still no ideal material specialized for the purpose of bioprinting. Most bioprinting process use the same biomaterials used in traditional bioengineering and sometimes in order to achieve the desired crosslinking and mechanical properties, it combines them [40]. Murphy and Atala, 2014 highlighted some factors that the materials should fulfill [14] in order to become an ideal material for bioprinting. The factors are printability which include the viscosity, gelation method and rheological properties, the compatibility with cell and also the printing system, the degradation rates and also the byproducts, the structural and mechanical properties, and also biomimicracy. In printing of tissue engineered human heart, besides design the materials to improve print fidelity, the material use also must capture the functional and physiological properties of native cardiac tissues [41].

Preparation of the materials for bioprinting process is the most challenging aspect, as the materials should satisfy both biological needs from the cell compatibility point of view and the physical and mechanical needs of the printing process itself [42]. The materials also have to be able to provide the desired mechanical and functional properties for tissue constructs [14]. Hydrogels are polymeric materials, which are the most commonly explored materials for fabricating the complex 3D cellular microenvironments [43]. Hydrogels are composed of polymer or peptide chains. Hydrogels are printed in a liquid precursor form, and then cross-linked to form a solidified macromolecular network. Hydrogels are commonly used in bioprinting because they have low cytotoxicity, structural similarity to the extracellular matrix (ECM) and can be tuned for ideal degradability and mechanics and the ability to incorporate biomolecules of interest [42].

Hydrogel can be classified based on their natural and synthetic origin [44, 45]. Natural hydrogel (including collagen, fibrin, hyaluronic acid, alginate, agarose and chitosan) are often derived from native ECM components and from non-mammalian sources such as algae and seaweed [43]. The advantages of natural hydrogel are they have high biocompatibility and degradability through natural enzymatic or chemical processes. They also have a similarity to human ECM, which contains natural ligands that allow for cellular adhesions that make it not require additional modifications to support cell growth and spreading [14, 43].

Synthetic hydrogel such as pluronic, poly(ethylene glycol) (PEG), poly(acrylamide) (PLA), poly-L-lactic acid (PLLA), poly(glycolic acid) (PGA), poly(caprolactone) (PCL) and poly(lactide-co-glycolide) (PLGA) are made through chemical synthesis and the chemical and mechanical properties are more controllable [46]. They are applicable for 3D bioprinting regenerative medicine applications because they are both hydrophilic and absorbent [14]. According to Malda et al. [30], printing using synthetic hydrogel like poly(ethyleneglycol) and pluronics can obtained a 3D printed structures with high shape fidelity. There are disadvantages of synthetic hydrogels including poor biocompatibility, toxic degradation products and loss of mechanical properties during degradation. However, the biocompatibality, and degradation properties synthetic hydrogel can be improved by facilitate the chemical manipulation of the structure [47]. This also will provide the ability to generate cross-linkable structures for tissue engineering. To meet the multiple requirements of the bioprinting process, hydrogels can also be combined as a composite material to achieve the desired properties [48].

4 Potential Cells Sources for Bioprinting of Tissue Engineered Human Heart

Choice of cells for bioprinting is crucial for correct functioning of the bioprinted construct. There are two main factors that should be considered in order to choose a suitable type of cell for bioprinting process. The factors are the bioprinted cells must closely resemble and mimic the physiological state of cells in vivo, and the cells must have the ability to maintain or develop in vivo functions under optimized microenvironments [14, 49]. Other than that, the cell chosen also should be able to expand into a sufficient numbers for printing. However, the cell should have precise control of the proliferation in vivo and in vitro as the bioprinted construct might lose the cell viability because off too low proliferation and causing hyperplasia and apoptosis if too much proliferation [14]. The bioprinting process can be done by either using a functional primary cells or progenitors or stem cells for further differentiation [16]. Printing with primary cells are challenging because it is difficult to isolate and culture the primary cells and have limited lifespan which make it not suitable for the long-term functionality of bioprinted constructs [50]. Stem cells are suitable to use in bioprinting because of its ability to expand to a large number of cells. Their ability to proliferate in an undifferentiated but multipotent state and capability to generate multiple functional tissue-specific cell phenotypes also make them suitable for use as a cell source in bioprinting [14].

To print the tissue engineered human heart, the bioprinted construct should be compatible with the native cardiac tissue where it comprise of heterogeneous cell population which are cardiomyocytes and non-cardiomyocytes [51]. Cardiomyocytes is the minority of cell population in heart where it give rise to the bulk volume of the heart, while fibroblast dominates the non-cardiomyocyte population and is in direct contact with cardiomyocytes [51]. Endothelial cells, adipocytes and neurons are other non-cardiomyocytes population. The coupling of cardiomyocytes and non-cardiomyocytes will improve the electrophysiology of engineered cardiac tissue and production of survival and trophic signals to cardiomyocytes [52, 53]. In bioprint of cardiac tissue, terminally differentiated cardiomyocytes and stem cells can be used. However, terminally differentiated cardiomyocytes have a limited ability to proliferate. They are also extremely sensitive to hypoxia and hardly to survive in a poorly vascularized environment [54]. Because of that, various progenitor and stem cell populations, including resident cardiac stem cells, embryonic stem cells, mesenchymal stem cells, or induced pluripotent stem cells have been used in bioprint of cardiac tissue instead of mature cardiovascular cells [54] as they can expand to a large number of cells and differentiate to vary cell type such as pacemaker cells, neural cells, fibroblasts, cardiomyocytes, endothelial cells and smooth muscle cells [55–58]. Apart from that, they are also more resistant to hypoxia than are mature cells. In addition, progenitor and stem cells can be isolated from autologous sources that limits the likelihood of rejection.

5 3D Bioprinting in Heart Tissue Engineering

Cardiomyocytes limitation in regenerative abilities, shortage of organs donor and problems of immune rejection and disadvantages of current intervention of CVD such as lifelong dependent on medication are the reasons why advanced therapeutic approach such as tissue engineering and bioprinting are extensively studied this recent years. Bioprinting as one facet of tissue engineering aims to provide biological solutions to restore the ill functioning heart and has the largest potential in regenerative medicine [59]. Followings are the previous work done by various researcher on bioprinting in heart tissue engineering.

5.1 Myocardium Tissue Bioprinting

Myocardial infarction happen when one of the coronary arteries have been block that will cause lacking in blood flow. These consequently results in ischemia and cause cardiomyocytes to die. After a vigorous inflammatory response is provoked, migration of fibroblasts and endothelial cells migrate to the site of injury and gradually form non-contracting fibrotic scars that ultimately leading to heart failure. 3D bioprinting for cardiac tissue regeneration can pattern and assemble cells with high density, defined organization and spatial distribution. It also enables the generation of multiple layered constructs with multiple cell types. In one study, Gaetani et al., bioprinted 3-D structures of a mixture of human cardiomyocyte progenitor cells (hCMPC) and alginate hydrogel using extrusion bioprinter. In their study, they found that printed construct had high cell viability, retained their commitment for the cardiac lineage and expressed the genes of the early cardiac transcription [60]. In their other work, they bioprinted the cardiac patches using hCMPC and a hyaluronic acid/gelatin-base matrix. They applied on in vivo model for the therapeutic potential evaluation. The results showed significant reduction in faulty remodeling when the patch is applied. The cardiac performances was also preserved and the long-term in vivo survival and implantation of hCMPCs is supported [61]. In another study, Gaebel et al., printed human umbilical vein endothelial cells (HUVEC) and human mesenchymal stem cell (hMSC) on polyester urethane urea cardiac patch by using laser-assisted bioprinting technique. The cardiac patch is then transplanted to infarction zone of rat heart and found that the vessel formation is successfully generated and transplanted human cells is integrated into the connected vessels of the host vascular system [62].

5.2 *Heart Valve Bioprinting*

Heart has 4 valves to control the unidirectional of the blood flow. The valves are mitral valve and tricuspid valve, which control blood flow from the atria to the ventricles and aortic valve and pulmonary valve, which control blood flow out of the ventricle. Each of the valve composed of leaflets that are attached to a fibrous root wall which they both are biomechanically and structurally anisotropic [63, 64]. Valve interstitial cells (VIC) made up the leaflets and smooth muscle cells (SMC) made up the root wall with the surface cover by valvular endothelial cells (VEC). Calcific aortic valve disease (CAVD) is one of the most common valve abnormalities and usually treated by surgical repair or replacement of the valve by either mechanical or bio-prosthetic valve. With complex geometry of heart valve, 3D bioprinting has great potential in producing the anatomically correct cardiac valves tailored for individual patient. Materials used in bioprinting of heart valve tissue must have strong mechanical properties for proper function of heart valve. Because of that, extrusion based bioprinter is often used in constructing the heart valve conduit as it suitable to be used with high strength of material. Used of extrusion based bioprinter also give an advantage as it able to generate anatomically accurate trileaflet valves, mechanically heterogeneous valve conduits and living engineered valves with spatial and temporal valve cells (VIC and SMC) distribution [65].

Currently, there are several bioprinting of heart valve models/designs have been reported. Duan et al., bioprinted a tri-leaflet heart valve conduit composed of combination hydrogel of methacrylated hyaluronic acid and methacrylated gelatin with encapsulation of human aortic VIC by using extrusion based bioprinter [66]. The results showed that the printed tri-leaflet heart valve conduit has high viability of cells and demonstrates a great potential for remodeling at 7 days. However this heart valve conduit model does not contain important structures to relieve abnormal stress and prevent blood back flow which are widenings between root wall and leaflets (sinus) and also joined places between leaflets (commissure). Another study use combination of alginate and gelatin hydrogel containing aortic root sinus smooth muscle cells and aortic VIC to build an anatomically complex living aortic valve conduit [67]. They found that the cells remained viable for more than a week, suggesting that complete fabrication of the aortic valve with viable cells may be feasible. These studies demonstrated the possibility that cellularized tissue valves can be created using the bioprinting technology for eventual clinical use. Hockaday et al. print aortic valve with present of key features such as the coronary ostium and sinuses [68]. They used combination of 700 and 8000 MW poly(ethylene glycol) diacrylate (PEGDA) hydrogel and seeded with porcine aortic valve interstitial cells where roots were formed by 700 MW PEGDA hydrogel while the leaflets were formed with 700/8000 MW PEG-DA hydrogels. The valve conduits constructs have biomechanical heterogeneity, where the leaflets were more flexible, while the root remained relatively rigid.

5.3 *Vascular Network Bioprinting*

Functional vascular network is necessary for removing metabolic waste, nutrients deliver, oxygen transfer, and promote the circulation of immune cells where play an important role in regeneration of cardiovascular tissue and other tissues such as pancreas, liver and bone, which are highly vascularized [69–71]. Approaches adapt from general tissue engineering and combining with 3D bioprinting own benefit able the 3D bioprinting to create in vitro vasculature and vascularized constructs. These approaches include generation of vascular constructs by self-assembly of cells (scaffold free), generation of microvasculatures by inkjet based bioprinting, generation of bioprinted constructs with growth factor delivery, coaxial nozzle assisted 3D bioprinting of vasculature and generation of channel based vascularized constructs [65].

Scaffold free approach usually use cell spheroids and make it as a building blocks which then the deposited spheroids will fused into vascular like constructs [72]. Norotte et al. build a scaffold-free small diameter vascular using cell spheroid of human smooth muscle cells and fibroblasts [73]. The cell spheroid were printed layer-by-layer by extrusion based bioprinter with agarose rods as a template. They were successful to build a tubular structures with controllable tube diameter, wall thickness and even branching pattern with double-layered vascular wall (smooth muscle cell and fibroblast). In other work, Kucukgul et al. printed an aortic vascular construct using an extrusion based bioprinter [74]. In their work, they use hydrogel called NovoGel as a support material and mouse embryonic fibroblast cell aggregates.

Chui et al. use a thermal inkjet based bioprinter to build a microvasculature by simultaneously deposit human microvascular endothelial cells and fibrin [75]. Printed fibrin scaffold retained proper shape after printing and endothelial cells proliferate to form a tubular structure. The fabricated microvasculature showed integrity after being cultured for 14 days. Growth factor such as vascular endothelial growth factor (VEGF) can be used to promote microvascularization in vitro. They can activate the endothelial (progenitor) cells, regulate their migration and promote cell assembly, vessel formation and maturation [76]. Poldervaart et al. build a 3D construct (Matrigel/alginate hydrogel with human endothelial progenitor cells (EPCs)) using extrusion based bioprinter and stimulates the vascularization by releasing of VEGF from gelatin microparticles (GMPs) [77].

Other than that, by using coaxial nozzle assisted 3D bioprinting, vasculature construct can be build. Gao et al. use an extrusion bioprinting method with coaxial nozzle system to fabricate hollow calcium alginate filaments [78]. The cell-laden alginate hydrogel (mouse fibroblasts cell) dispensed through the outer tube of the coaxial nozzle get crosslinked when contact with calcium chloride solution in the inner of the coaxial nozzle which consequently form the filament with a hollow channel. The results image from scanning electron microscopy (SEM) confirmed the formation of hollow structure and the uniform fusion section between adjacent hollow filaments. Yu et al. also fabricate alginate with encapsulation of human umbilical vein smooth muscle based vasculatures using similar bioprinting setup [79]. The deposition of collagen was observed after 6 week culture. Despite the successful result of using coaxial nozzle bioprinting system, however, it still has disadvantage. The type of bioink that can incorporated with this coaxial nozzle-assisted 3D bioprinting is limited. Currently, only alginate based bioink is reported to be used as it has fast ionic crosslinking capacity.

Another approach to build vasculature construct is by generating the channel. Usually the channel network is generate by printing sacrificial materials, using water soluble material, within engineered tissue constructs, typically a cell-laden construct. Then the soluble channel network is dissolved either by solution or changing temperature. Kolesky et al. bioprinted a perfusable networks using sacrificial materials using Pluronic F127 and human neonatal dermal fibroblast cell-laden methacrylated gelatin hydrogels by extrusion based bioprinting [80]. After removal of Pluronic by exposed to 22 °C temperature, microchannels were endothelialized with human umbilical vein endothelial cells (HUVEC) (Table 2).

6 Challenges of Bioprinting Heart Tissue Engineering

3D bioprinting shows a promising technique for the construction and development of cardiovascular tissues as it able to print heterogeneous and clinical relevant sized heart tissue constructs. However, despite of all the advantages and promising results, there are still many challenges to go through that will interfere its further applications and translations. It is a great challenge to develop an ideal material or bioinks to bioprint anatomically correct, biologically functional and mechanically robust tissue constructs although many bioinks have been implemented. According to Ozbolat and Yu, ideal bioink should be bioprintable, have comparable mechanical properties to native tissue/organ after crosslink, support cardiovascular cell functions, and affordable and commercially available with appropriate regulatory guidelines for clinical use [10]. Hydrogel is the most commonly explored materials for fabricating the complex 3D cellular microenvironments. However, it is still difficult to balance the cell functions and mechanical properties of crosslinked hydrogels. Hydrogels with high concentration will cause high mechanical properties, low mobility and less spreading of the encapsulated cells.

Table 2 Three dimensional bioprinting in heart tissue engineering

Method/Technique	Material	Cell type	Model/design	References
Extrusion	Alginate hydrogel	Human cardiomyocyte progenitor cells	Myocardium tissue construct scaffold	[60]
Extrusion	Combination of hyaluronic acid and gelatin	Human cardiomyocyte progenitor cells	Cardiac patch	[61]
Laser-assisted	Polyester urethane urea	Human umbilical vein endothelial cells and human mesenchymal stem cell	Cardiac patch	[62]
Extrusion	Methacrylated hyaluronic acid and methacrylated gelatin	Human aortic valve interstitial cells	Tri-leaflet heart valve conduit	[66]
Extrusion	Combination of alginate and gelatin hydrogel	Aortic root sinus, smooth muscle cells and aortic valve interstitial cells	Aortic valve conduit	[67]
Extrusion	Poly(ethylene glycol) diacrylate (PEGDA)	Porcine aortic valve interstitial cells	Aortic valve	[68]
Extrusion	Agarose rod as template	Human smooth muscle cells and fibroblasts	Scaffold-free small diameter vascular	[73]
Extrusion	NovoGel hydrogel	Mouse embryonic fibroblast	Aortic vascular	[74]
Thermal inkjet	–	Human microvascular endothelial cells and fibrin	Microvasculature	[75]
Extrusion	Matrigel/alginate hydrogel and gelatin microparticles with vascular endothelial growth factor	Human endothelial progenitor cells	Vascularized 3D construct	[77]
Extrusion	Alginate hydrogel	Mouse fibroblasts cell	Fabricate hollow calcium alginate filaments	[78]

(continued)

Table 2 (continued)

Method/Technique	Material	Cell type	Model/design	References
Extrusion	Alginate hydrogel	Human umbilical vein smooth muscle	Vascular	[79]
Extrusion	Pluronic F127 and methacrylated gelatin hydrogels	Human neonatal dermal fibroblast	Vascular network	[80]

7 Conclusions

Bioprinting as one facet of tissue engineering which has been gained more and more attention and interest in these recent years to fabricate human cell based tissue models and biological products, especially cardiovascular tissues. Wide range of material or bioinks with various type of cells can be used in bioprinting using different printing mechanisms such as extrusion, inkjet and laser-assisted. Cardiovascular constructs, such as myocardium tissue, heart valve and vasculature constructs have been successfully bioprinted with similar architecture to the native tissue and certain functions and decent resolution. However, despite all the advantages of 3D bioprinting, there are still have challenges for building a tissue construct or artificial organ with complex microarchitecture and fully biological functions. Therefore, more research efforts should be dedicated to develop more high-performance bioinks and high-resolution bioprinters.

References

1. World Health Organization.: Cardiovascular diseases (CVDs). http://www.who.int/mediacentre/factsheets/fs317/en/ (2017). Retrieved Dec 2017
2. World Health Organization.: Global atlas on cardiovascular disease prevention and control. 3–18 (2011)
3. Abubakar, I.I., Tillmann, T., Banerjee, A.: Global, regional, and national age-sex specific all-cause and cause-specific mortality for 240 causes of death, 1990–2013: a systematic analysis for the Global Burden of Disease Study 2013. Lancet **385**(9963), 117–171 (2015)
4. Eschenhagen, T., Zimmermann, W.H.: Engineering myocardial tissue. Circ. Res. **97**(12), 1220–1231 (2005)
5. Harris, C., Croce, B., Cao, C.: Tissue and mechanical heart valves. Ann. Cardiothorac. Surg. **4**(4), 399 (2015)
6. Tillquist, M.N., Maddox, T.M.: Cardiac crossroads: deciding between mechanical or bioprosthetic heart valve replacement. Patient Prefer. Adherence **5**, 91 (2011)
7. Atala, A., Yoo, J.J.: Bioprinting of Cardiac Tissue. Essentials of 3D Biofabrication and Translation, pp. 351–370. Academic Press, London (2015)
8. Wang, F., Guan, J.: Cellular cardiomyoplasty and cardiac tissue engineering for myocardial therapy. Adv. Drug Deliv. Rev. **62**(7), 784–797 (2010)
9. Georgiadis, V., Knight, R.A., Jayasinghe, S.N., Stephanou, A.: Cardiac tissue engineering: renewing the arsenal for the battle against heart disease. Integr. Biol. **6**(2), 111–126 (2014)

10. Ozbolat, I.T., Yu, Y.: Bioprinting toward organ fabrication: challenges and future trends. IEEE Trans. Biomed. Eng. **60**(3), 691–699 (2013)
11. Langer, R., Vacanti, J.P.: Tissue engineering. Science **260**(5110), 920–926 (1993)
12. Zhang, L.G., Fisher, J.P., Leong, K.: 3D Bioprinting and Nanotechnology in Tissue Engineering and Regenerative Medicine, p. 61. Academic Press, London (2015)
13. Lee, V.K., Dai, G.: Printing of three-dimensional tissue analogs for regenerative medicine. Ann. Biomed. Eng. **45**(1), 115–131 (2017)
14. Murphy, S.V., Atala, A.: 3D bioprinting of tissues and organs. Nat. Biotechnol. **32**(8), 773–785 (2014)
15. Atala, A., Yoo, J.J.: Bioprinting essentials of cells and protein viability. Essentials of 3D biofabrication and translation, pp. 1–14. Academic Press, London (2015)
16. Mandrycky, C., Wang, Z., Kim, K., Kim, D.H.: 3D bioprinting for engineering complex tissues. Biotechnol. Adv. **34**(4), 422–434 (2016)
17. Smith, C. M., Stone, A. L., Parkhill, R. L., Stewart, R. L., Simpkins, M. W., Kachurin, A. M., … Williams, S. K.: Three-dimensional bioassembly tool for generating viable tissue-engineered constructs. Tissue Eng. **10**(9–10), 1566–1576 (2004)
18. Khalil, S., Sun, W.: Biopolymer deposition for freeform fabrication of hydrogel tissue constructs. Mater. Sci. Eng. **27**(3), 469–478 (2007)
19. Chang, C.C., Boland, E.D., Williams, S.K., Hoying, J.B.: Direct-write bioprinting three-dimensional biohybrid systems for future regenerative therapies. J. Biomed. Mater. Res. B Appl. Biomater. **98**(1), 160–170 (2011)
20. Cohen, D.L., Malone, E., Lipson, H.O.D., Bonassar, L.J.: Direct freeform fabrication of seeded hydrogels in arbitrary geometries. Tissue Eng. **12**(5), 1325–1335 (2006)
21. Visser, J., Peters, B., Burger, T.J., Boomstra, J., Dhert, W.J., Melchels, F.P., Malda, J.: Biofabrication of multi-material anatomically shaped tissue constructs. Biofabrication **5**(3), 035007 (2013)
22. Keriquel, V., Guillemot, F., Arnault, I., Guillotin, B., Miraux, S., Amédée, J., Fricain, J.C., Catros, S.: In vivo bioprinting for computer-and robotic-assisted medical intervention: preliminary study in mice. Biofabrication **2**(1), 014101 (2010)
23. Xu, T., Kincaid, H., Atala, A., Yoo, J.J.: High-throughput production of single-cell microparticles using an inkjet printing technology. J. Manuf. Sci. Eng. **130**(2), 021017 (2008)
24. Singh, M., Haverinen, H.M., Dhagat, P., Jabbour, G.E.: Inkjet printing process and its applications. Adv. Mater. **22**(6), 673–685 (2010)
25. Cui, X., Boland, T., DD'Lima, D., K Lotz, M.: Thermal inkjet printing in tissue engineering and regenerative medicine. Recent Pat. Drug Delivery Formulation **6**(2), 149–155 (2012)
26. Tekin, E., Smith, P.J., Schubert, U.S.: Inkjet printing as a deposition and patterning tool for polymers and inorganic particles. Soft Matter **4**(4), 703–713 (2008)
27. Bohandy, J., Kim, B.F., Adrian, F.J.: Metal deposition from a supported metal film using an excimer laser. J. Appl. Phys. **60**(4), 1538–1539 (1986)
28. Barron, J.A., Ringeisen, B.R., Kim, H., Spargo, B.J., Chrisey, D.B.: Application of laser printing to mammalian cells. Thin Solid Films **453**, 383–387 (2004)
29. Schiele, N.R., Corr, D.T., Huang, Y., Raof, N.A., Xie, Y., Chrisey, D.B.: Laser-based direct-write techniques for cell printing. Biofabrication **2**(3), 032001 (2010)
30. Malda, J., Visser, J., Melchels, F.P., Jüngst, T., Hennink, W.E., Dhert, W.J., Groll, J., Hutmacher, D.W.: 25th anniversary article: engineering hydrogels for biofabrication. Adv. Mater. **25**(36), 5011–5028 (2013)
31. Dababneh, A.B., Ozbolat, I.T.: Bioprinting technology: a current state-of-the-art review. J. Manuf. Sci. Eng. **136**(6), 061016 (2014)
32. Billiet, T., Vandenhaute, M., Schelfhout, J., Van Vlierberghe, S., Dubruel, P.: A review of trends and limitations in hydrogel-rapid prototyping for tissue engineering. Biomaterials **33**(26), 6020–6041 (2012)
33. Chang, R., Nam, J., Sun, W.: Effects of dispensing pressure and nozzle diameter on cell survival from solid freeform fabrication–based direct cell writing. Tissue Eng. Part A **14**(1), 41–48 (2008)

34. Nair, K., Gandhi, M., Khalil, S., Yan, K.C., Marcolongo, M., Barbee, K., Sun, W.: Characterization of cell viability during bioprinting processes. Biotechnol. J. **4**(8), 1168–1177 (2009)
35. Nakamura, M., Kobayashi, A., Takagi, F., Watanabe, A., Hiruma, Y., Ohuchi, K., ... Takatani, S.: Biocompatible inkjet printing technique for designed seeding of individual living cells. Tissue Eng. **11**(11–12), 1658–1666 (2005)
36. Ozbolat, I.T., Yu, Y.: Bioprinting toward organ fabrication: challenges and future trends. IEEE Trans. Biomed. Eng. **60**(3), 691–699 (2013)
37. Mironov, V., Reis, N., Derby, B.: Bioprinting: a beginning. Tissue Eng. **12**(4), 631–634 (2006)
38. Pataky, K., Braschler, T., Negro, A., Renaud, P., Lutolf, M.P., Brugger, J.: Microdrop printing of hydrogel bioinks into 3D tissue-like geometries. Adv. Mater. **24**(3), 391–396 (2012)
39. Culver, J.C., Hoffmann, J.C., Poché, R.A., Slater, J.H., West, J.L., Dickinson, M.E.: Three-dimensional biomimetic patterning in hydrogels to guide cellular organization. Adv. Mater. **24**(17), 2344–2348 (2012)
40. Skardal, A., Atala, A.: Biomaterials for integration with 3-D bioprinting. Ann. Biomed. Eng. **43**(3), 730–746 (2015)
41. Lee, J.M., Sing, S.L., Tan, E.Y.S., Yeong, W.Y.: Bioprinting in cardiovascular tissue engineering: a review. Int. J. Bioprinting **2**(2), 136–145 (2016)
42. Chung, J.H., Naficy, S., Yue, Z., Kapsa, R., Quigley, A., Moulton, S.E., Wallace, G.G.: Bio-ink properties and printability for extrusion printing living cells. Biomater. Sci. **1**(7), 763–773 (2013)
43. Bajaj, P., Schweller, R.M., Khademhosseini, A., West, J.L., Bashir, R.: 3D biofabrication strategies for tissue engineering and regenerative medicine. Annu. Rev. Biomed. Eng. **16**, 247 (2014)
44. Ahmed, E.M.: Hydrogel: preparation, characterization, and applications: a review. J. Adv. Res. **6**(2), 105–121 (2015)
45. Zhao, W., Jin, X., Cong, Y., Liu, Y., Fu, J.: Degradable natural polymer hydrogels for articular cartilage tissue engineering. J. Chem. Technol. Biotechnol. **88**(3), 327–339 (2013)
46. Zhu, J., Marchant, R.E.: Design properties of hydrogel tissue-engineering scaffolds. Expert Rev. Med. Devices **8**(5), 607–626 (2011)
47. Carrow, J. K., Kerativitayanan, P., Jaiswal, M. K., Lokhande, G., Gaharwar, A. K. (2015). Polymers for bioprinting. In: Essentials of 3D Biofabrication and Translation (pp. 229–248)
48. Wüst, S., Godla, M.E., Müller, R., Hofmann, S.: Tunable hydrogel composite with two-step processing in combination with innovative hardware upgrade for cell-based three-dimensional bioprinting. Acta Biomater. **10**(2), 630–640 (2014)
49. Ambesi-Impiombato, F.S., Parks, L.A., Coon, H.G.: Culture of hormone-dependent functional epithelial cells from rat thyroids. Proc. Natl. Acad. Sci. **77**(6), 3455–3459 (1980)
50. Dimri, G.P., Lee, X., Basile, G., Acosta, M., Scott, G., Roskelley, C., Medrano, E.E., Linskens, M., Rubelj, I., Pereira-Smith, O.: A biomarker that identifies senescent human cells in culture and in aging skin in vivo. Proc. Natl. Acad. Sci. **92**(20), 9363–9367 (1995)
51. Kohl, P., Camelliti, P., Burton, F.L., Smith, G.L.: Electrical coupling of fibroblasts and myocytes: relevance for cardiac propagation. J. Electrocardiol. **38**(4), 45–50 (2005)
52. Camelliti, P., Green, C.R., LeGrice, I., Kohl, P.: Fibroblast network in rabbit sinoatrial node: structural and functional identification of homogeneous and heterogeneous cell coupling. Circ. Res. **94**(6), 828–835 (2004)
53. Hirsch, E., Nagai, R., Thum, T.: Heterocellular signalling and crosstalk in the heart in ischaemia and heart failure. Cardiovasc. Res. **102**(2), 191 (2014)
54. Taylor, D.A., Sampaio, L.C., Gobin, A.: Building new hearts: a review of trends in cardiac tissue engineering. Am. J. Transplant. **14**(11), 2448–2459 (2014)
55. Ott, H.C., Matthiesen, T.S., Goh, S.K., Black, L.D., Kren, S.M., Netoff, T.I., Taylor, D.A.: Perfusion-decellularized matrix: using nature's platform to engineer a bioartificial heart. Nat. Med. **14**(2), 213 (2008)
56. Bagó, J. R., Soler-Botija, C., Casaní, L., Aguilar, E., Alieva, M., Rubio, N., ... Blanco, J.: Bioluminescence imaging of cardiomyogenic and vascular differentiation of cardiac and subcutaneous adipose tissue-derived progenitor cells in fibrin patches in a myocardium infarct model. Int. J. Cardiol. **169**(4), 288–295 (2013)

57. Shin, S. R., Jung, S. M., Zalabany, M., Kim, K., Zorlutuna, P., Kim, S. B., ... Wan, K. T.: Carbon-nanotube-embedded hydrogel sheets for engineering cardiac constructs and bioactuators. ACS nano. **7**(3), 2369–2380 (2013)
58. Ye, X., Wang, H., Zhou, J., Li, H., Liu, J., Wang, Z., ... Zhao, Q.: The effect of Heparin-VEGF multilayer on the biocompatibility of decellularized aortic valve with platelet and endothelial progenitor cells. PloS one. **8**(1), e54622 (2013)
59. Tillman, B., Hardin-Young, J., Shannon, W., Russell, A.J., Parenteau, N.L.: Meeting the need for regenerative therapies: translation-focused analysis of US regenerative medicine opportunities in cardiovascular and peripheral vascular medicine using detailed incidence data. Tissue Eng. Part B Rev. **19**(2), 99–115 (2012)
60. Gaetani, R., Doevendans, P.A., Metz, C.H., Alblas, J., Messina, E., Giacomello, A., Sluijter, J.P.: Cardiac tissue engineering using tissue printing technology and human cardiac progenitor cells. Biomaterials **33**(6), 1782–1790 (2012)
61. Gaetani, R., Feyen, D.A., Verhage, V., Slaats, R., Messina, E., Christman, K.L., Giacomello, A., Doevendans, P.A., Sluijter, J.P.: Epicardial application of cardiac progenitor cells in a 3D-printed gelatin/hyaluronic acid patch preserves cardiac function after myocardial infarction. Biomaterials **61**, 339–348 (2015)
62. Gaebel, R., Ma, N., Liu, J., Guan, J., Koch, L., Klopsch, C., Gruene, M., Toelk, A., Wang, W., Mark, P., Wang, F.: Patterning human stem cells and endothelial cells with laser printing for cardiac regeneration. Biomaterials **32**(35), 9218–9230 (2011)
63. Butcher, J.T., Mahler, G.J., Hockaday, L.A.: Aortic valve disease and treatment: the need for naturally engineered solutions. Adv. Drug Deliv. Rev. **63**(4–5), 242–268 (2011)
64. Hasan, A., Ragaert, K., Swieszkowski, W., Selimović, Š., Paul, A., Camci-Unal, G., Mofrad, M.R., Khademhosseini, A.: Biomechanical properties of native and tissue engineered heart valve constructs. J. Biomech. **47**(9), 1949–1963 (2014)
65. Duan, B.: State-of-the-art review of 3D bioprinting for cardiovascular tissue engineering. Ann. Biomed. Eng. **45**(1), 195–209 (2017)
66. Duan, B., Kapetanovic, E., Hockaday, L.A., Butcher, J.T.: Three-dimensional printed trileaflet valve conduits using biological hydrogels and human valve interstitial cells. Acta Biomater. **10**(5), 1836–1846 (2014)
67. Duan, B., Hockaday, L.A., Kang, K.H., Butcher, J.T.: 3D bioprinting of heterogeneous aortic valve conduits with alginate/gelatin hydrogels. J. Biomed. Mater. Res., Part A **101**(5), 1255–1264 (2013)
68. Hockaday, L.A., Kang, K.H., Colangelo, N.W., Cheung, P.Y.C., Duan, B., Malone, E., Wu, J., Girardi, L.N., Bonassar, L.J., Lipson, H., Chu, C.C.: Rapid 3D printing of anatomically accurate and mechanically heterogeneous aortic valve hydrogel scaffolds. Biofabrication **4**(3), 035005 (2012)
69. Sun, X., Altalhi, W., Nunes, S.S.: Vascularization strategies of engineered tissues and their application in cardiac regeneration. Adv. Drug Deliv. Rev. **96**, 183–194 (2016)
70. Horch, R.E., Kneser, U., Polykandriotis, E., Schmidt, V.J., Sun, J., Arkudas, A.: Tissue engineering and regenerative medicine–where do we stand? J. Cell Mol. Med. **16**(6), 1157–1165 (2012)
71. Rouwkema, J., Rivron, N.C., van Blitterswijk, C.A.: Vascularization in tissue engineering. Trends Biotechnol. **26**(8), 434–441 (2008)
72. Mironov, V., Visconti, R.P., Kasyanov, V., Forgacs, G., Drake, C.J., Markwald, R.R.: Organ printing: tissue spheroids as building blocks. Biomaterials **30**(12), 2164–2174 (2009)
73. Norotte, C., Marga, F.S., Niklason, L.E., Forgacs, G.: Scaffold-free vascular tissue engineering using bioprinting. Biomaterials **30**(30), 5910–5917 (2009)
74. Kucukgul, C., Ozler, S.B., Inci, I., Karakas, E., Irmak, S., Gozuacik, D., Taralp, A., Koc, B.: 3D bioprinting of biomimetic aortic vascular constructs with self-supporting cells. Biotechnol. Bioeng. **112**(4), 811–821 (2015)
75. Cui, X., Boland, T.: Human microvasculature fabrication using thermal inkjet printing technology. Biomaterials **30**(31), 6221–6227 (2009)

76. Novosel, E.C., Kleinhans, C., Kluger, P.J.: Vascularization is the key challenge in tissue engineering. Adv. Drug Deliv. Rev. **63**(4–5), 300–311 (2011)
77. Poldervaart, M.T., Gremmels, H., van Deventer, K., Fledderus, J.O., Öner, F.C., Verhaar, M.C., Dhert, W.J., Alblas, J.: Prolonged presence of VEGF promotes vascularization in 3D bioprinted scaffolds with defined architecture. J. Controlled Release **184**, 58–66 (2014)
78. Gao, Q., He, Y., Fu, J.Z., Liu, A., Ma, L.: Coaxial nozzle-assisted 3D bioprinting with built-in microchannels for nutrients delivery. Biomaterials **61**, 203–215 (2015)
79. Yu, Y., Zhang, Y., Ozbolat, I.T.: A hybrid bioprinting approach for scale-up tissue fabrication. J. Manuf. Sci. Eng. **136**(6), 061013 (2014)
80. Kolesky, D.B., Truby, R.L., Gladman, A., Busbee, T.A., Homan, K.A., Lewis, J.A.: 3D bioprinting of vascularized, heterogeneous cell-laden tissue constructs. Adv. Mater. **26**(19), 3124–3130 (2014)

Printed by Printforce, the Netherlands